Systems Engineering of Software-Enabled Systems

Systems Engineering of Software-Enabled Systems

Richard E. Fairley
Software and Systems Engineering Assoc.
CO, US

This edition first published 2019
© 2019 John Wiley & Sons, Inc.

All rights reserved. No part of this publication may be reproduced, stored in a retrieval system, or transmitted, in any form or by any means, electronic, mechanical, photocopying, recording or otherwise, except as permitted by law. Advice on how to obtain permission to reuse material from this title is available at http://www.wiley.com/go/permissions.

The right of Richard E. Fairley to be identified as the author of this work has been asserted in accordance with law.

Registered Office
John Wiley & Sons, Inc., 111 River Street, Hoboken, NJ 07030, USA

Editorial Office
111 River Street, Hoboken, NJ 07030, USA

For details of our global editorial offices, customer services, and more information about Wiley products visit us at www.wiley.com.

Wiley also publishes its books in a variety of electronic formats and by print-on-demand. Some content that appears in standard print versions of this book may not be available in other formats.

Limit of Liability/Disclaimer of Warranty
While the publisher and authors have used their best efforts in preparing this work, they make no representations or warranties with respect to the accuracy or completeness of the contents of this work and specifically disclaim all warranties, including without limitation any implied warranties of merchantability or fitness for a particular purpose. No warranty may be created or extended by sales representatives, written sales materials or promotional statements for this work. The fact that an organization, website, or product is referred to in this work as a citation and/or potential source of further information does not mean that the publisher and authors endorse the information or services the organization, website, or product may provide or recommendations it may make. This work is sold with the understanding that the publisher is not engaged in rendering professional services. The advice and strategies contained herein may not be suitable for your situation. You should consult with a specialist where appropriate. Further, readers should be aware that websites listed in this work may have changed or disappeared between when this work was written and when it is read. Neither the publisher nor authors shall be liable for any loss of profit or any other commercial damages, including but not limited to special, incidental, consequential, or other damages.

Library of Congress Cataloging-in-Publication Data

Names: Fairley, R. E. (Richard E.), 1937- author.
Title: Systems engineering of software-enabled systems / Richard E. Fairley, Software Engineering Management Assoc., CO, US.
Description: Hoboken, NJ, USA : Wiley, 2019. | Includes bibliographical references and index. |
Identifiers: LCCN 2019004173 (print) | LCCN 2019009630 (ebook) | ISBN 9781119535027 (Adobe PDF) | ISBN 9781119535034 (ePub) | ISBN 9781119535010 (hardback)
Subjects: LCSH: Software engineering. | Systems engineering.
Classification: LCC QA76.758 (ebook) | LCC QA76.758 .F355 2019 (print) | DDC 005.1–dc23
LC record available at https://lccn.loc.gov/2019004173

Cover image: © Henrik5000/E+/Getty Images
Cover design by Wiley

Set in 10/12pt WarnockPro by SPi Global, Chennai, India

Printed in the United States of America

10 9 8 7 6 5 4 3 2 1

Contents

Preface

Modern physical systems would be inoperable without software. Software elements provide interfaces among physical elements, coordinate the interactions among software elements and physical elements, enable connections to system environments, and provide some (or most) of the functionality, behavior, and quality attributes of software-enabled physical systems.

It is not an exaggeration to state that software-enabled physical systems are ubiquitous throughout modern society, including but not limited to systems in the domains of transportation, communication, health care, energy, aerospace, military defense, manufacturing, ecology, agriculture, and intelligent structures such as buildings and bridges. Software-enabled physical systems include real-time embedded systems (navigation, communication), consumer products (cell phones, microwave ovens), health care devices (pacemakers, heart and lung machines), transportation systems (automobiles, light rail), energy systems (solar and wind farms, power grids), military defense systems (Tomahawk missiles, F-35 aircraft), earth orbiting satellites (GPS, weather), and interplanetary missions (Cassini, Rosetta).

Software-enabled physical systems have evolved so rapidly in size, complexity, and number of deployments that the systems engineering processes, methods, and techniques for gracefully coordinating development of the software elements and physical elements of software-enabled physical systems have not evolved at a comparable pace.

This text presents processes, methods, and techniques that can be used to bridge the gaps between systems engineering and software engineering that will enable members of both disciplines to perform their tasks in a more cooperative and coordinated manner, which will permit more efficient system development and higher quality systems.

The text is presented in three parts. Part I includes Chapters 1–3. Chapter 1 provides an introduction to and overview of systems engineering and software engineering. Chapter 2 presents similarities and differences in the professional disciplines of systems engineering and software engineering.

Chapter 3 presents issues that can facilitate and inhibit successful execution of software-enabled physical systems development projects.

Part II includes Chapters 4–9 and covers the technical aspects of systems engineering for software-enabled physical systems from system definition to system delivery. Chapter 4 presents traditional approaches to analysis, design, and development of complex systems. Chapters 5–9 present an approach for improving the linkages among systems engineering, software engineering, and the other engineering disciplines that collaborate to build software-enabled physical systems.

Part III (Chapters 10–12) is concerned with technical management of systems engineering activities: planning and estimating, assessing and controlling, and organizing and leading project teams that develop software-enabled systems.

Emphasis of this text is on systems engineering of software-enabled physical systems but the material in the text can be applied to development of other kinds of complex systems.

Explanatory sidebars are included throughout the text. Each chapter of the text concludes with key points, exercises, and references. Two running examples are used to illustrate various processes, methods, and techniques for system engineering of software-enabled physical systems: automated teller machines and a driving system simulator for ground-based vehicles. The examples are large enough to illustrate key issues in systems engineering of software-enabled physical systems but small enough to be treated within the page limitations of the text.

The intended audiences for the text include advanced undergraduate students, graduate students, practitioners of systems engineering and software engineering, and others who desire to know more about systems engineering and software engineering of software-enabled physical systems.

The following references provide guidance for the material in this text:

- *ISO/IEC/IEEE Standard 15288:2015 Systems and software engineering – System life-cycle processes;*
- *ISO/IEC/IEEE Standard 12207:2017 Systems and software engineering – Software life-cycle processes*;
- *The Guide to the Systems Engineering Body of Knowledge* (SEBoK);
- The *Guide to the Software Engineering Body of Knowledge* (SWEBOK);
- The *Software Engineering Competency Model* (SWECOM);
- *NASA's Systems Engineering Competency Model*; and
- The *INCOSE Systems Engineering Handbook.*

Citations for accessing these documents are provided throughout the text and in the bibliography at the end of the text. Access to these documents will be helpful in understanding the material in this text but is not essential.

I am indebted to the students and practitioners from whom I have learn as much and perhaps more than they have learned from me. I am doubly indebted to my wife and best friend Mary Jane for her thoughtful advice on the structure and content of the text.

15 January 2019

Dick Fairley
Teller County, Colorado

Part I

Systems Engineering and Software Engineering

This text is presented in three parts. Part I includes Chapters 1–3. Chapter 1 provides an introduction and overview of systems engineering and software engineering. Chapter 2 presents attributes of the professional disciplines of systems engineering and software engineering. Chapter 3 includes issues that can facilitate and inhibit successful execution of software-enabled system development projects.

Systems Engineering of Software-Enabled Systems, First Edition. Richard E. Fairley.
© 2019 John Wiley & Sons, Inc. Published 2019 by John Wiley & Sons, Inc.

1

Introduction and Overview

1.1 Introduction

This book is about the life cycle processes, methods, and techniques used by systems engineers, software engineers, and other engineers to develop and modify software-enabled physical systems. Software-enabled systems (SESs) are sometimes termed “software-intensive,” “cyber-physical,” “embedded,” or “Internet of Things (IOT).” The term “software-enabled system” is used throughout this text to denote these and other kinds of systems for which software provides functionality, behavior, quality attributes, interfaces among system elements, and connections to entities in external environments.

An SES includes physical elements and software elements. The physical elements may be naturally occurring (e.g. wind, water, sun), may have been purposefully engineered by humans (e.g. solar and wind farms, hydroelectric dams), or may include a combination of naturally occurring and engineered elements. A hydroelectric system for example includes flowing water, the physical structure of the dam, and turbine/generator machines; all are sensed and controlled by software in digital devices. A dam or a flowing river provides the operational environment. The natural force of gravity provides the energy that rotates the turbine/generator machines. A case study of the Northwest hydroelectric system in the United States is presented in Appendix A of this text. Alternatively, the physical elements of an SES may be entirely engineered. These systems operate in natural and engineered environments. A case study of modern software-enabled automobiles is presented in Appendix B.

The digital elements of an SES include digital hardware and the software that senses, measures, regulates, and controls the physical elements, including the digital hardware. In addition, the digital elements of an SES may include analog/digital and digital/analog converters for interconnecting analog system elements and digital devices. Software may also provide data management capabilities and communication among internal system elements and to external entities. Communication may be provided by direct linkage or by Internet-enabled or Intranet-enabled software links.

Systems Engineering of Software-Enabled Systems, First Edition. Richard E. Fairley.
© 2019 John Wiley & Sons, Inc. Published 2019 by John Wiley & Sons, Inc.

SESs range from smartphones to household appliances to pacemakers to automobiles to military systems to the International Space Station. They are deployed in every domain of modern society, including but not limited to aerospace, agriculture, communication, consumer products, defense, ecology, energy, health care, intelligent buildings, manufacturing, and transportation.

It is not an exaggeration to state that SESs are ubiquitous throughout modern society; they are constantly growing in size, complexity, and number of deployments.

Part 1 of the *Guide to the Systems Engineering Body of Knowledge* (Engineering Disciplines Other than Systems Engineering) includes the following statement, attributed to Dr Barry Boehm, that describes the relationship between system engineering and software engineering (Boehm 1994):

> SwE and SE are not just allied disciplines, they are intimately intertwined. Most functionality of commercial and government systems is now implemented in software, and software plays a prominent, or dominant role in differentiating competing systems in the marketplace. Software is usually prominent in modern systems architectures and is often the "glue" for integrating complex system components.
>
> (SEBoKwiki.org, Part 1, Systems Engineering and Other Disciplines)

The intimate intertwining is reflected in the close relationships between systems engineering (SE) and software engineering life cycle processes in ISO/IEC/IEEE Standards 15288 and 12207 (ISO 2015, 2017). However, the methods and processes of SE do not always match those needed to accommodate development of the intimate intertwined physical elements and software elements of SESs. This text is concerned with bridging the gap between SE and software engineering and bridging the gap between the working relationships of physical systems engineers (PhSEs) and software engineers.

This chapter provides the opportunity for readers to learn about (or review) some background for the coverage of SES development in the remainder of the text. Topics in this chapter include the following:

- The evolution of engineering;
- The nature of systems, SE, and software engineering;
- Related disciplines;
- The product, service, enterprise, and systems of systems (SoSs) perspectives; and
- The roles played by PhSEs and SwSEs in developing SESs.

References and exercises are provided to enable further investigation of the topics presented in this chapter.

1.2 The Evolution of Engineering

Engineers develop systems, processes, and tools that improve the safety, security, ease, and convenience of human life and other life forms. Engineers also develop methods and tools to exploit natural resources. In recent times, engineers have become increasingly involved in developing mechanisms to protect and preserve natural resources and the ecological environment.

It is said that the first engineers developed structures to protect humans from animals and the environment, build crude bridges across streams, and exploit the lever and the wheel. Notable engineering achievements in ancient times included construction of the pyramids in Egypt, the Great Wall of China, the temples of ancient Greece, and the aqueducts that brought water to Rome.

In the early 1800s, the term "civil engineering" was coined to distinguish construction of bridges and buildings from development of military devices. During the mid-1800s, development of machine tools, and invention of the steam engine and the machines powered by it, fostered the discipline of mechanical engineering. An early (stationary) steam engine, known as a beam engine, is illustrated in Figure 1.1.

During the late 1800s, electrical engineering evolved based on experiments with electricity and invention of the electric motor and generator. Civil engineering emerged from ancient practices; mechanical and electrical engineering were initially thought of as applied physics in the later 1800s and early 1900s. In modern times, mechatronics has emerged as an engineering discipline that combines mechanical and electrical engineering; biology provides the basis for biological and biomedical engineering.

Engineers purposefully build and sustain systems to provide beneficial consequences for one or more segments of society. Bridges are built to allow safe

Figure 1.1 A stationary steam engine. Source: Attribution: Nicolás Pérez.

passage across rivers and streams, commercial airplanes are built to enable air travel, and hydroelectric dams are built to generate electricity from the energy in flowing water.

An example of a software-enabled physical system, the smart factory, is presented in "Smart Factory."

Smart Factory

Smart factories are SESs. A seminal paper "Smart Factory Systems" presents an overview of enabling technologies for smart manufacturing factories that integrate the IoT with computer networks, data integration, and data analytics (Lee 2015).

The Smart Factory System includes a virtual model of a factory that monitors and controls the physical factory. The physical factory provides data to the virtual model using a network that connects the physical devices to the software-based model that resides in one or more computers.

The referenced paper identifies three sections of a smart factory: components, machines, and production systems. Components contain software-enabled smart sensors that provide data on performance of the components, including degradation monitoring, prediction of remaining useful life, and fault detection. Machine controllers use sensor data to generate monitoring and diagnostics information that includes predictive uptime and failure prevention. Production systems include control systems that use data collected from the machines that allows them to self-configure and self-organize.

Figure 1.2 (figure 2 from Lee 2015) illustrates the architecture of the "five C" layers of a smart factory: connection, conversion, digital, cognition, and configure.

Level 1: Connection – The connection level connects sensors to components that are elements of machines and may connect sensors into networks. Some sensors may generate analog signals and some may generate digital signals, either directly or by way of analog to digital converters.

Level 2: Conversion – Level 2 converts input data to information for signal processing, feature extraction, Prognostics and Health Management (PHM) algorithms, and predictive analytics to monitor and predict component and machine performance.

Level 3: Digital – Level 3 provides the digital image of the physical system. Data from individual machines is recorded and analyzed over time to determine machine utilization and machine health. Data is shared among machines to permit self-comparison of each machine to collective performance of the machines in the production factory.

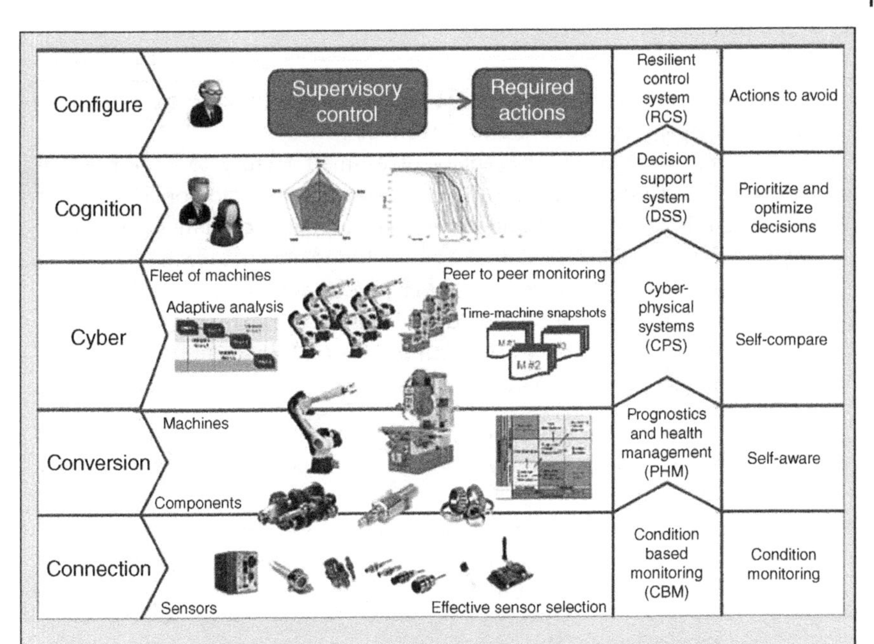

Figure 1.2 The 5C levels of a smart factory architecture. Source: Springer Verlag.

Level 4: Cognition – The cognition level provides information concerning the overall monitored system and correlates the effects of different components and machines, which is presented in a form that can be used by expert users to make decisions concerning components, machines, and the production system.

Level 5: Configure – Level 5 provides feedback from the digital model to the physical system and supports actions taken by humans or automated supervisory control to apply corrective and preventive decisions made at the cognition level.

A second paper provides a comprehensive description of smart factories. In the paper, Burke et al. (2017) identify five characteristics for consideration in making the transition from a conventional factory to a smart factory.

1. *Technology*: Digital technology will enable various elements of plant equipment to communicate with one another.
2. *Data and algorithms*: Algorithmic analysis of data will enable the manufacturing processes and detection of operations errors, provide user feedback, and predict operational inefficiencies and failures.
3. *People*: People will be key elements of a smart factory but changes in processes and governance may result in changes to the roles to be played by

(Continued)

(Continued)

operators, users, and maintainers, as well as to the supporting information technology (IT) organizations and personnel.

4. *Process and governance*: The ability of a smart factory to self-optimize, self-adapt, and autonomously run production will alter traditional models of manufacturing processes. Connectivity with suppliers, customers, and other factories may result in redesign of manufacturing governance models.
5. *Digital security*: Digital security risks will be increased for a smart factory, as compared with a conventional factory because of the increased interconnectivity of, and digital interfaces among, machines. Digital security will be a priority consideration for a smart factory.

Oborski describes an integrated monitoring system for manufacturing processes that includes a multilayer hardware and software reference model (Oborski 2016).

The above considerations make it apparent that software-based digital elements play a central role in development and deployment of smart sensors, smart machines, and smart factories.

1.3 Characterizations of Systems

A system is a collection of interconnected elements that exist within and interact with an environment. There are several definitions of "system," including those provided by the International Council on Systems Engineering (INCOSE), the ISO 15288 standard for SE processes, and the Software and Systems Engineering Vocabulary.

The INCOSE definition of "system" is as follows:

> A system is a construct or collection of different elements that together produce results not obtainable by the elements alone. The elements, or parts, can include people, hardware, software, facilities, policies, and documents; that is, all things required to produce systems-level results.
>
> (Walden et al. 2015)

According to ISO/IEC/IEEE Standard 15288 a "system element" is

> a discrete part of a system that can be implemented to fulfill specified requirements. A system element can be hardware, software, data, humans, processes (e.g., processes for providing service to users), procedures (e.g., operator instructions), facilities, materials, and naturally occurring entities (e.g., water, organisms, minerals), or any combination.
>
> (ISO 2015)

The System and Software Engineering Vocabulary defines a "complete system" as follows:

> A *complete system* includes all of the associated equipment, facilities, material, computer programs, firmware, technical documentation, services, and personnel required for operations and support to the degree necessary for self-sufficient use in its intended environment.
>
> (SEVOCAB 2017)

Although not explicitly mentioned, natural elements may be part of a system; for example, the flow rate of water that energizes a hydroelectric dam must be controlled and excess water must be routed to a spillway.

Systems can be characterized and categorized in various ways. Some of the characterizations follow.

1.3.1 Open and Closed Systems

Physicists, chemists, and thermodynamicists distinguish between open and closed physical systems. Open systems exchange energy, physical matter, and/or information across one or more system boundaries. Closed physical systems limit the kinds of transfers that can occur, depending on the physical discipline.

In engineering, a closed system is one for which all inputs are known and all outputs or outcomes can be determined within a bounded time. Some engineered systems are closed systems; for example, a closed loop control system (such as a bimetal thermostat) is a closed system, but most engineered systems are open systems.

1.3.2 Static and Dynamic Systems

System elements have relationships with other system elements that, in combination, provide the system features. Some systems such as traditional bridges and buildings are *static*. The physical elements of a bridge include abutments, pilings, substructure, girders, and deck. Relationships among these elements are fixed or change slowly over time by use and deterioration (draw bridges are engineered to permit slow raising and lowering of the deck but the structural relationships of the deck to other elements do not change, or change slowly over time).

Some systems such as aircraft navigation and control systems and IT systems are *dynamic*. The behavioral relationships among elements of dynamic system typically change frequently and rapidly. Thus, static systems *are* and dynamic systems *do*.

Static systems have structural, functional, and quality attributes. For example, a bridge has structural properties that include functionality and quality attributes: the function of a bridge is to provide safe passage of humans, animals, and vehicles across rivers and streams; the quality attributes of a bridge include load rating, stability, durability, adaptability, and aesthetics. Some modern bridges are software-instrumented because they have embedded sensors and digital elements that are used to periodically, and when needed, measure and report the physical conditions of a bridge.

Dynamic systems have *behavior,* in addition to structure, functionality, and quality attributes. For example, an automobile is structurally, functionally, and behaviorally engineered. Structural quality attributes include those in the bridge example. The primary behavioral attribute of an automobile is transportation of people and physical material from one geographic location to another. Safety is the primary quality attribute. Modern automobiles that have Internet connections also have security concerns (e.g. connections to the automaker's web site, GPS navigation, or Internet radio).

Behavioral attributes of a dynamic system include performance and reliability, in addition to safety and security. The *state* of a dynamic system is characterized by the attributes of and relationships among the system elements at a particular point in time. Taken together, the states of all system elements and their interactions at a particular point in time constitute the *system state* at that instant. Behavior of a dynamic system is exhibited as a sequence of states and state changes over time. For example, the behavior of an automobile can be characterized as a time sequence of concurrent states and state changes of attributes such as position of the steering wheel, accelerator, brakes, turn signals, radio, and other system elements. In addition, systems of all kinds have boundaries.

1.3.3 System Boundaries

An open engineered system is a collection of elements that exist within and interact with one or more environments. Some open system boundaries are precisely delineated; for example, an office computer is an open system; the boundary between the computer and a human user is clear. Other boundaries are inexact; there is no precisely defined boundary between the Earth's atmosphere and outer space – the atmosphere is densest near the Earth and gradually fades with distance from the earth.

Context diagrams are used to illustrate the entities in a system's environment that interact with the system across the system boundary. Figure 1.3 is a context diagram depicting entities that will exploit data generated by an earth-orbiting NOAA satellite (i.e. the National Polar-Orbiting Operational Environmental Satellite System, NPOESS, in Figure 1.3).

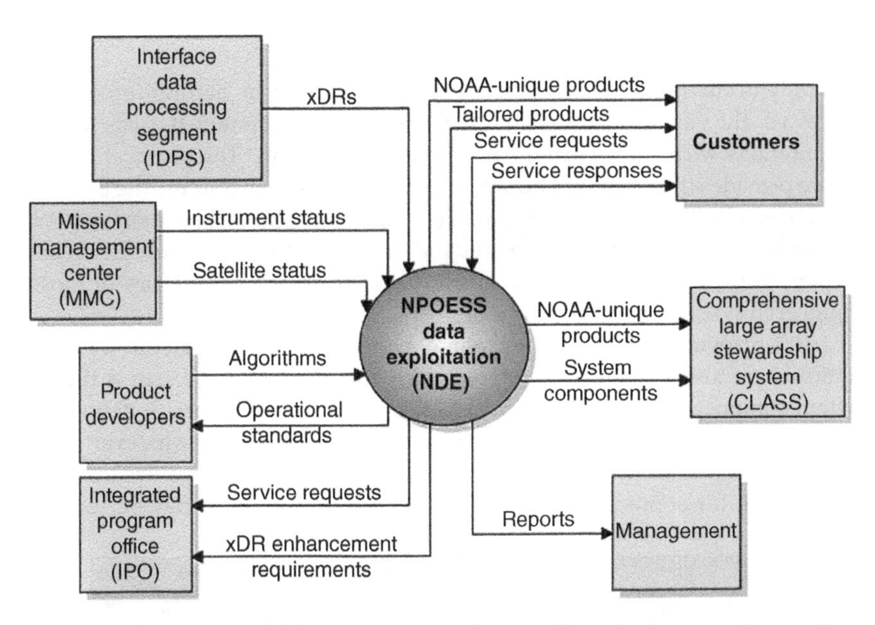

Figure 1.3 The NOAA NPOESS context diagram. Source: Wikimedia Commons free media repository.

NOAA is the U.S. National Oceanic and Atmospheric Administration; NOAA satellites provide weather data to facilitate weather forecasting and also to monitor the environment and climate. NDE is the National Data Exploitation project. NPOESS, illustrated in Figure 1.3, evolved into Joint Polar Satellite System-1 (JPSS-1), which was launched on 18 November 2017 and became operational as NOAA-20 with a lifetime requirement of seven years. More information can be obtained from the JPSS Level 1 Requirements Document (JPSS 2016).

Some humans may be elements of a system and others may be external users that interact across a system boundary. Medical doctors may be characterized as internal elements of a Health Maintenance Organization (HMO), where they are employed, and they may also be external users of hospital facilities, where they care for their patients.

The determination of external entities versus internal system elements is often influenced by the degree of control exerted by one or more regulatory agencies. There are no regulations governing the behavior of automated teller machine users who are external to an automated teller system, other than laws regarding criminal behavior. However, air traffic controllers are internal elements of an air traffic control system because government regulations require that they be trained and certified; their performance is monitored and they can be dismissed for unacceptable performance.

1.3.4 Naturally Occurring Systems

Naturally occurring systems are those that have arisen by natural forces and processes. The Earth, illustrated in Figure 1.4, is a natural system. It exists within and interacts with the environment of our solar system. The physical atmosphere provides the boundary between Earth and our solar system. The boundary between Earth and our solar system (i.e. the atmosphere), like many natural system boundaries, is a "fuzzy boundary" because the atmosphere becomes thinner at greater distances from Earth. In contrast, the boundaries of most engineered systems are precise.

Earth's primary elements (atmosphere, continents, oceans, global weather patterns) are also systems that interact with one another. The Earth is thus a system composed of interacting systems (i.e. an SoS). This recursive definition of the Earth's systems can be hierarchically decomposed to whatever level is needed for the purpose at hand. Identifying and naming the oceans may be sufficient for some purposes but the definition of oceans may be further

Figure 1.4 A complex natural system (the Blue Marble). Source: Public Domain File.

decomposed to permit identification of and differentiation among the biological forms that exist in different oceans. Further decomposition might be needed to study the effect of various ocean environments on the reproduction rates of blue whales.

Thus, it can be said

Every system is an element of one or more larger systems.

With one possible exception: We do not currently know if our universe is embedded in a larger system, i.e. whether it is an open system, a partially closed system, or a totally closed system.

1.3.5 Engineered Systems

Some engineered systems include naturally occurring elements and purposefully engineered elements. A hydroelectric dam, for example, is illustrated in Figure 1.5.

The rectangle in the figure indicates (approximately) the system boundary. Water and gravity are naturally occurring elements in Figure 1.5. Some natural elements may be reconfigured when included as part of a system; for example, the banks and depth of a river might be altered to facilitate construction and

Figure 1.5 A software-enabled natural/engineered system. Source: Used with permission of Missoulian newspaper, Missoula, MT, 18 April 2013.

operation of a hydroelectric dam. Engineered elements of a hydroelectric dam include the concrete structure, a spillway (used to provide controlled release of water that bypasses the dam), a penstock (a water-flow channel with a flow regulator that routes water into the turbines that rotate the generators), and the turbines and generators. The software-enabled digital elements are used to monitor the dam structure, measure the level of and regulate the flow rate of water (both upstream and downstream), control the turbines and generators and the amount of energy extracted from the flowing water, deliver electrical energy to an electrical transmission system, and coordinate operation with other dams in a hydroelectric system.

The U.S. Northwest Hydro System (NwHS) presented in Appendix A to this text is a case study of a large complex natural/engineered SES. The NWHS, like many SESs, also includes human, regulatory, and societal elements.

An automobile is an example of an SES that includes no natural elements; it is a totally engineered system. The partial structure of an automotive electronics system for a modern automobile that includes several software-enabled subsystems is illustrated in Figure 1.6.

Modern automobiles have 100 and more electronic control units (ECUs); each is a microprocessor that contains software modules that respond to inputs, interact with other ECUs, and generate signals through local area networks. Automobiles are increasingly becoming mobile software platforms that include millions of lines of software code. The evolution of self-driving vehicles further increases the central role of software in these SESs. As illustrated in Figure 1.6, modern vehicles also have increasing connectivity to the Internet, which indicates the need for digital security in addition to physical safety. A case study of modern automobiles as SESs is presented in Appendix B to this text.

"The Apollo" describes one of the first modern applications of SE processes and methods to a large-scale interdisciplinary project.

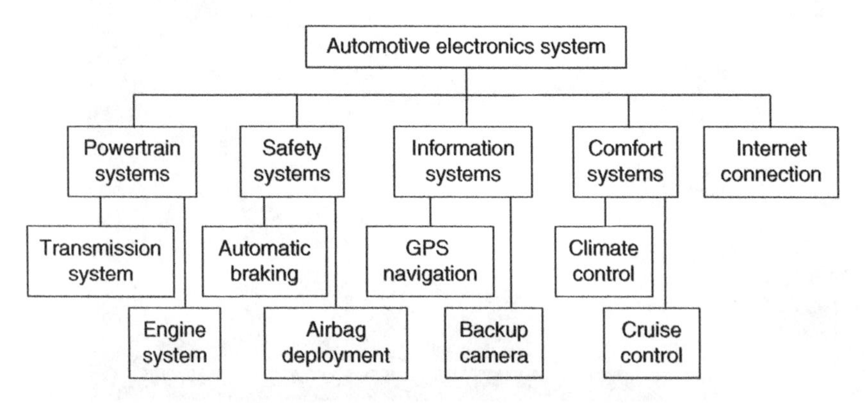

Figure 1.6 A complex software-enabled engineered system.

The Apollo

In 1961, U.S. President John Kennedy addressed a special session of the U.S. Congress to propose a Moon exploration program. During that speech, he told Congress: "First, I believe that this nation should commit itself to achieving the goal, before this decade is out, of landing a man on the moon and returning him safely to the earth" (Kennedy 1961a). A recording of President Kennedy's speech to the U.S. Congress can be heard at Kennedy (1961b). In 1969, the Apollo 11 mission achieved that goal.

Technical achievements of the Apollo program included development of the Saturn launch vehicle, the Apollo spacecraft, and the lunar module plus integration of the launch vehicle, spacecraft, and lunar module into a cohesive system. Diverse technologies needed for communication, navigation, launch, and reentry/recovery were also developed. Images of the Saturn V launch and the Apollo spacecraft are provided in Figures 1.7 and 1.8.

Figure 1.7 Launch of Saturn V and Apollo spacecraft. Source: NASA.

(Continued)

(Continued)

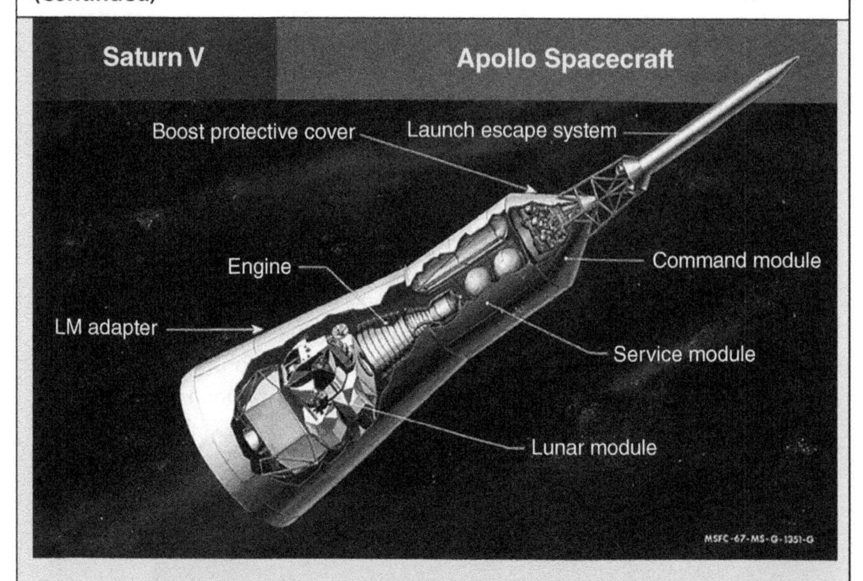

Figure 1.8 The Apollo spacecraft. Source: Wikimedia Commons. Public Domain image.

A NASA staff of thousands plus more than 500 contractor organizations in prime contractor/subcontractor relationships worked on various aspects of the Apollo program (Bilstein 1996).

A program management office (PMO) was established to provide centralized authority over design, engineering, procurement, testing, construction, manufacturing, spare parts, logistics, training, and operations. System engineering was the primary emphasis of the program management concept. The Apollo PMO was based on earlier experience with the Minuteman ICBM System Program Office that was established to provide for acquisition, SE, and logistical support for silo-based ICBM systems (SPO 2016).

In the 15 November 1968 issue of *Science* magazine, a publication of the American Association for the Advancement of Science, Dael Wolfle wrote the following:

> In terms of numbers of dollars or of men, NASA has not been our largest national undertaking, but in terms of complexity, rate of growth, and technological sophistication it has been unique … It may turn out that the [space program's] most valuable spin-off of all will be human rather than technological: better knowledge of how to plan, coordinate, and monitor the multitudinous and varied activities of the organizations required to accomplish great social undertakings.
>
> (Wolfle 1968)

> Since the Apollo success, system engineering has evolved into a logical and consistent approach to dealing with any situation that involves conceiving, developing, deploying, operating, and sustaining a complex system. Today, system engineering processes and methods are being applied to provisioning of products and services, and engineering of enterprises in every domain of modern society.

Figure 1.9 illustrates an aircraft carrier system in a natural environment. Aircraft carrier systems (ship, aircraft, infrastructure) are among the most complex of engineered systems. They may include as many as 5000 personnel and many carry as many as 60 aircraft. Carrier systems are self-sufficient and provide all of the facilities and services of a small city. They have many complex subsystems, some of which may not have been originally engineered to be elements of an aircraft carrier system. An aircraft carrier is thus an SoS.

1.3.6 Systems of Systems

The Software and Systems Engineering Vocabulary defines an SoS as follows:

> a large system that delivers unique capabilities, formed by integrating independently useful systems.
>
> (SEVOCAB 2017)

Figure 1.9 A complex engineered system in a natural environment. Source: www.navy.mil/viewGallery.asp?id=10.

An SoS is typically a large complex system consisting of standalone systems that were not designed to be integrated with other systems. An SoS provides capabilities not provided by any of the individual systems.

SoSs are developed to meet crisis situations or to provide needed capabilities without the time and expense that would be required to develop a new system. While many subsystems of a complex system such as an aircraft carrier system are designed to be compatible, legacy systems and systems designed for multipurpose use may not be compatible with other carrier subsystems.

Issues to be addressed in developing an integrated SoS include the following:

- Developing additional capabilities needed in the SoS but not provided by any of the constituent systems;
- Modifying and extending constituent systems without altering the standalone capabilities of those systems because the systems may continue to be used in standalone mode, in addition to becoming member systems of an SoS;
- Providing resources needed by the set of individual constituent systems in excess of what is needed by each system individually;
- Providing interfaces and communication links among the constituent systems;
- Ensuring interoperability among the systems; and
- Eliminating or masking undesired behaviors that result from interactions of the combined systems.

In addition to being open or closed, natural, natural/engineered, engineered, or an SoS, many systems can also be categorized as one or more of social, economic, political, ethical, legal, and/or regulatory. Some systems span across multiple domains. Hydroelectric dams, automobiles, and aircraft carriers are examples of systems that span natural, physical engineered, software engineered, service, enterprise, regulatory, and social (human) domains.

These examples of complex SESs illustrate the need for SE and software engineering expertise. Each discipline is discussed in turn.

1.4 Systems Engineering

The first modern use of the term "systems engineering" can be traced to Bell Laboratory employees in the 1940s (Liu 2015). Other important dates in the evolution of SE as a discipline are listed in table 1.2 of the *Systems Engineering Handbook* (Walden et al. 2015). The terms "system engineering" and "systems engineering" are used interchangeably in this text depending on the context of usage.

The INCOSE definition of SE is as follows:

> Systems engineering (SE) is an interdisciplinary approach and means to enable the realization of successful systems. It focuses on defining

> customer needs and required functionality early in the development cycle, documenting requirements, and then proceeding with design synthesis and system validation while considering the complete problem: operations, cost and schedule, performance, training and support, test, manufacturing, and disposal. SE considers both the business and the technical needs of all customers with the goal of providing a quality product that meets the user needs.
>
> (Walden et al. 2015)

The NASA definition of SE is as follows:

> Systems engineering is a methodical, disciplined approach for the design, realization, technical management, operations, and retirement of a system. A "system" is a construct or collection of different elements that together produce results not obtainable by the elements alone.
>
> (NASA 2016)

The term "interdisciplinary" as used in the INCOSE definition indicates that SE teams usually include engineers from various engineering disciplines and engineering specialties, including software engineering. The NASA term "technical management" indicates that systems engineers, in addition to engaging in technical activities, also plan and coordinate the work activities of the interdisciplinary team, or teams, engaged in a systems project.

1.4.1 The Systems Engineering Profession

A profession is an occupation that requires specialized knowledge and skill that are usually obtained by advanced education and work experience. Many professions share common attributes, including postsecondary education, peer-reviewed publications, a code of ethics, and a not-for-profit organization that represents the interests of the profession's practitioners. In some cases, individuals must be licensed to perform designated activities in their profession.

A fundamental aspect of the engineering professions, as opposed to science, is practical application of scientific theories to find acceptable solutions to problems that are constrained by time, effort, cost, available technology, quality attributes, and cultural norms. Engineers apply mathematics, science, economics, and management skills to development of systems that make human life easier, safer, and/or more secure. Engineering professions are based on education, training, and experience. An engineer typically has a college degree in an engineering discipline (e.g. biomedical, chemical, civil, electrical, mechanical, or software engineering). Systems engineers typically have an engineering college degree and work experience in an engineering discipline.

Figure 1.10 illustrates the primary elements of an engineering profession; they include professional practice, self-governance, preparatory education,

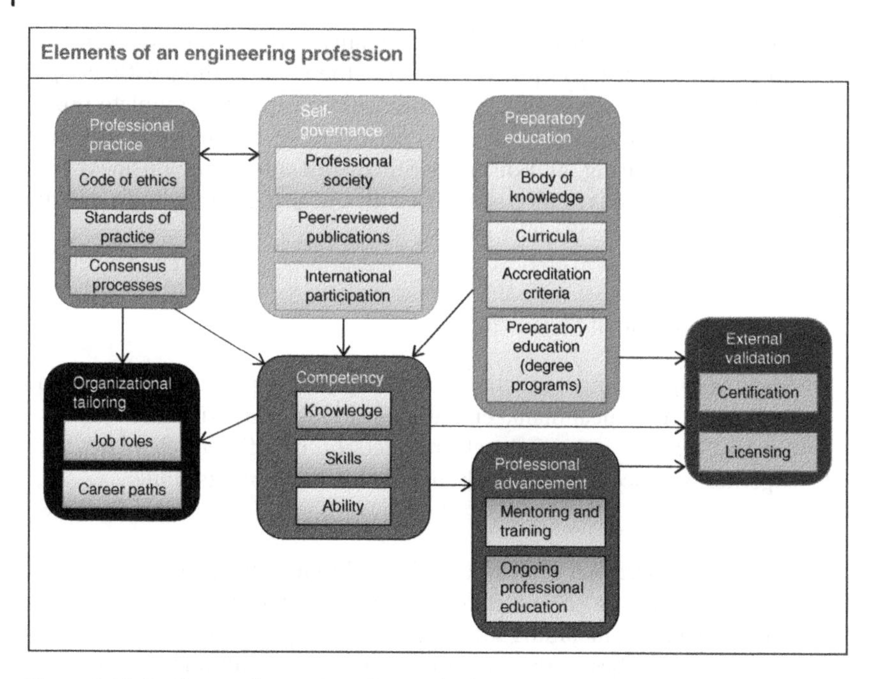

Figure 1.10 Attributes of an engineering profession.

organizational tailoring, competency, professional advancement, and external validation. The links between elements are bidirectional relationships that indicate mutual influence. Elements not directly connected in Figure 1.10 may be influenced by transitive relationships with other elements.

Professional practice in an engineering discipline involves having consensus processes that are generally practiced by engineers in that discipline; consensus processes are documented in standards, guidelines, and engineering handbooks that are developed and approved by practitioners of the discipline. A code of ethics provides guidance for ethical business practices, ethical conduct, and ethical professional practice.

Self-governance for an engineering discipline typically includes a professional society that sponsors development of standards, guidelines, handbooks, codes of ethics, and publications such as newsletters and magazines that provide practical information; peer-reviewed publications including archival journals that publish advances in research; and services such as tutorials, workshops, and conferences.

Postsecondary education for an engineering discipline is based on a body of knowledge, which includes the totality of textbooks, journals, conference proceedings, and all other sources of publicly available information relevant to the discipline – a guide to the body of knowledge is sometimes prepared.

International standards may be developed for the processes of an engineering discipline – the International Standards Organization (ISO) and the Institute of Electrical and Electronic Engineers (IEEE) are the primary standards organizations for SE and software engineering standards. Bodies of knowledge provide the basis for developing model curricula, degree programs, and accreditation criteria for academic programs.

Professional practice, governance, and postsecondary education provide the basis for developing competency, which involves having the knowledge, skills, and ability to perform assigned tasks at various levels of competency. Organizational adaptation involves mapping competency and professional practice into job roles and career paths within individual organizations; job roles vary among organizations depending on an organization's goals, structure, mission, and vision.

Professional advancement involves improving the competencies of individuals in selected skill areas by providing mentoring, training, apprenticeship, and advanced education. Postsecondary education, competency, and professional advancement can prepare an engineering professional for certification and licensing.

Intellectual foundations provide the basis for engineering professions, which are applied disciplines based on fundamental principles of science, mathematics, economics, and management; different engineering disciplines are based on different foundations: electrical engineering is based on the science of electromagnetism, mechanical engineering is based on physics and material science, and chemical engineering is based on chemistry. The intellectual foundations of software engineering are discrete mathematics and computer science. The intellectual foundations for SE are systems science and systems thinking. Table 1.1 is the instantiation of Figure 1.10 for the SE profession.

Not all attributes in Figure 1.10 are covered in Table 1.1. For example, "Organizational Tailoring" is not included in Table 1.1 because tailoring is unique to each organization.

1.5 Applications of Systems Engineering

Part 4 of the *Guide to the Systems Engineering Body of Knowledge* (Applications of Systems Engineering) indicates that a system may be a product and/or a set of processes that (SEBoK 2017):

- Satisfies business or mission requirements;
- Provides a service;
- Enhances an enterprise; or
- Composes an SoS.

Table 1.1 Attributes of the systems engineering profession.

Attributes	Systems engineering
Professional society	INCOSE, IISE, IEEE SMC
Code of ethics	NSPE Code of Ethics for Engineers (NSPE)
	INCOSE Code of Ethics (INCOSE1)
Guide to professional practice	*INCOSE Systems Engineering Handbook* (INCOSE2)
Peer-reviewed publications	Systems Engineering (INCOSE3)
	IEEE Systems Journal (IEEE1)
	IEEE SMC (SMC)
Consensus processes	ISO/IEC/IEEE Standard 15288:2015 (ISO 2015)
	IEEE Standard 15288.1-2014 (IEEE2)
Codified body of knowledge	*Guide to the System Engineering Body of Knowledge* (SEBoK)
Model curricula	Graduate Reference Curriculum for Systems Engineering (GRCSE)
Program accreditation	ABET Engineering Accreditation (ABET)
Competency framework	Systems Engineering Competency Framework (COMP)
Certification and licensure	INCOSE SEP Certification (SEP)
	Professional Engineer License (NCEES)

ABET, Incorporated as the Accreditation Board for Engineering and Technology, Inc.; IEEE, Institute of Electrical and Electronics Engineers; IISE, Institute of Industrial and Systems Engineers; INCOSE, International Council on Systems Engineering; NCEES, National Council of Examiners for Engineering and Surveying®; SEP, INCOSE Systems Engineering Professional; SMC, Systems, Man, and Cybernetics.
ABET, http://www.abet.org/accreditation; COMP, https://www.incose.org/products-and-publications/competency-framework; GRCSE, https://www.bkcase.org/grcse; IEEE, https://www.ieee.org; IEEE1, http://ieeexplore.ieee.org/xpl/aboutJournal.jsp?punumber=4267003; IEEE2: 15288.1-2014, https://standards.ieee.org/findstds/standard/15288.1-2014.html; IEEE SMC, http://www.ieeesmc.org; IISE, http://www.iise.org/Home; INCOSE, https://www.incose.org; INCOSE1, https://www.incose.org/about/leadershiporganization/codeofethics; INCOSE2, https://www.incose.org/ProductsPublications/PapersProceedings/sehandbook; INCOSE3, http://onlinelibrary.wiley.com/journal/10.1002/(ISSN)1520-6858; ISO 2015, https://www.iso.org/standard/63711.html; NCEES, https://ncees.org; NSPE, https://www.nspe.org/resources/ethics/code-ethics; SMC, http://www.ieeesmc.org; SEBoK, www.sebokwiki.org; SEP, https://www.incose.org/docs/default-source/certification/20151008-incose-sep-overview.pdf?sfvrsn=4; IEEE2, https://standards.ieee.org/findstds/standard/15288.1-2014.html; ISO, https://www.iso.org/standard/63711.html; SEBoK, http://sebokwiki.org/wiki/Guide_to_the_Systems_Engineering_Body_of_Knowledge_(SEBoK).

System engineering can thus be characterized as a discipline that develops products, provisions services, enhances development and operation of enterprises, and composes SoSs.

According to SEBoK, the Part 4 applications of SE will be expanded over time to include a series of domain application knowledge areas. The healthcare knowledge area in Part 4 is the first of these domain-specific extensions of SEBoK at the time of publication of this text.

1.5.1 Systems Engineering of Products

Systems engineering of products is concerned with planning and conducting work processes (analysis, design, implementation, verification, transition, validation) for the technical aspects of projects that may require multiple engineering disciplines and specialty disciplines, such as constructing, operating, and maintaining a hydroelectric dam (civil, mechanical, electrical, hydrological, ecological, and software engineering). Specialty disciplines such as safety, security, reliability, availability, ruggedness, and maintainability may also be applied.

1.5.2 Systems Engineering of Service Provision

Provisioning of services applies the processes and methods of SE needed to satisfy stakeholders' expectations, users' needs, and support for system operators and maintainers. For example, one way to develop and support the software-enabled elements of a health care facility is to focus on delivery of the services to be provided to patients. In a similar manner, the safety and convenience features to be provided by a software-enabled automobile can be developed by focusing on the service needs and desires of automobile drivers and passengers.

1.5.3 Systems Engineering for Enterprises

According to SEBoK Part 4, systems engineering "involves application of SE principles, concepts, and methods to the planning, design, improvement, and operation of an enterprise." Enterprises usually include multiple organizations, multiple interacting SESs, and multiple stakeholders; examples include health care enterprises, automobile manufacturing enterprises, and enterprise information systems (EISs).

1.5.4 Systems Engineering for Systems of Systems

A discussion of SE for SoSs was provided in Section 1.3.6. An EIS is an SoS that provides support for the member organizations and collects data from them that are used to make business decisions, both strategic and tactical.

An example of an EIS is provided in Figure 1.11, which depicts an EIS as a collection of IT systems used to support the various semi-autonomous organizations that form the enterprise.

Each organization may have a different computing infrastructure to support the local business needs of that organization (i.e. different computing hardware

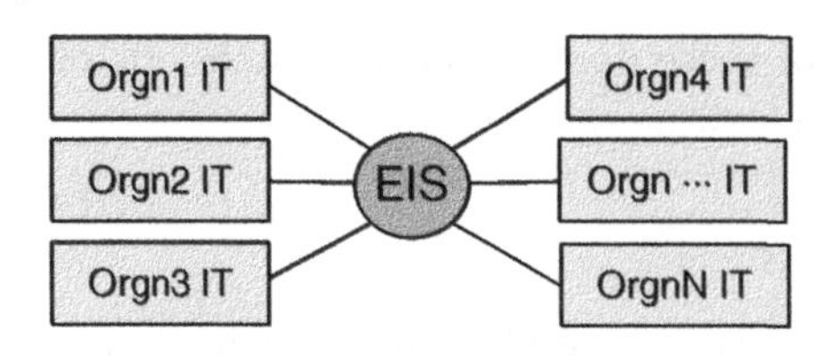

Figure 1.11 An enterprise information system (EIS) in context.

and a tailored operating system plus software applications). The EIS must provide software interfaces to connect each organization's IT unit to the EIS. The SoS issues presented in Section 1.3.6 may have to be addressed for each of the semi-autonomous IT systems, for example handling of behaviors that emerge from the interactions of the IT systems and masking unwanted EIS behaviors while maintaining the integrity of the local IT systems.

1.6 Specialty Engineering

Specialty engineering includes engineering disciplines that are not normally central to development or sustainment of a system but may be needed to address specific concerns that require more than normal emphasis. According to SEBoK Part 6,

> Specialty engineering requirements are often manifest as constraints on the overall system design space. The role of system engineering is to balance these constraints with other functionality in order to harmonize total system performance.
>
> (SEBoK 2017)

Specialty disciplines for SE included in Part 6 of SEBoK are the following:

- Reliability, availability, and maintainability;
- Human systems integration;
- Safety engineering;
- Security engineering;
- System assurance;
- Electromagnetic interference/electromagnetic compatibility
- Manufacturing and producibility;
- Affordability; and
- Environmental engineering.

Specialty engineers participate as members of SE teams to increase confidence that overall system performance will satisfy the appropriate specialty requirements; their involvement may be continuous, periodic, or as needed.

They contribute their expertise by advising on issues of design and implementation that will satisfy requirements in a technically satisfactory and cost-effective manner (e.g. safety, security, reliability). Specialty engineering requirements, plans, and resources are, or should be, included in the systems engineering management plan (SEMP) by direct incorporation or by reference. The form and content of an SEMP is presented in Chapter 10.

1.7 Related Disciplines

Industrial engineering (IE) and project management are two disciplines that are closely related to SE, in addition to software engineering.

1.7.1 Industrial Engineering

According to the definition provided by the Institute of Industrial and Systems Engineers, IE is

> concerned with the design, improvement and installation of integrated systems of people, materials, information, equipment and energy. It draws upon specialized knowledge and skill in the mathematical, physical, and social sciences together with the principles and methods of engineering analysis and design, to specify, predict, and evaluate the results to be obtained from such systems.
>
> (IISE 2018)

According to Part 6 of SEBoK IE encompasses several aspects of SE (i.e. production planning and analysis, continuous process improvement) and also many elements of the engineered systems domain (production control, supply chain management, operations planning and preparation, operations management, etc.). Descriptions of the following elements of the IE body of knowledge are provided in Part 6 of SEBoK (2017):

- Operations engineering;
- Operations research;
- Production engineering/work design;
- Facilities engineering and engineering management;
- Ergonomics;
- Engineering economic analysis;
- Quality and reliability;
- Engineering management; and
- Supply chain management.

1.7.2 Project Management

Systems engineering and project management are closely related disciplines. Project managers are responsible for delivering acceptable systems within the constraints of the schedule, budget, and resources. System engineers are responsible for developing the acceptable systems within the constraints imposed by the system requirements, technology, and the resources and schedule allocated by project managers and perhaps directly by customers.

Project managers prepare and administer project management plans (PMPs) that address the above constraints (Fairley 2009). System engineers prepare and administer SEMPs to coordinate the technical activities of developing a system (see Chapter 10).

On small projects, the roles of project manager and systems engineer may be played by a single individual or by two individuals, each assigned on a part-time basis. For larger projects, the workload of duties, activities, and needed skills for the project management and the SE require different individuals; each may head a team: one for project management and one for SE.

The project manager or the systems engineer may be the designated project leader with the other playing a subordinate role. In either case, it is essential that a healthy working relationship, clearly defined responsibilities, and shared decision-making are established and practiced.

The relationships and responsibilities of the project manager and of the systems engineer depend on many factors including the nature of the product, service, or enterprise to be provisioned; the customers and other stakeholders; the organizational structure of the parent organization; corporate and contractual policies and procedures; and relationships with affiliate contractors and subcontractors, if any. The relationships between SE and project management are further explored in Chapter 12 of this text.

Systems engineering and software engineering are closely related disciplines as evidenced by the close relationships between the ISO standards for SE processes and software engineering processes (ISO 2015, 2017). Systems engineering and software engineering have been described as being "intimately intertwined" (Boehm 1994).

1.8 Software Engineering

The term "software engineering" was coined as the descriptor for NATO-sponsored conferences held in Munich, Germany in 1968 and in Rome, Italy in 1969 (Randell et al. 2001). These first software engineering conferences were held in recognition of the increased size and complexity of software that occurred in the 1960s. Development of transistorized computers in the late 1950s and early 1960s resulted in more powerful and more reliable computers. Software became larger and more complex to support the increasing applications of the more powerful computers. Those who convened the first conferences saw the need for an engineering approach to software development.

The *Guide to the Software Engineering Body of Knowledge* defines software engineering as follows:

> Application of a systematic, disciplined, quantifiable approach to the development, operation, and maintenance of software; that is, the application of engineering to software.
>
> (Bourque and Fairley 2014)

Continuing advances in the technology of computing hardware (decreased cost, smaller size, larger memory, increased computing power, better reliability) has resulted in increasing application of digital hardware and software to measurement and control of physical systems, i.e. to development of SESs.

1.8.1 The Software Engineering Profession

Elements of the software engineering profession, based on Figure 1.3, are illustrated in Table 1.2. A URL link for each of the software engineering attributes in Table 1.2 is provided in the references at the end of this chapter.

The SE contributions that can be made by competent SwSEs is sometimes missing because software engineers are often treated as disciplinary engineers who apply their skills to develop system elements that satisfy requirements allocated by PhSEs, in the same way that the mechanical, electrical, and other disciplinary engineers accomplish similar tasks. However, some mechanical engineers and other physical engineers acquire the skills to become PhSEs (i.e. systems engineers whose expertise is in engineering of physical systems). In the same way, some software engineers have acquired the skills to become SwSEs (i.e. systems engineers whose expertise is in SE of software systems). Allowing qualified SwSEs to participate as peers of PhSEs in a coordinated and cooperative manner will help to bridge the gap between these “intimately intertwined” disciplines.

1.9 Applications of Software Engineering

Software engineers develop, modify, and maintain software that is used in different ways for different purposes. The following categories indicate some of the ways software is used; however, there are no standardized definitions for these categories.

Table 1.2 Attributes of the software engineering profession.

Attributes	Software engineering
Professional society	IEEE Computer Society (IEEE CS)
Code of ethics and professional practice	Software Engineering Code of Ethics (SwCE)
Peer-reviewed publications	Transaction on Software Engineering (TSE)
	International Conference on Software Engineering (ICSE)
Consensus processes	ISO/IEC/IEEE Standard 12207 (ISO 2017)
	IEEE Standards based on 12207 (IEEE SA)
Body of knowledge	*Guide to the Software Engineering Body of knowledge* (Bourque and Fairley 2014)
Model curricula	Software Engineering 2014 (SwE14)
	Graduate Software Engineering (GswE 2009)
Accredited education programs	ABET Accredited Software Engineering Programs (ABET)
Competency model	Software Engineering Competency Model (SWECOM)
Certification and licensure	IEEE Computer Society Certification and Credential Program (SWCERT)
	Professional Engineer License (NCEES)

ABET 2017, http://www.abet.org/accreditation/accreditation-criteria/criteria-for-accrediting-engineering-programs-2016-2017; GswE 2009, https://www.acm.org/binaries/content/assets/education/gsew2009.pdf; ICSE, http://www.icse-conferences.org; IEEE CS, https://www.computer.org; IEEE SA, https://standards.ieee.org/findstds/standard/software_and_systems_engineering.html; ISO 2017, https://www.iso.org/standard/63712.html; NCEES, https://ncees.org; SwCE, https://ethics.acm.org/code-of-ethics/software-engineering-code; SwCERT, https://www.computer.org/web/education/certifications; SwE14, https://www.acm.org/binaries/content/assets/education/se2014.pdf; SWEBOK 2014, https://www.computer.org/web/swebok/v3-guide; SWECOM, https://www.computer.org/web/peb/swecom; TSE, https://www.computer.org/web/tse.

1.9.1 Application Packages

Software application packages are software products that are developed to facilitate users' interactions with computers and multimedia devices. They include the software that provides Internet browsers, games, spreadsheets, word processors, smartphone apps, business software, project management packages, and software tools used by engineers to support analysis and design of the systems they develop.

1.9.2 System Software and Software Utilities

A computer operating system is the primary element of system software; it provides the interfaces between application packages and the hardware.

Operating systems also manage the scheduling of concurrently executing software programs and the memory used by them in a computing device. Software utilities are low-level adjuncts to an operating system; they include screen savers, encryption/decryption software, file managers, network utilities that support Internet browsers, and many other low-level functions. In some domains, such as business data processing, a database management system is an important element of system software. Virtual machine software is an element of system software that allows multiple operating systems to run concurrently on a computing device. Virtualization is the basis for cloud computing.

1.9.3 Software Tools

Software tools are software that are used to develop other software; they include software construction tools such as text editors, compilers, and debuggers. Other software tools support analysis, design, integration, verification, validation, and configuration control of software. A software engineering environment is an integrated set of software tools that support activities across the various phases of a software development life cycle. Individual software tools and software engineering environments are used to develop and maintain all kinds of software.

1.9.4 Software-Intensive Systems

A software-intensive system is a collection of integrated software application packages or special-built software that are developed on a stable, preexisting platform of computing hardware and system software. IT systems that support business operations and software engineering environments used to develop other software are software-intensive systems.

1.9.5 Software-Enabled Systems

The software in an SES provides interconnections among the physical elements, coordinates their interactions, provides connections to entities in the system environment, and provides some or most of a system's functionality, behavior, and quality attributes. The intertwining of software engineering with SE and other engineering disciplines makes development of software for complex SESs the most challenging form of software engineering.

1.10 Physical Systems Engineers and Software Systems Engineers

Most systems engineers have college degrees and work experience in a traditional engineering discipline (e.g. aerospace, mechanical, electrical, civil, chemical, industrial engineering). Some by virtue of interest, opportunity, or corporate necessity become systems engineers. Some are registered professional engineers (see Section 1.4.1). We refer to these systems engineers as PhSEs because their expertise is in developing physical systems.

Traditionally, software engineers have been regarded as disciplinary engineers who develop the software elements of systems that will satisfy the requirements allocated to them by systems engineers. But qualified software engineers need to be recognized as the peers of PhSEs because of the increasing size and complexity of software-enabled physical systems and the increasing reliance on software in those systems. These qualified software engineers are termed "software systems engineers" in this text (see Section 1.10).

In a similar manner, most software engineers have college degrees and work experience in a computing discipline (computer science, IT, software engineering). As in the case of PhSEs, some software engineers have the skill, ability, experience, and interest to become systems engineers. We refer to these systems engineers as SwSEs because their expertise is in developing software systems.

PhSEs and SwSEs engage in the following activities:

- Eliciting stakeholders' needs, wants, and desires;
- Defining stakeholders' requirements, system capabilities, and system requirements;
- Planning for system development and life cycle sustainment;
- Developing the system architecture definition and system design definition;
- Establishing interfaces to the environments in which systems will operate;
- Organizing, coordinating, and leading interdisciplinary engineering teams;
- Facilitating system realization;
- Measuring progress and take corrective actions as needed; and
- Facilitating verification, transition, validation, and acceptance of systems.

Both kinds of systems engineers (PhSEs and SwSEs) apply the methods, tools, and techniques of SE within the realms of their expertise – physical or software systems. But, unfortunately, most PhSEs and SwSEs do not have the expertise to apply SE methods to development of the opposite kinds of systems – physical or software systems. Development of complex software-enabled physical systems can be most effectively accomplished when one or more PhSEs and one or more SwSEs work together as peers in a collaborative and cooperative manner.

Some systems engineers may also continue to apply their disciplinary engineering skills to projects (e.g. as mechanical or software engineers), but when they do so, they are playing different roles than PhSE or SwSE.

A major theme of this text is the ways PhSEs and SwSEs can work together in a synergistic manner to bridge the gap that often exists between the two kinds of systems engineers to the benefit of their projects, the project stakeholders, and their personal job satisfaction.

Throughout this text, the term "systems engineer" is used to denote both PhSEs and SwSEs. The distinction between PhSEs and SwSEs is made when it is appropriate.

1.11 Key Points

- A system is a collection of interacting elements that exist within and interact with an environment.
- Systems can be categorized in various ways, including natural or engineered, open or closed, and static or dynamic.
- Engineered systems are purposefully built to satisfy business needs and support missions; they are open and dynamic and may include natural elements.
- SESs are dynamic engineered systems in which software provides significant amounts of functionality, behavior, and quality attributes, coordinates and controls physical elements, and provides interfaces to one or more external environments.
- Large, complex SESs are found in most domains of modern society, including but not limited to aerospace, agriculture, communication, consumer products, defense, ecology, energy, health care, intelligent buildings, manufacturing, and transportation.
- Systems engineering and software engineering are professional engineering disciplines; they are closely related (i.e. intertwined) disciplines, particularly for development of SESs.
- Some systems engineers are PhSEs and some are SwSEs.
- Complex SESs are most effectively developed when PhSEs and SwSEs work together as peers in a collaborative and cooperative manner.
- Specialty engineering disciplines support the mainstream processes and activities of SE; they include safety, security, reliability, availability, robustness, and ergonomics engineering.
- Disciplines closely related to SE include project management, IE, and software engineering.

Part I of this text continues with Chapter 2, where attributes of the professional disciplines of SE and software engineering are presented. Issues that can inhibit and facilitate successful execution of SES development are presented in Chapter 3.

Exercises

1.1. Systems engineering has evolved from the 1940s to the present time and continues to evolve. Find and briefly describe a system that was developed using systems engineering processes and methods in each of the following decades. Briefly describe your four chosen systems and how systems process and methods were used to develop each of them.
(a) The 1960s
(b) The 1980s
(c) The 2000s
(d) The 2010s

1.2. The technical activities of systems engineering projects are constrained by requirements, schedule, resources, infrastructure, and technology. Provide an example of each kind of constraint when applied to a systems project of interest to you.
(a) Briefly describe the system.
(b) Provide an example of each kind of constraint for system development.

1.3. A system of systems is an integrated set of systems that each operates independently and when integrated provides capabilities that exceed the capabilities of the systems operating independently.
(a) Find and briefly describe an example of a system of systems.
(b) Briefly describe each of the constituent systems in your chosen system of systems.
(c) Describe some capabilities of your chosen system of systems that cannot be provided by the constituent systems when they are operating independently.

1.4. Software-enabled systems include physical elements and software elements.
(a) Briefly describe a software-enabled system of interest to you.
(b) List and briefly describe three software elements and three physical elements for your chosen system.
(c) Briefly describe the ways in which the software and physical elements interact.

1.5. Some software-enabled systems include natural elements, engineered physical elements (e.g. hardware), and software elements.
(a) Briefly describe an example of a software-enabled system that includes natural and engineered physical elements.

(b) Describe some of the interfaces between the natural and engineered physical elements.
(c) Describe the roles played by software in the system.

1.6. Specialty engineering disciplines include those that address safety, security, reliability, availability, maintainability, and other quality attributes. For each of the following attributes, briefly describe a system for which a specialty engineer would likely be involved in system development.
(a) Safety engineering
(b) Security engineering
(c) Reliability engineering
(d) Availability engineering
(e) Maintainability engineering

1.7. Section 1.9 describes different kinds of software: application packages, system software and software utilities, software tools, software-intensive systems, and software-enabled systems.
(a) Provide a specific example of each of the five kinds of software.
(b) Briefly explain the difference between a software-intensive system and a software-enabled system.

1.8. Software-enabled systems include systems that are variously named: software-intensive, cyber-physical, embedded real-time, and the Internet of Things. Briefly describe the characteristics of each of these four kinds of systems that distinguish each of them from the other three.
(a) Software-intensive
(b) Cyber-physical
(c) Embedded real-time
(d) Internet of Things

1.9. The NASA systems engineering competency model includes 10 areas of systems engineering competency with four levels of proficiency for each. The systems engineering competency model can be found at https://www.nasa.gov/pdf/303747main_Systems_Engineering_Competencies.pdf.
(a) Scan through the model to gain an understanding of systems engineering competencies.
(b) Briefly describe the four levels of competency.
(c) Choose three of the competencies and briefly describe each of them.

1.10. The software engineering competency model includes 15 areas of software engineering competency with five levels of proficiency for each. Choose three of the competencies and briefly describe the competency and the nature of the levels of proficiency. The software engineering competency model can be found at https://www.computer.org/web/peb/swecom.

(a) Scan through the model to gain an understanding of software engineering competencies.
(b) Briefly describe the five levels of competency.
(c) Choose three of the competencies and briefly describe each of them.

References

Bilstein, R. (1996). Stages to Saturn: a technological history of the Apollo/Saturn launch vehicles. National Aeronautics and Space Administration, NASA History Office, Washington, D.C. https://history.nasa.gov/SP-4206/sp4206.htm (accessed 6 September 2018).

Boehm, B. (1994). Integrating software engineering and systems engineering. *The Journal of NCOSE* 1 (1): 147–151.

Bourque, P. and Fairley, R. (eds.) (2014). *Guide to the Software Engineering Body of Knowledge V3.0.* IEEE (Institute of Electrical and Electronic Engineers) https://www.computer.org/web/swebok/v3-guide (accessed 6 September 2018).

Burke, R., Mussomeli, A., Laaper, S. et al. (2017). *The Smart Factory.* Deloitte Insights https://www2.deloitte.com/insights/us/en/focus/industry-4-0/smart-factory-connected-manufacturing.html (accessed 6 September 2018).

Fairley, R. (2009). *Managing and Leading Software Projects.* Wiley.

IISE (2018). Institute of Industrial and Systems Engineers. http://www.iise.org/Home/ (accessed 6 September 2018).

ISO (2015). ISO/IEC/IEEE Standard 15288:2015 – Systems and software engineering – System life cycle processes. https://www.iso.org/standard/63711.html (accessed 20 August 2018).

ISO (2017). ISO/IEC/IEEE Standard 12207:2017 – Systems and software engineering – Software life cycle processes. https://www.iso.org/standard/63712.html (accessed 20 August 2018).

JPSS (2016). Joint Polar Satellite System (JPSS) Level 1 Requirements Document – Final Version: 2.0, JPSS-REQ-1001, 3 March 2016. https://www.jpss.noaa.gov/assets/pdfs/technical_documents/level_1_requirements.pdf (accessed 29 January 2019).

Kennedy, J. F. (1961a). Special Message to Congress on Urgent National Needs, 25 May 1961. https://www.nasa.gov/vision/space/features/jfk_speech_text.html#.VwU3qhIrLBI (written text accessed 6 September 2018).

Kennedy, J. F. (1961b) Special Message to Congress on Urgent National Needs, 25 May 1961. http://www.jfklibrary.org/Asset-Viewer/Archives/JFKWHA-032.aspx (audio recording accessed 6 September 2018).

Lee, J. (2015). Smart factory systems. *Informatik-Spektrum* 38 (3): 230–235. https://www.researchgate.net/publication/276145281_Smart_Factory_Systems (accessed 6 September 2018).

Liu, D. (2015). *Systems Engineering: Design Principles and Models,* section 1.4.2. CRC Press, Taylor and Francis.

NASA (2016). *NASA Systems Engineering Handbook, Revision 2,* 21. NASA, NASA/SP-2016-6105 Rev 2, 2016. https://www.nasa.gov/connect/ebooks/nasa-systems-engineering-handbook (accessed 29 January 2019).

Oborski, P. (2016). Integrated monitoring system of production processes. *Management and Production Engineering Review* 7 (4): 86–96. http://mper.org/mper/images/archiwum/2016/nr4/9-oborski.pdf (accessed 6 September 2018).

Naur, P. and Randell, B. (2001). *The NATO Software Engineering Conferences.* Brussels (1969) 231 pp. and (1970) 164 pp. http://homepages.cs.ncl.ac.uk/brian.randell/NATO/: Scientific Affairs Division, NATO (accessed 22 August 2018).

SEBoK (2017). Guide to the Systems Engineering Body of Knowledge V1.9. www.sebokwiki.org (accessed 6 September 2018).

SEVOCAB (2017). Software and systems engineering vocabulary, IEEE Computer Society and ISO/IEC JTC 1/SC7. https://pascal.computer.org/sev_display/index.action (accessed 6 September 2018).

SPO (2016). *ICBM System Program Office (SPO).* GlobalSecurity.org http://www.globalsecurity.org/wmd/agency/icbm-spo.htm (accessed 6 September 2018).

Walden, D., Roedler, G., Forsberg, K. et al. (2015). *System Engineering Handbook: A Guide for System Life Cycle Processes and Activities,* 4e, INCOSE-TP-2003-002-04 2015. International Council on Systems Engineering (INCOSE).

Wolfle, D. (1968). The administration of NASA. *Science* 162, Number 3855. http://science.sciencemag.org/content/162/3855/753: (accessed on 6 September 2018).

2

Systems Engineering and Software Engineering

2.1 Introduction

This text is about bridging the gap between systems engineering and software engineering and increasing the synergy when physical systems engineers (PhSEs) and software systems engineers (SwSEs) work together to develop complex software-enabled systems that include physical elements (natural and/or engineered) plus digital devices and the associated software. In this text, synergy is the result of interactions between PhSEs and SwSEs that produce more and better outcomes than would result from the sum of the outcomes if the two kinds of systems engineers were working with less-effective interactions, which, unfortunately, is often the case.

This chapter provides the opportunity for readers to review or to learn about the following:

- Categories of systems;
- Common attributes of PhSEs and SwSEs;
- Ten things PhSEs need to know about software and software engineering; and
- Ten things SwSEs need to know about physical systems and physical systems engineering.

A summary of key points concludes the chapter. References and exercises are included for further study.

Chapter 3 continues the themes of this chapter by presenting issues that inhibit better working relationships among PhSEs and SwSEs; opportunities for improving working relationships are presented.

2.2 Categories of Systems

In their paper, “Exploring the Relationship Between Systems Engineering and Software Engineering,” Pyster et al. (2015b) distinguish three kinds of

Systems Engineering of Software-Enabled Systems, First Edition. Richard E. Fairley.
© 2019 John Wiley & Sons, Inc. Published 2019 by John Wiley & Sons, Inc.

Table 2.1 Three categories of systems.

Category	Characteristics	Examples	Roles played
Older physical systems	Relatively "dumb"	Bridges, buildings, vehicles, roads	Predominantly systems engineers
Computational systems	Computational algorithms, behaviors, and software representations	Information systems, operating systems, middleware	Predominantly software engineers
Cyber-physical systems	Complex configurations of physical and computational elements	Self-driving vehicles, Internet of things, robotic manufacturing systems	Complex interactions of systems engineers and software engineers

systems: physical, computational, and cyber-physical; they also present the corresponding roles played by systems engineers and software engineers for each kind of system. Table 2.1 provides characteristics and examples of the three categories of systems and the roles played by systems engineers and software engineers for each category.

As indicated, software engineers have had little or no involvement in development and sustainment of "dumb" physical systems, although newer version of previously dumb systems are including more software, for example, instrumented smart building and vehicles.

The role of PhSEs is minimal for computational systems. One or more SwSEs typically play the role of systems engineer for computational systems because computing hardware and other system infrastructure elements are available or can be procured as needed; i.e. these systems are "software-intensive" (see Section 1.9.4).

Cyber-physical systems include physical and computational elements that are combined in complex configurations that require complex interactions of PhSEs and SwSEs.

In this text, the term "software-enabled" is used rather than "cyber-physical" to describe the third category of systems. Cyber-physical is typically used to mean "Internet-connected" while other systems that are described as software-intensive, embedded, or real-time may have no Internet connectivity. These systems are also software-enabled (see Section 1.9). PhSEs and SwSEs play, or should play, complementary and cooperative roles when developing and modifying software-enabled systems.

Over time, PhSEs have adapted many of the methods and techniques developed by SwSEs and, conversely, SwSEs have adapted many of the methods

Table 2.2 Mutual adaptation of methods by PhSEs and SwSEs.

PhSE methods adapted by SwSEs	SwSE methods adapted by PhSEs
Stakeholder analysis	Model-driven development
Requirements engineering	UML
Functional decomposition	Use cases
Design constraints	Object-oriented design
Architectural design	Iterative development
Design criteria	Agile methods
Design tradeoffs	Continuous integration
Interface specification	Incremental V&V
Traceability	Process modeling
Configuration management	Process improvement
Systematic verification and validation	

and techniques developed by PhSEs. Table 2.2 itemizes some of the methods developed by PhSEs adapted for use by SwSEs and, conversely, methods developed by SwSEs adapted for use by PhSEs.

Each discipline adapts the methods developed by the other discipline to accommodate the needs of that discipline, the domains of application, and the current project.

2.3 Common Attributes of PhSEs and SwSEs

PhSEs and SwSEs share many common attributes; there are also many differences, and there are many things they typically don't know about the work of their counterparts.

The following indicates some of the common attributes shared by PhSEs and SwSEs. Some of these attributes are from the cited paper by Pyster et al. Differences are addressed subsequently.

Some common attributes of PhSEs, SwSEs, and their professional disciplines are the following:

1. Most who will become PhSEs or SwSEs receive "silo" educations in an engineering discipline or a computing discipline that includes minimal exposure to the counterpart discipline.
2. Most PhSEs and SwSEs acquire their system-level skills through on-the-job training, mentoring, and work experiences.
3. Most who become competent PhSEs and SwSEs, by whatever means, receive little or no on-the-job exposure to the counterpart discipline.

4. Some PhSEs and SwSEs specialize in applying systems-level technical skills and some specialize in applying systems-level engineering management skills.
5. PhSEs and SwSEs may, at times, perform work activities in their engineering disciplines (e.g. mechanical design or software development).
6. The tools used by PhSEs and SwSEs lag in maturity and capability when compared with the tools used by other traditional engineers, such as mechanical engineers and electrical engineers.
7. Successful PhSEs and SwSEs often rely on interpersonal communication and leadership skills to influence others rather than attempting to use positional power.
8. PhSEs and SwSEs work, or should work, closely with project managers and other relevant personnel.
9. Communication and coordination between PhSEs/SwSEs and project managers/others is often inhibited because neither group understands the incentives and concerns of their counterparts.
10. Project managers and others often emphasize cost and schedule metrics over measures of technical progress, to the frustration of PhSEs and SwSEs.

2.4 Ten Things PhSEs Need to Know About Software and Software Engineering

To better understand software engineering, PhSEs do not need to know how to write computer programs (i.e. the details of software construction) any more than they need to know how to fabricate a special purpose computer chip or design a power supply. But in order to effectively communicate and work with SwSEs, they do need to understand the nature of software, the resulting processes and procedures used by software engineers, and the constraints under which software engineers design and construct software.

The following 10 items are significant aspects of software and software engineering that PhSEs need to understand to improve the synergy between PhSEs and SwSEs.

These 10 items are based on a paper by Fairley and Willshire; additional information about the things systems engineers need to know about software and software engineering is provided by the topic "Key Points a Systems Engineer Needs to Know About Software Engineering" in the Part 6 knowledge area of the *Guide to the Systems Engineering Body of Knowledge*: "Systems Engineering and Software Engineering" (Fairley and Willshire 2011a; Adcock 2017).

More information about software engineering can be found in IEEE Standard 12207, the *Guide to the Software Engineering Body of Knowledge*, and the *Software Engineering Competency Model* (ISO 2017; Bourque and Fairley 2014; Fairley et al. 2014).

2.4.1 Systems Engineering and Software Engineering are Distinct Disciplines

Engineering disciplines are concerned with applying mathematics, scientific principles, economics, and management skills to development of systems that make life easier, more pleasurable, safer, and more secure. It might appear that software engineering is an application of systems engineering because both disciplines apply similar processes of analysis, design, realization, verification, transition, and validation to develop systems of interest. And systems engineers and software engineers, like all engineers, seek to provide effective solutions to technical problems within the constraints of time, effort, cost, available technology, necessary quality attributes, and cultural norms.

However, most PhSEs have traditional engineering educations and work experiences; hence, they apply their problem-solving skills (based on continuous mathematics and the physical sciences) to develop physical systems. In contrast, most software engineers and SwSEs apply their problem-solving skills (based on discrete mathematics and computer science) to develop computational systems.

While software engineering is a key discipline for implementing and modifying the software elements of physical systems, software engineers also develop other kinds of software, including application packages, software tools, and operating systems and utilities, as explained in Sections 1.9 and 2.2, so their skills are not limited to participating in development of complex software-enabled systems.

The correspondingly different approaches to problem solving have resulted in issues that inhibit effective synergy between PhSEs and SwSEs. Some of these issues and the opportunities for improving them are presented in Chapter 3.

The second thing PhSEs need to know about software and software engineering is the logical, as opposed to physical, nature of software.

2.4.2 Software Is a Logical Medium

The software elements of software-enabled systems are logical constructions expressed in algorithmic form, in contrast to the physical elements of systems that are realized in mechanical, electrical, chemical, biological, and other physical media. Software is said to be intangible because it has no physical properties and malleable because of the relative ease with which software code can be modified. Obtaining the desired effect by modify software code may not be easy to achieve but modifications of code, per se, are straightforward when compared with modifying physical elements.

A software program (i.e. source code) is a written representation of software but it is not the executable code because the source code is transformed in various ways by other software programs (i.e. software that transforms source code into executable machine code). One of the most significant developments in the history of technology occurred when Alan Turing, in the 1930s, developed the theory of computability and demonstrated that software machines (i.e. Turing machines) could be written to manipulate and transform other software (Turing 1937).

Machine code enables discrete voltage levels that activate digital hardware elements; the voltage levels are typically 0 and 5 V, which can be interpreted as logical "false" and "true" or as **B**inary dig**IT**s 0 and 1 (i.e. "bits"). The source code translators are also software programs that have been transformed into machine code (e.g. by software tools called compiler-compilers).

Humans write source code and, being humans, sometimes make mistakes. Because source code is written as logical statements, it is easy to make mistakes that may be hard to detect. For example, a human programmer might intend to write a logical decision in the source code as follows:

if $x < y$ then do A otherwise do B.

But the programmer might mistakenly type the "less than" symbol as "greater than":

if $x > y$ then do A otherwise do B.

Simple mistakes such as this one have resulted in destruction of systems and property and, in some cases, have resulted in loss of human life (Leveson and Turner 1993).

The software that translates source code into machine code can analyze the string of input symbols that constitute a software program and detect some kinds of programming mistakes. Typing "«" is easily detected and reported when "«" is not a valid construct in the programming language being used (by a software program called a syntax analyzer).

However, semantic errors such as typing ">" when "<" was intended are much more difficult, and sometimes impossible to detect until the corresponding machine code is executed during testing. But testing may not expose the programmer's mistake because the logical paths taken by the set of test cases may not follow a path that included the offending statement (see Section 2.4.3).

In general, the set of logical paths in a computer program are determined by nesting, interleaving, and recursion of three basic constructs:

- *Sequencing*: Logical statements are executed in sequential order;
- *Branching*: A sequence is altered based on a logical test (true, false); and
- *Repetition*: Repeated execution of nested and interleaved statements that can include sequences, branches, and nested repetitions.

The infeasibility of exhaustive testing is the third thing systems engineers need to know about software and software engineering.

2.4.3 Exhaustive Testing of Software is Not Feasible

The undesired effect of the ">" mistake cited above may, or may not, be easy to detect in a small program but it may go undetected until later for programs of even moderate size if manual reviews and the logical execution paths taken by the tests do not expose it. One of the reasons exhaustive testing is not possible is because of the very large number of logical paths that exist in even moderately sized software programs; it would take an inordinate amount of time to execute all of the combinations of logical paths, given the many ways the logical segments can be intermixed and traversed.

And some of the paths may be difficult to access during testing because they are traversed to execute exception conditions that are difficult to replicate during testing. A program having one hundred thousand lines of source code (a moderate size) with an average of 10 symbols per line will contain one million symbols; one incorrect symbol, as above, in one of the logical segments can cause system failure when it is encountered for the first time during system operation.

Another of the several reasons exhaustive testing of software is not feasible is because of the very large numbers of data values that would be required to exhaustively test even a single program variable for each data value.

Suppose, for example, that a single mathematical variable in a program is encoded in a 32-bit representation of 0s and 1s; there are thus 2^{32} encodings of the variable. Suppose, also, that running a full test of the program that contains the variable, for each representation of the variable (each combination of 0s and 1s) within the program where it is embedded, requires one second of computing time. The time to exhaustively test the program for all values of this variable is thus 2^{32} seconds, or approximately 10^9 seconds, which is approximately 10^7 hours (i.e. 10 million hours). This is, of course, a worst case because an encoding that exposes a programming mistake might be encountered early during testing, but without exhaustive testing, it is not known if another undetected error remains in the code that would be detected by another of the 10^{32} data values.

This single variable could be 1 of 100 or more variables in a moderately sized program. Exhaustive testing of each variable in combinations with the other variables would result in an astronomical number of tests and prohibitive testing time, even if each test of the combinations required only 1 nanosecond.

An attempt to reduce the number of tests to be executed can be made by partitioning the input space for the variables into equivalences classes so that one successful test in each class would provide assurance that all values of the variables within a partition would execute without error – but this is an inexact science.

The next four things systems engineers need to know about software and software engineering are presented in the book *The Mythical Man-Month*, written by Fred Brooks, in which he (among other things) described four essential properties of software that differentiate it from other kinds of engineering artifacts; the essential properties are complexity, conformity, invisibility, and changeability (Brooks 1995). Some information about the book is provided in "The Mythical Man-Month".

The Mythical Man-Month

Fred Brooks titled his book *The Mythical Man-Month* because people and time are not interchangeable for intellect-intensive work. A project requiring 60 engineer-months of effort could likely be completed by 6 engineers in 10 months (6 × 10) but it would probably be difficult for 10 engineers to complete the project in 6 months (10 × 6) and impossible for 60 engineers to complete it in 1 month. The latter schedules (6 months and 1 month) are difficult or impossible because of the sequential nature of the work activities; some things have to be partially or totally completed before subsequent activities can be started.

Also, software development requires closely coordinated work among team members. It has been observed that the productivity of individual software engineers decreases as the number of team members increases. Compressing the example schedule from 10 months to 6 months might require as many as 15 software engineers (instead of 10) to compensate for the loss of individual productivity caused by the increase in communication overhead among team members.

In general, the number of communication links among N team members is $N(N-1)/2$, which is the number of links in a fully connected graph; the number of communication links thus grows on the order of N^2. Ten team members working in close coordination would have 45 communication paths. Fifteen team members would have 105 communication paths.

2.4.4 Software Is Highly Complex

Fred Brooks observed that, for the amount of effort and other resources invested, the resulting software elements are more complex than most other system elements. He stated, "Software entities are more complex *for their size* (emphasis added) than perhaps any other human construct, because no two parts are alike (at least above the statement level)."

It is difficult to visualize the size of a software program because software has no physical attributes; however, if one were to print one million lines of software code at 50 lines per page the stack of paper would be a few feet high (one million lines of source code is not an excessively large software system). Humans are biological entities of similar volume and they are far more complex than computer software. However, there are few, if any, physical engineered artifacts of comparable size that are as complex as software.

Few software parts in a program are alike (beyond the statement level) because software elements are encapsulated (i.e. as functions, subroutines, and objects), stored in a library, included once each in a program as needed, and invoked rather than being replicated throughout a program. Another way to state this is most software elements are unique. A large software-enabled system may have thousands of encapsulated software parts that are combined in myriad ways, which results in complex systems (and further inhibits exhaustive testing).

The complexity of software also arises from the large numbers and kinds of interactions among the software elements, including serial and concurrent invocations, state transitions, data couplings, and interactions with databases and external systems. Depiction of a software program often requires several different representations to portray the static structures, dynamic couplings, behaviors, and modes of interaction that exist in computer software.

2.4.5 Software Conformity Must Be Exact

Conformity is the second issue cited by Brooks that makes software engineering difficult. Software must conform to exacting specifications in the representation of each part, in the interfaces to other internal parts, and in the connections to the environment in which a system operates.

Tolerances among the interconnections of physical entities are the foundation of manufacturing and construction; no two physical parts that are joined together are required to match exactly, nor could they. The machining industry, for example, uses the following standard tolerances (Kibbe et al. 2009):

1 decimal place ($y.x$)	±0.2″
2 decimal places ($y.0x$)	±0.01″
3 decimal places ($y.00x$)	±0.005″
4 decimal places ($y.000x$)	±0.0005″

The first known practical demonstration of machine tolerances occurred in 1785, when a French gunsmith name Honoré Blanc demonstrated the interchangeability of mechanical parts. He disassembled 50 muskets locks, mixed the parts, and reassembled the locks by randomly choosing the parts for each lock. He was not the first person to think about interchangeable parts but was the first to demonstrate it on a practical scale. The myth that Eli Whitney, in the United States in 1798, was the first to demonstrate interchangeable musket parts has been widely discredited (Woodbury 1960).

There are no corresponding tolerances in the interfaces among software entities; software interfaces must agree exactly in the numbers and kinds of parameters and the types of couplings. There are no interface specifications for software stating that an interface parameter can be "an integer number ±2%."

Also, lack of conformity can cause problems when an existing software element cannot be reused as planned because it does not conform to the needs of the system under development. Lack of conformity might not be discovered until late in a project, thus necessitating development and integration of an acceptable element to replace the one that cannot be used. This requires unplanned allocation of resources that can delay system completion. Complexity may have made it difficult to determine that the element to be used lacked the necessary conformity until the elements with which it would interact were available.

Mismatched interfaces (software–software and software–hardware) are responsible for many system failures. "Ariane 5" describes the result of a well-known software interface error. A mismatched interface between new and reused system elements was not discovered until the Ariane 5 was launched, which produced a catastrophic result.

Ariane 5

In June 1996, a new version of the Ariane (the Ariane 5), illustrated in Figure 2.1, veered off course and exploded shortly after launch.

Figure 2.1 The Ariane 5 rocket at Le Bourget Air and Space Museum, Paris. Source: Wikimedia.

Failure was caused by an interface error between the altitude and guidance elements of the launcher's software. Numbers encoded in 64-bit floating-point format were converted to 16-bit integer-encoded numbers and passed across the interface. The interface was correctly connected and the conversion algorithm was correctly coded. The problem was that a value of the floating-point number being converted was so large it could not be accurately represented as a 16-bit integer. This caused the guidance system to shut down. The rocket went off course and exploded (Gleick 1996).

The root cause of the failure was that the guidance software from the previous version of the Ariane rocket (Ariane 4) was used without modification. The new Ariane 5 altitude software generated a 64-bit floating-point value larger than the reused Ariane 4 guidance software could process. The failure was not detected in system testing.

The problem was corrected after the explosion and subsequent versions of the Ariane 5 performed well but not before loss of two years of work and $500 million for the European Space Agency.

A YouTube video of the launch and explosion of Ariane 5 can be seen at https://www.youtube.com/watch?v=PK_ygulapgA.

2.4.6 Software Is an Invisible Medium

Invisibility is the third of Brooks' essential properties of software. Software is said to be invisible because it has no physical properties. The functional and behavioral effects of machine code execution can be observed but software itself cannot the seen, tasted, smelled, touched, or heard. Source code, as written, is of course visible but the source code is not the voltage levels in the digital devices that are controlled by the machine code.

In contrast, progress (or lack thereof) in fabricating physical entities is supported by observing the physical presence of the entity. It is not credible to claim that construction of a bridge, building, or satellite launch facility is 95% complete when the groundwork for the site is being prepared. Unfortunately, the corresponding claim is sometimes made for software because it is invisible.

Well-known sayings based on software invisibility are the following:

> Developing the first 50% of the software takes 95% of the schedule; developing the remaining 50% of the software takes the other 95% of the schedule.

and

> Software is always 95% complete until the delivery date approaches; it then becomes 50% complete.

Unfortunately, the effects of these facetious sayings are often observed.

For this reason, emphasis is placed on surrogate indicators of progress, including

- Number of software requirements designed;
- Number of lines of source code written;
- Number of software modules constructed and unit tested;
- Number of system test cases prepared; and
- Number of system test cases executed.

Process metrics such as effort expended and backlog of rework to be completed are also used.

These surrogate measures attempt to make progress visible when developing inherently invisible software but they may or may not be accurate indicators of progress. Another saying is as follows:

> Activity does not equal progress.

Overcoming the problems of using surrogate indicators to measure progress accounts for the popularity of iterative software development processes, such as agile software development that emphasizes frequent demonstrations of executable increments of software code – a more exact measure of progress.

2.4.7 Software Appears to Be Easily Changed

Changeability is Brooks' fourth essential property that makes software engineering difficult. Software coordinates the operation of physical elements; establishes interconnection to the environment; and provides functionality, behavior, and quality attributes in software-enabled systems. As compared with modifying hardware elements, software elements are comparatively easy to change, particularly in the late stages of a project – comparatively easy to change but it is not necessarily easy to obtain the desired effect.

Changes to a system may occur because stakeholders change their minds; mission objectives change; laws, regulations, and business practices change; and underlying hardware and software technology changes. In addition, the operational environment of the system may change, necessitating changes to a system. Also, incremental demonstrations or installation of an early version of a software-enabled system in the operational environment may result in new requirements that will necessitate changes to the system (i.e. now that the evolving system enables me to do A and B, I would like for it to also allow me to do C or to do C instead of B).

Complexity within software elements and in the connections among them, the need for exact conformity, and the invisibility of the software may result in a large amount of rework for a "small" change, thus disrupting the ability to make progress according to plan. Invisibility of the 0s and 1s (as opposed to the design and source code representations of the software) may result in an unforeseen ripple effect of undesired consequences in other parts of a system.

An additional consideration: Conformity of software can produce unwanted side effects when making changes. For example, changing the order of sorting a list of data items from ascending order to descending order to facilitate processing of the list in one part of the software may disrupt the functionality and behavior of other parts of the software or entities in the system environment that depend on conformity of the ascending order.

For these reasons, many experienced software engineers say the following:

> There are no small changes to software.

Additional things system engineers need to understand about software and software engineering follow.

2.4.8 Software Development Is a Team-Oriented, Intellect-Intensive Endeavor

In addition to the essential properties of software described by Brooks (complexity, conformity, changeability, and invisibility), an additional factor distinguishes software projects from other kinds of projects:

> Software projects are team-oriented, intellect-intensive endeavors.

All large engineering projects are team-oriented, intellectual endeavors but the level of communication, interaction, and coordination required of members of a software development team are different in kind. A relevant simile would be to compare a team effort to concurrently write a set of integrated software elements to a team effort to concurrently write the chapters of a book and integrate them into a seamless whole. (*Note*: A single misspelled word could make the book unreadable if it were a software program.) "Book Writing" outlines the processes that would be used if writing a book were a software engineering project that included a team of writers.

Book Writing

To illustrate the problems of intellect-intensive teamwork, consider the difficulties that would arise if a group of individuals were to write a book as a team effort, on a predetermined schedule and within a specified budget of staff-hours that could be devoted to the project.

(Continued)

(Continued)

Determining the kind of book to write and documenting the requirements for the book in a clear, complete, and unambiguous manner is analogous to specifying the requirements for a complex system. Deciding on the structure of and relationships among chapters and (perhaps) sections of the book is the analog of architectural design. Specifying writing style, voice, tense, and page layout for each chapter is similar to detailed design.

Writing the chapters and checking spelling, syntax, and grammar is analogous to coding, reviewing, and testing system elements. Merging the chapters corresponds to system integration. When the book is completed, the integrated text should flow smoothly, as if written by a single individual in a single session. An editor performs independent verification and makes suggestions for improvements to the system, both during writing and upon completion of the book. The perceived value of the finished product is largely determined by reviews of critics and work of mouth among book buyers. If the book is popular, it may be updated and released through several editions following initial release – the simile to system development is apparent.

A team project to write a book could include the following systems/software engineering processes:

- Investigating the idea of writing a book (business analysis process)
- Deciding what kind of book would likely be successful (stakeholder needs and requirements definition process)
- Deciding on a title, theme, and style for the book (systems requirements definition process)
- Conducting a market analysis and making a go/no-go decision (system analysis process)

Assuming a "go" decision:

- Preparing an initial plan for the writing project to include estimated schedule, progress milestones to be achieved, and resources – including number of writers and their skill levels for the kind of book to be written – plus project inhibitors and risk factors (project planning process)
- Identifying infrastructure needs and prepare the writing environment; procuring infrastructure tools and preparing the development environment (project planning process)
- Deciding how many sections, chapters, and appendices to include in the book (architecture definition process)
- Preparing an abstract for each chapter (design definition process)
- Specifying the relationships among the chapters (interface design definition process)
- Deciding the style of prose, font to be used, form of reference citations, and other composition details (design definition process)

- Refining the initial project plan to include detailed plans for writing the book chapters; recruiting writers and assign chapters to them (project planning process)
- Preparing a detailed schedule for writing draft versions of chapter increments, periodically merging individual drafts of chapters, and conducting team reviews (project planning process)
- Reviewing all of the preceding steps to identify and confront project inhibitors and risk factors (system analysis process)
- Authorizing the writing project (kickoff process)
- Periodically integrating chapter increments, conducting reviews, and planning the details of the next writing cycle (integration, project planning verification and validation; architecture definition and design definition revisited as necessary)
- Conducting a final review of the manuscript for the book prior to delivering it to the publisher (verification process)
- Delivering the manuscript to the publisher (transition process)
- Assisting the publisher with problems encountered in publishing the manuscript to achieve a satisfactory result (validation process)

Each of the steps indicated in parentheses is a system engineering/software engineering process documented in the ISO/IEC/IEEE Standards 15288 and 12207 (ISO 2015, 2017).

Because there are no physical laws that govern book writing and no mathematical theories of how to write a book, a successful outcome for the team effort would depend on clearly defined goals, common understanding and acceptance of those goals, a common approach, adequate resources and calendar time, the skills of the individual contributors, shared ownership of the book, and their ability to work as a cohesive team.

As mentioned in "Mythical Man-Month" the number of communication paths among N workers engaged in closely coordinated work is $N(N-1)/2$ (the number links in a fully connected graph where each team member is a node in the graph). Analysis has shown that individual productivity of software developers decreases with increasing team size because of the need for increased communication and coordination among the increasing number of team members; the same decrease in individual productivity probably occurs for any team engaged in intellect-intensive work, including teams of PhSEs.

Increasing the size of a programming team from 5 to 10 members might, for example, increase the production rate of the team by 75%, but not by 100%, as would occur if each team member could work effectively in isolation from all of the other team members. Increasing the size of a software team soon reaches the level of diminishing returns, where the overall quantity and quality of generated work products declines with addition of more team members.

In *The Mythical Man-Month,* Fred Brooks described this phenomenon as follows (Brooks 1995):

> Oversimplifying outrageously, we state Brooks' Law:
> Adding manpower to a late software project makes it later.

Brooks' law (so called) is based on three factors:

1. The nonproductive time required for existing team members to indoctrinate new team members;
2. The learning curve for the new members; and
3. The increased communication overhead that results from the new and existing members working together.

Brooks' law would not be true if the work assigned to the new members did not invoke any of these three conditions, for example, if some of the work could be "spun off" for implementation in relative isolation by individuals who are "up to speed" on the project but this situation is unlikely to occur.

For these reasons, software development teams are usually limited to a few members (e.g. three to seven) to minimize the overhead and disconnects of constant communication among the team members. Multiple small teams are employed to accomplish larger projects.

Continuity of team members is also important to avoid factors 1 and 2 above when turnover occurs.

2.4.9 Software Developers Use Mostly Iterative Processes

The malleable nature of software allows iteration among and interleaving of the phases of the development process to a much greater degree than is possible when developing physical artifacts. Software engineers favor iterative development processes having cycle times that range from a few days to two weeks. Most iteration cycles produce a next demonstrable increment of working software; occasionally, an iteration cycle may be a "timeout" to regroup, redesign, and pay accumulated technical debt (Fairley and Willshire 2017). Iterative software development processes include the evolutionary, agile, and spiral processes that are described in Chapter 2 of the text *Managing and Leading Software Projects* (Fairley 2009).

A widely held but incorrect perception is that iterative software development does not include an analysis phase, an architecture phase, or a design phase. The backlog of prioritized capabilities illustrated in Figure 2.2 is provided by these systematic phases of software development, which may also be accomplished in an iterative manner.

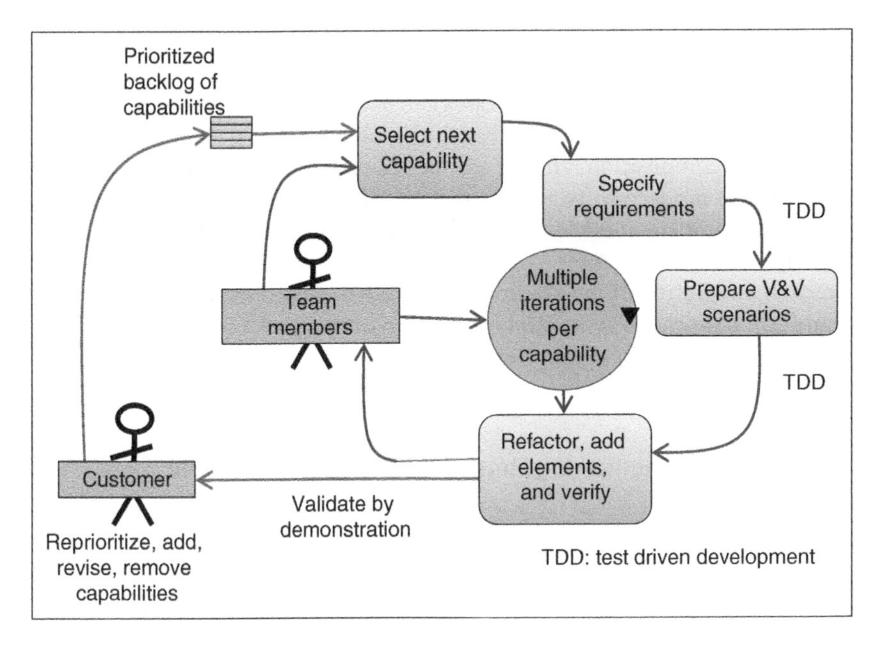

Figure 2.2 Iterative software development.

The customer illustrated in Figure 2.2 is a knowledgeable representative of the system acquirer.

Each of the outer iteration cycles illustrated in the figure can include some or all of incremental analysis, design, and construction. As indicated, frequent internal iterations may occur during each outer iteration cycle. Test-driven development (TDD) is often used, whereby test cases are written before the corresponding software is written. TDD is widely used, although there is some controversy as to its effectiveness (Karac and Turhan 2018).

Refactoring involves rearranging the structure of the cumulatively implemented software to facilitate testing of it or to facilitate construction of additional software. However, refactoring must not alter the behavior of the implemented software increments. Refactoring is possible because of the malleable nature of software and the modularized constructs of object-oriented software development.

Regression testing of the currently implemented software capability combined with the previously implemented capabilities provides verification that the software satisfies the requirements implemented to date (or not). Demonstrations for a knowledgeable customer validates that the software provides the capabilities needed to satisfy the operational requirements for the software implemented to date (or not). A detailed description of the process illustrated in Figure 2.2 can be found in Section 4.5.2, Chapters 5 and 9 of this text, and in Section 2.4 of the *Software Extension to the PMBOK® Guide* (Fairley et al. 2013).

While it is true that iterative processes can be, and are, used to develop physical system elements, it is also true that the nature of physical elements involves functional decomposition and linear development processes to a greater degree than in modern software engineering practice. The Vee and incremental Vee development models favored by systems engineers are presented in Chapter 4.

Smooth integration of the development processes used in systems engineering and software engineering, and integration of the work products generated, are addressed by the integrated-iterative-incremental (I^3) development process introduced in Chapter 5 of this text and continued in Chapters 6–9.

2.4.10 Software Engineering Metrics and Models Are Different in Kind

Software engineering metrics and models are different in kind from the metrics and models of traditional engineering because of the logical, versus physical, nature of software. During the twentieth century, great advances were made in the traditional engineering disciplines based on development of quantitative measures and analytical models. It is often the case that a physical system element can be characterized by a few parameters such as voltage level, current flow, and heat dissipation or mass, volume, and strength of materials. Metrics of interest can often be determined from mathematical models of the artifact in question. These metrics then guide design and fabrication of physical elements.

Because software has no physical properties, these kinds of metrics and models are not possible. There are no mathematical models that can accurately predict, for example, the failure-rate reliability of software with the degree of precision that is possible for physical artifacts. This is, in part, because the goal of manufacturing physical entities is to produce multiple units that satisfy objectively specified functional and quality attributes. Sampling individual units and generalizing the results to statistics of production runs can determine the reliability of manufactured components. In contrast, the goal of software development is to develop perfect single copies of individual units that can be stored in libraries and used repeatedly.

This is not to imply that there are no models in software engineering (examples include complexity metrics for source code and queuing models for throughput of a software network) but the models and metrics in software engineering are fewer and different in nature than the models and metrics of traditional engineering (Kan 2002).

The following section presents 10 things software engineers need to know about systems engineering.

2.5 Ten Things Software Engineers Need to Know About Systems Engineering

As indicated in Table 2.2 and Section 2.3, PhSEs and SwSEs have each adapted methods from the other discipline and thus have many commonalities. But there are many things systems engineers need to know about software engineering and software engineers (addressed in Section 2.4), and there are many things software engineers need to know about systems engineering of physical systems. The latter topic is address in this section.

More info on the following topics can be found in the paper "Teaching Systems Engineering to Software Engineering Students" (Fairley and Willshire 2011b).

2.5.1 Most PhSEs Have Traditional Engineering Backgrounds

Most PhSEs are educated and trained in a traditional engineering discipline (e.g. electrical, mechanical, civil) that is based on continuous mathematics, quantitative models, and design of physical artifacts. They usually start their careers within their engineering discipline. Opportunity, personal interest, and mentoring have traditionally provided entry points to physical systems engineering.

Increasingly, PhSEs are obtaining undergraduate and master's degrees in systems engineering. However, there is no clear consensus as to the form and content of systems engineering degree programs, as can be seen from the program names in the Wikipedia list of systems engineering degree programs offered by universities (Wikipedia 2018).

The *Graduate Reference Curriculum for Systems Engineering* (*GRCSE®*) provides a template for developing master's programs in systems engineering, but at the present time, there is no counterpart to GRCSE for undergraduate systems engineering degree programs (Pyster et al. 2015a). ABET criteria for certification of undergraduate systems engineering degree programs are being developed but the criteria are a work in progress at the time of writing of this text (see www.ABET.org for updates).

The current state of systems engineering degree programs is reminiscent of how computer science programs emerged from applied mathematics and other academic disciplines and how software engineering emerged from computer science, computer engineering, and related academic disciplines. As often happens in emerging engineering disciplines, degree programs start at the graduate level. Undergraduate programs evolve as the discipline matures and educational materials become available.

Some individuals contend that systems engineering education should occur only at the graduate level, based on an undergraduate degree in an engineering discipline plus some "real-world" experience. This also occurred in the evolution of software engineering education.

2.5.2 Systems Engineers' Work Activities Are Technical and Managerial

Systems engineers lead the technical work of analyzing, designing, building, integrating, verifying, validating, installing, delivering, maintaining, and retiring complex systems. Systems engineers also engage in managerial activities to plan and coordinate the work activities of the disciplinary and specialty engineers for a systems project. They thus plan, estimate, measure, and control the work products and work processes. In addition, they coordinate and lead the system development personnel. Some systems engineers specialize in the technical aspects of systems engineering while others specialize in technical management, which is a specialization of project management tailored to managing the technical aspects of developing and sustaining complex systems.

Systems engineers typically apply their skills within a domain such as health care, automotive systems, aerospace, defense, civil infrastructure, or electrical power. They are thus systems engineering domain experts who acquire their domain expertise through a combination of education, training, and on-the-job experience.

Developing a large system may require a team of systems engineers who are led by a senior systems engineer; systems engineering team members may be assigned to facilitating development of complex subsystems, such as the launch, navigation, telemetry, and satellite elements of a space mission. Some systems engineers may have staff members to supplement their technical and/or managerial work activities. Alternatively, some systems engineers may contribute their expertise in their engineering discipline, in addition to their systems engineering work activities, for example, contributing to mechanical design of system or subsystem elements.

2.5.3 The Scope of Systems Engineering Work Is Diverse

Systems engineers provide technical and managerial leadership to develop and sustain products, provision services, enable enterprises, and integrate independent systems to provide system-of-systems capabilities – all within application domains such as health care and transportation – see SEBok Part 4 (Adcock 2017).

A *product* is an engineered system that operates in standalone mode within an environment. The *service* sector of an economy, as the name implies, is concerned with providing services. Retail sales, banks, hotels, and education are examples of service sector businesses. *Enterprises* are organizations that include one or more local/regional offices and include, for example, insurance companies and financial institutions. A *system-of-systems* is a collection of standalone systems that are integrated to cooperatively provide capabilities that cannot be provided by independent operation of each system.

The particular roles played by systems engineers vary according to the domains in which they specialize and the kinds of applications within those domains. Systems engineering methods and techniques are tailored to satisfy stakeholders' needs when developing and maintaining products, services, enterprises, and systems-of-systems in different domains (e.g. medical, automotive, electrical power).

2.5.4 Systems Engineers Apply Holistic and Reductionist Thinking

Systems engineers apply holistic thinking (i.e. systems thinking) to incorporate stakeholders' needs, wants, and desires into an envisioned pattern of elements and relationships that will form a complete system, the environment in which the system will exist, and the impact a system will have on the operational context and the entities that affect or will be affected by the system (humans, other life forms, and natural and engineered entities). In addition, systems engineers apply holistic thinking to envision the entire life cycle of support needed for sustainment, upgrading, retiring, and replacing complex systems. A complex system is best understood, and perhaps can only be understood, by applying holistic thinking.

Systems engineers apply reductionist thinking to decompose an envisioned system into subsystems and elements to determine the role each subsystem or element must play within the context and environment of the other subsystems and elements and to consider how the performance of the system will impact its context and environment.

Additional information on systems thinking is provided in Part 2 of the *Guide to the Systems Engineering Body of Knowledge* (Adcock 2017).

2.5.5 Systems Engineering Covers the Full Spectrum of Life-Cycle Processes

ISO/IEC/IEEE Standards 15288 and 12207 for system life-cycle processes and software life-cycle processes include four areas: agreement processes, organizational project-enabling processes, technical management processes, and technical processes (ISO 2015, 2017).

Agreement processes include acquisition and supply processes that are used to establish agreements with organizational entities external to the system development organization, such as acquiring products and services from vendors and subcontractors and supplying a deliverable system to an acquirer (i.e. a customer). Acquisition and supply agreements may also exist between organizational entities internal to the development organization, such as an agreement with an internal configuration management or quality assurance (QA) group that is organizationally distinct from the development projects or an agreement with an internal supplier to fabricate some system elements.

External agreements are often documented in Statements of Work (SOWs) that are legally binding. Internal agreements are less formal and are often documented in Memos of Understanding (MOUs). SOWs and MOUs are described in Chapter 10.

Organizational project-enabling processes include life-cycle model management, portfolio management, human resources management, quality management (QM), and knowledge management. These processes, as the name implies, are best accomplished at the organizational level in which systems engineering projects are housed – they provide resources and infrastructure to support the projects.

Technical management processes include project planning, project assessment and control, decision management, risk management, configuration management, information management, measurement, and QA processes.

These processes are concerned with managing the technical aspects of a systems project – they are subordinate to and complementary to overall project management as conducted by a project manager. It is perhaps self-evident that the project manager (and his/her staff) and the lead systems engineer (and her/his staff) must work together in a cooperative manner. Technical management processes are covered in Chapters 10 and 11 of this text.

QM, QA, and quality control (QC) are sometimes confused; they are addressed in "QM/QA/QC."

QM/QA/QC

According to 15288 and 12207, QM is an organizational project-enabling process. Subclause 6.2.5.1 in both standards states that the purpose of the QM process is to develop and enforce policies and procedures intended to assure uniform QM for all projects within an organization.

> to assure that products, services and implementations of the quality management process meet organizational and project quality objectives and achieve customer satisfaction.

QA is a technical management process in 15288 and 12207. Subclause 6.3.8.1 in each standard states that the purpose of the QA process is

> To help ensure the effective application of the organization's Quality Management process to the project.

The role of QA is to provide confidence that quality requirements will be fulfilled by each project. As illustrated in Figure 2.3, QA personnel observe and analyze the project life-cycle processes and the outputs of each project to assure that organizational and project-level policies and procedures are followed so that the system being developed will have the desired quality attributes

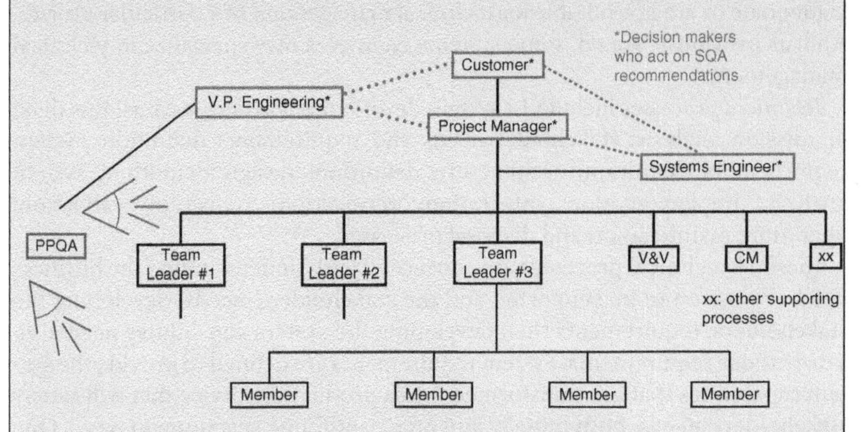

Figure 2.3 Quality assurance observation of a systems project.

As indicated, QA personnel may make recommendations to decision makers for improvement of a project's development processes, the QA practices applied to all projects, and the policies and procedures developed by QM personnel. The proper role of QA is to observe and make recommendations. A QA group or organization should not have authority over projects; otherwise a police mentality will disrupt cooperation and coordination of systems/software projects with QA.

QA and QC are often confused, particularly in software engineering where "testing" is often used interchangeably with "quality assurance" rather than correctly regarding software testing as a QC mechanism. QA is an engineering management process used to assure that regulations, policies, procedures, guidelines, and plans that impact the quality of system elements, subsystems, system increments, and the final system are being followed.

QA should not to be confused with QM and QA should not be confused with QC. The purpose of QA is to ensure that the QM policies and procedures are applied uniformly to each project by observing projects and making recommendations for improvements to individual projects. QA is sometimes confused with the QC processes of verification and validation (V&V). And, "QA" is sometimes used to erroneously denote the testing activities of V&V. QA personnel monitor development processes, including V&V, to assure adherence to QM policies.

Separate organizational units may conduct some of the technical management processes; for example, separate organizational units sometimes conduct configuration management and QA. Systems engineers must establish close working relationships with the separate organizational units.

In other cases, systems engineers may conduct most or all of the technical management processes, as for example, when the organizational processes are inadequate or are not suitable for technical management of a particular project. And, as previously stated, some systems engineers may specialize in technical management.

Technical processes include 14 system development processes: business need or mission analysis; stakeholder needs and requirements definition; system requirements definition; architecture definition; design definition; system analysis; implementation; integration; verification; transition; validation; operation; maintenance; and disposal processes.

These 14 technical processes are concerned with understanding the business need or mission to be supported and the stakeholders' needs, developing the stakeholders' requirements then developing the system capabilities needed to satisfy those requirements. System requirements are defined to provide the system capabilities that are transformed into a product or service that will satisfy stakeholders' needs, both initially and on a continuing sustainment basis. One of the organization's life-cycle models, perhaps tailored to fit the needs of the project, is typically used to develop a system, a service, or a system-of-systems or to enable an enterprise. The technical processes of system development are presented in Chapters 4–9 of this text.

2.5.6 System Engineers Plan and Coordinate the Work of Interdisciplinary Teams

Modern complex systems are typically developed by various kinds of disciplinary engineers (mechanical, electrical, civil, software, and so forth) as needed to realize a total system. In addition, specialty engineers in areas such as security, reliability, availability, and maintainability participate as needed.

A systems engineer or a team of systems engineers develop the system requirements, participate in developing the system architecture, and determine the kinds of disciplinary and specialty engineers that will be needed, how many of each kind will be needed, when they will be needed, for how long they will be needed, and how they will be acquired. Systems engineers then coordinate the work activities of the disciplinary and specialty engineers who design, implement, integrate, verify, and validate individual system elements, system increments, and the deliverable system.

2.5.7 Current Systems Engineering Development Processes Are Mostly Incremental

The nature of physical artifacts constrains the development processes used by PhSEs; most prefer the linear and incremental development processes presented in Chapter 4 of this text. Some software engineers view these processes as outdated. However, the constraints imposed by procurement, development, and modifications of physical artifacts do not allow the comparatively rapid evolution and demonstration of partially completed work products afforded by iterative software development. Compensating for the constraints of physical system development is a driving force behind the movement toward model-based systems engineering (MBSE).

2.5.8 Systems Engineering Is Transitioning to a Model-Based Approach

Systems engineering is transitioning to MBSE wherein domain models for system requirements, architecture, and design are created and analyzed to reduce the use of traditional documents and to facilitate simulation and, in some cases, partial realization of system elements, subsystems, and systems. Section 9.2.1 of the *Systems Engineering Handbook*, Fourth Edition, defines MBSE as follows (Walden et al. 2015):

> The formalized application of modeling to support system requirements, design, analysis, verification, and validation activities beginning in the conceptual design phase and continuing throughout development and later life cycle phases.

The *Systems Engineering Handbook* lists the following benefits of MBSE as compared with traditional document-based approaches to system development:

- Improved communications among system development stakeholders;
- Increased ability to manage system complexity;
- Improved product quality;
- Enhanced knowledge capture; and
- Improved ability to teach and learn systems engineering fundamentals.

MBSE methods and tools are being created and deployed. The term modeling and simulation-based systems engineering (M&SBSE) is used to describe digital simulation of executable MBSE systems models.

MBSE and M&SBSE approaches may support more iterative processes for system development. Chapter 5 of this text presents a capability-based M&SBSE process for system development that facilitates concurrent development and integration of the iteratively developed software elements and incrementally developed physical elements of a software-enabled system.

Changes in systems development processes can thus be anticipated as MBSE and M&SBSE mature but the traditional development processes presented in Chapter 4 of this text will be the prevalent approaches to systems development during the near-term transition period.

2.5.9 Some Who Perform Systems Engineering Tasks Are Not Called Systems Engineers

Some who lead development and sustainment of complex systems have job titles other than "systems engineer," although they perform systems engineering tasks in various domains when developing products, delivering services, provisioning enterprises, or integrating collections of standalone systems to form systems-of-systems. As examples, those who provide technical leadership for developing electronic systems for automobiles (hardware and software) and those who develop and sustain complex enterprise information systems are engaged in systems engineering endeavors regardless of their titles.

These individuals often succeed (or fail) on the basis of their innate ability (or lack thereof) and may be unaware of the benefits that could be gained from knowledge of systems engineering methods and techniques.

2.5.10 Most PhSEs View Software Engineers as Disciplinary Engineers or Specialty Engineers

Most PhSEs fail to take advantage of involving qualified SwSEs at the system level because they don't understand that SwSEs can offer design options for system architecture; interfaces among system elements; connections to the system environment; and system functionality, behavior, and quality attributes. Instead, some PhSEs regard software engineers as programmers who write code, as needed, to support the efforts of other engineers who are working to develop the various physical elements of a system.

Alternatively, some PhSEs may regard software engineers as specialists who provide their services on an as-needed basis, similar to engineers in specialty disciplines such as safety, security, and reliability.

These views are reinforced by separate allocation of software requirements to the software engineers during the system implementation phase of system development. Inefficient system development and reduced quality are the results of this approach.

Chapter 3 presents some of the factors that inhibit close cooperation and coordination of PhSEs and SwSEs and indicates some of the ways that the inhibitors can be ameliorated.

2.6 Key Points

- Systems engineering and software engineering are distinct but intimately intertwined disciplines.
- Most systems engineers have traditional engineering backgrounds. They are termed PhSEs.
- Qualified software engineers can effectively participate at the systems level. They are termed SwSEs.
- Traditional systems engineers (PhSEs) have adapted methods and techniques developed by SwSEs and, conversely, SwSEs have adapted methods and techniques developed by PhSEs.
- The work activities of PhSEs and SwSEs are technical and managerial.
- PhSEs and SwSEs rely on interpersonal communication and leadership skills rather than positional authority to convince others.
- Most PhSEs and SwSEs have minimal exposure to the other discipline.
- PhSEs prefer incremental system development processes.
- SwSEs prefer iterative system development processes.
- Systems engineering is transitioning to a model-based approach to system development and simulation-based models.

Chapter 3 continues the exploration of systems engineering and software engineering. Issues that can arise when PhSEs, SwSEs, disciplinary engineers, and specialty engineers attempt to work together in a collaborative manner are presented, along with opportunities for improvements.

Exercises

2.1. Table 2.1 lists three kinds of systems: older physical systems, computational systems, and cyber-physical systems. Find an example of each kind of system and briefly describe it. Do not use the examples in Table 2.1.
   (a) An older physical system
   (b) A computational system
   (c) A cyber-physical system

2.2. Table 2.2 lists software engineering methods that have been adopted and adapted by systems engineers and systems engineering methods that have been adopted and adapted by software engineers.

(a) Choose three software engineering methods that have been adopted and adapted by systems engineers. For each of the three you have chosen, briefly explain why you think systems engineers have found them to be useful.

(b) Choose three systems engineering methods that have been adopted and adapted by software engineers. For each of the three you have chosen, briefly explain why you think software engineers have found them to be useful.

2.3. Section 2.3 lists 10 common attributes of physical systems engineers and software systems engineers. Item 1 states that most students who will become PhSEs or SwSEs receive "silo" educations in an engineering discipline or a computing discipline that includes minimal exposure to the counterpart discipline.

(a) Briefly explain what is meant by a "silo" education.

(b) List and briefly explain three reasons why PhSEs and SwSEs receive silo educations.

2.4. Section 2.4.7 includes the statement: "there are no small changes to software." Hardware engineers could, and perhaps do, make the same claim but for different reasons. Briefly describe the different reasons that there are no small changes to software and no small changes to hardware.

(a) There are no small changes to software because:

(b) There are no small changes to hardware because:

2.5. The box about "Book Writing" in Section 2.4.8 describes how a team effort to write software is like a team effort to write a book (i.e. a simile). List and briefly describe three ways in which a team effort to write software is not like a team effort to write a book.

2.6. Section 2.5.3 indicates that systems engineers facilitate development and sustainment of products, provisioning of services, enabling enterprises, and integrating independent systems to provide system-of-systems capabilities. Briefly explain how systems engineering is applicable to each of the following endeavors.

(a) Development and sustainment of products

(b) Provisioning of services

(c) Enabling of enterprises

(d) Integrating independent systems to provide system-of-systems capabilities

2.7. Systems engineers typically specialize in an application domain such as health care or transportation. Briefly explain why most systems engineers specialize in an application domain.

2.8. Section 2.5.10 indicates that most physical systems engineers view software engineers as disciplinary engineers or specialty engineers. Briefly explain why physical systems engineers do not regard qualified software engineers (i.e. software systems engineers) as peer systems engineers, who can make significant contributions to systems engineering of software-enabled systems.

References

Adcock, R. (EIC) (2017). *The Guide to the Systems Engineering Body of Knowledge (SEBoK) V1.9*. BKCASE Editorial Board https://www.sebokwiki.org (accessed 8 June 2018).

Ardis, M., Fairley, D., Hilburn, T. et al. (2013). *Software Extension to the PMBOK® Guide*, 5e. Project Management Institute. https://www.pmi.org/pmbok-guide-standards/foundational/pmbok/software-extension-5th-edition (accessed 9 June 2018).

Ardis, M., Fairley, D., Hilburn, T. et al. (2014). *Software Engineering Competency Model*, Version 1.0 (SWECOM). IEEE Computer Society. https://www.computer.org/volunteering/boardsand-committees/professional-educational-activities/software-engineering-competency-model (accessed 20 Sept 2018).

Bourque, P. and Fairley, R. (eds.) (2014, 2014). *Guide to the Software Engineering Body of Knowledge (Version 3.0)*. IEEE https://www.computer.org/web/swebok/v3-guide (accessed 8 June 2018).

Brooks, F. (1995). *The Mythical Man-Month: Essays on Software Engineering*, Anniversary Edition, 2e. Addison Wesley.

Fairley, R. (2009). *Managing and Leading Software Projects*, Chapter 2. Wiley.

Fairley, R. and Willshire, M. (2011a). Teaching software engineering to undergraduate systems engineering students. In: *Proceedings of the 2011 American Society for Engineering Education Annual Conference and Exposition*, 26–29. Vancouver, BC.

Fairley, R. and Willshire, M. (2011b). Teaching systems engineering to software engineering students. In: *Proceeding of the 24th IEEE-CS Conference on Software Engineering Education and Training*, 219–226. Honolulu, HI.

Fairley, R. and Willshire, M. (2017). Better now than later: managing technical debt in systems development. *IEEE Computer* 6: 80–87.

Gleick, J. (1996). A bug and a crash. https://www.around.com/ariane.html (accessed 8 June 2018).

ISO (2015). ISO/IEC/IEEE 15288:2015, Systems and software engineering – System life cycle processes, ISO/IEEE, 2015.

ISO (2017). ISO/IEC/IEEE 12207:2017, Systems and software engineering – Software life cycle processes, ISO/IEEE, 2015.

Kan, S. (2002). *Metrics and Models in Software Quality Engineering*, 2e. Pearson Education.

Karac, I. and Turhan, B. (2018). What do we (really) know about test driven development?, IEEE Software, July/August 2018. *IEEE Computer Society* 35 (4): 81–85.

Kibbe, R. et al. (2009). *Machine Tool Practices*, 9e. Pearson.

Leveson, N. and Turner, C. (1993). An investigation of the Therac-25 accidents. *IEEE Computer* 26 (7): 18–41.

Pyster, A., Olwell, D.H., Ferris, T.L.J. et al. (eds.) (2015a). Graduate Reference Curriculum for Systems Engineering (GRCSE™) V1.1. Hoboken, NJ: Trustees of the Stevens Institute of Technology http://www.bkcase.org/grcse (accessed 8 September 2018).

Pyster, A., Adcock, R., Ardis, M. et al. (2015b). *Exploring the Relationship between Systems Engineering and Software Engineering*, vol. 44, 708–717. Procedia Computer Science https://www.sciencedirect.com/science/article/pii/S1877050915002525 (accessed 8 June 2018).

Turing, A. (1937). On computable numbers with an application to the Entscheidungsproblem. *Proceedings of the London Mathematical Society* s2-42 (1): 230–265.

Walden, D., Roedler, G., Forsberg, K. et al. (eds.) (2015). *Systems Engineering Handbook: A Guide for System Life Cycle Process and Activities*, 4e), International Council on Systems Engineering. Wiley.

Wikipedia (2018). List of systems engineering universities. https://en.wikipedia.org/wiki/List_of_systems_engineering_universities (accessed 9 June 2018).

Woodbury, R. (1960). The legend of Eli Whitney and interchangeable parts. *Technology and Culture* 1 (3): 235–253.

3

Issues and Opportunities for Improvements

3.1 Introduction

Given the close coupling of software and the other elements in a software-enabled system (the "intimate intertwining" cited in Section 1.1), it is desirable that physical systems engineers (PhSEs), software systems engineers (SwSEs), and other engineers who develop system elements work in a synergistic manner so that their combined contributions will be greater than the contributions that would result if each group worked in relative isolation.

The other engineers, in addition to PhSEs and SwSEs, are the disciplinary engineers (e.g. electrical, mechanical, and software) and the specialty engineers (e.g. safety, security, and reliability) who follow the guidance of PhSEs and SwSEs to develop software-enabled systems.

This chapter presents some of the issues that can arise when PhSEs, SwSEs, and other engineers attempt to work together in a collaborative manner. We do not provide definitive solutions for the issues that can arise. Instead we explore underlying reasons for the issues and offer some observations on opportunities for improvement.

Effective interactions among engineers improve project schedules and budgets, the features and quality attributes of the deliverable system, and the collective engineers' morale and productivity.

We start with some background that recaps the separate evolutionary histories of systems engineering and software engineering presented in Chapter 1.

3.2 Some Background

As discussed in Chapter 1, systems engineering evolved in response to the need to coordinate the work of interdisciplinary engineering teams that developed large complex systems consisting of physical elements. The process that evolved was "divide and conquer," namely, to decompose large complex systems into functional subsystems, allocate requirements to each subsystem, and specify

Systems Engineering of Software-Enabled Systems, First Edition. Richard E. Fairley.
© 2019 John Wiley & Sons, Inc. Published 2019 by John Wiley & Sons, Inc.

the interfaces among the subsystems and to the system environment. Complex subsystems were recursively decomposed into subordinate parts as necessary to gain control of size and complexity. This "divide and conquer" strategy continues to be a commonly used approach in systems engineering.

Interdisciplinary teams were, and are, assigned to develop the subsystems, develop interfaces, and work with other teams whose subsystems interacted with their subsystems. The roles of systems engineers were, and are, to

- Do the "upstream work" of business or mission analysis, requirements development, functional decomposition, architectural design, and system design;
- Recursively decompose subsystems and allocate requirements to subsystem elements;
- Plan the technical work to be done, recruit and organize interdisciplinary teams, assign them to the subsystems, and coordinate their work activities while working with the project manager;
- Provide needed resources, track progress, and anticipate and solve problems; and
- Coordinate the "downstream" activities of integrating subsystems and verifying, transitioning, and validating the resulting system.

In some cases, one or more specialty engineers were, and are, designated to provide expertise for critical quality attributes (safety, security, reliability, and so forth).

And in many cases, systems were, and are, developed by multiple organizations contracted to provide the needed expertise, meet staffing requirements, and accommodate political considerations for systems projects that may be part of larger programs. Within organizations, engineers were, and are, often "matrixed" to the teams that develop system elements for which their skills were, and are, needed. These practices continue today.

Software engineering emerged as a professional discipline in the late 1960s and early 1970s when it became apparent that the increasing size and complexity of computing hardware required increasing size and complexity of the corresponding software. System software was developed to control operation of the computer hardware. Software applications were developed using the interfaces provided by the system software. This resulted in increasing size and complexity of software development teams, which led to the development of process models used to develop the software and the management techniques used to plan and control the growing complexity of the work (Randell et al. 2001).

Computers and associated software were increasingly incorporated into large and complex physical systems in the 1980s. The versatility of software resulted in allocation of functional, behavioral, and quality requirements to software. The malleability of software made it comparatively easy to change systems during development and in operation as compared with modifying hardware elements and hardware interfaces.

Initially, software was developed using the traditional engineering approach of sequential analysis, design, and development phases with rework accomplished as needed. Over time, the malleability of software has facilitated various approaches to software development including incremental, iterative, and agile development processes (Fairley 2009; Fairley et al. 2013).

The following issues have resulted from the separate evolutionary paths of systems engineering and software engineering:

- Lack of reciprocal workplace respect;
- Differences in terminology used;
- Differences in styles of problem solving;
- Application of holistic and reductionist thinking;
- Different approaches to logical design and detailed design;
- Different process models;
- Different application of technical processes; and
- Reciprocal workplace respect.

These issues and opportunities for making improvements are presented here.

3.3 Professional Literacy

The traditional definition of literacy is the ability to read and write, but more generally, literacy is knowledge of a specific subject. Many of those who engage in systems engineering and software engineering do not have sufficient literacy in their respective disciplines or sufficient literacy in the other discipline.

3.3.1 System Engineering Literacy

Lack of systems engineering literacy presents three kinds of issues:

First, some traditionally educated engineers (electrical, mechanical, civil, and so forth) become systems engineers without sufficient training or opportunities to gain experience by shadowing or being mentored. They are left to "sink or swim." Some become expert swimmers but some sink to the detriment of their projects, organizations, and customers.

Second, not all who perform systems engineering tasks have job titles of "systems engineer." Many who perform systems engineering tasks do not have training or experience in systems engineering. It is apparent that a great deal of system-level planning and coordination must be applied to orchestrate multiple disciplines within the constraints of requirements, schedule, budget, resources, and technology when observing the building site for an apartment complex, a shopping center, or a high-rise building regardless of the job titles of those involved. In a similar manner, scientists and engineers of various kinds plan and execute complex projects, as do many information technologists, service providers, and business professionals.

Unfortunately, these "systems engineers" may fail to be as effective as they might be if they had better understanding of system engineering processes and methods. Many scientists, engineers, information technologists, service providers, and business professionals are increasingly involved in developing complex social and technological systems. Their need for system engineering literacy is becoming increasingly important.

The third issue is systems engineering literacy for software engineers. Many competent software engineers receive educations based on computer science, which concentrates on computational applications of discrete mathematics, software theory and practice, algorithms, and software development processes. These software engineers do not receive orientation to or education in traditional engineering approaches to problem solving that are based on continuous mathematics, physical sciences, and quantitative approaches to designing and developing physical artifacts, nor do software engineers receive much, if any, orientation to engineering of systems that include physical artifacts.

Some software engineers have traditional engineering backgrounds and acquire their knowledge of software and software engineering by virtue of education, training, mentoring, and on-the-job experience. However, they, like software engineers whose backgrounds are based in computer science, often typically receive little, if any, exposure to systems engineering competencies.

3.3.1.1 Opportunities for Improving Systems Engineering Literacy of Systems Engineers and Software Engineers

Some succeed at systems engineering activities by native intelligence and some learn about systems engineering from workshops, seminars, and webinars. Others may have opportunities to be trained and mentored and to gain on-the-job experience sponsored by their employers. Directed readings, academic classes, industrial training courses, mentoring, and shadowing also provide opportunities to gain or increase knowledge of systems engineering processes, methods, and tools. Section 2.5 of this text provides background information on systems engineering (10 things software engineers [and others] need to know about systems engineering).

Sources of foundation knowledge include ISO/IEC/IEEE Standard 15288, the *Guide to the Systems Engineering Body of Knowledge* (SEBoK), the *INCOSE Systems Engineering Handbook*, and the *INCOSE Systems Engineering Competency Framework* (ISO 2015, Adcock 2017, Walden et al. 2015, Beasley et al. 2018). A description of the handbook follows; the competency framework is described in "The INCOSE Systems Engineering Competency Framework."

The *INCOSE Systems Engineering Handbook* is a product of the International Council for Systems Engineering (INCOSE). It is written for students and practicing professionals; the latter group includes new systems engineers, product engineers, engineers in another profession who perform systems engineering, and experienced systems engineers who will use it as a reference document. Fledgling and experienced PhSEs and SwSEs will find the handbook to be a valuable resource.

The handbook's fourth edition contains a chapter for each of the four process groups of ISO/IEC/IEEE Standard 15288:2015 (ISO 2015):

- Agreement processes;
- Organizational project-enabling processes;
- Technical management processes; and
- Technical processes.

Each of the 15288 processes in each of the four groups is covered (25 in all). The handbook additionally contains chapters on tailoring, crosscutting methods, and specialty engineering activities. It includes seven appendices.

The handbook is a recommended reference for those preparing to take the exam to become an INCOSE associate systems engineering professional (ASEP), which is the first level of systems engineering certification (INCOSE 2018a).

The *INCOSE Systems Engineering Handbook* is available as a free download for INCOSE members and can be purchased from various sources, including the INCOSE Store, which also has many other useful resources. The store inventory is available for purchase and some items are free for INCOSE members (INCOSE 2018b).

The INCOSE Systems Engineering Competency Framework

The *INCOSE Systems Engineering Competency Framework* (ISECF) provides a framework for a set of systems engineering competencies that describes knowledge, skills, abilities, and behaviors for each competency (ISECF 2018). ISECF includes five areas of competency with multiple competencies in each area. The five areas are core competencies, professional competencies, management competencies, technical competencies, and integrating competencies.

Competencies in each of the five areas include, for example, the following:

- *Core competencies*: Systems thinking and lifecycles;
- *Professional competencies*: Communication, and ethics and professionalism;
- *Management competencies*: Planning, and monitoring and controlling;
- *Technical competencies*: Requirements definition and system architecting; and
- *Integrating competencies*: Project management and quality.

Indicators are included that can be used to verify attainment of the various competencies on five levels for each competency: aware, supervised practitioner, practitioner, lead practitioner, and expert.

(Continued)

(Continued)

Table 1 of ISECF includes some intended users and how they might use the framework. Table 2 describes who might use the framework: individuals for self-assessment, managers for employee assessment, and outsiders for independent assessment.

Guidance is provided for tailoring the framework to fit the needs of organizations when defining job roles for their workers.

ISECF can be obtained from the INCOSE Store at https://connect.incose.org/Pages/Product-Details.aspx?ProductCode=ISECFv1. It is a free download for INCOSE members.

3.3.2 Software Engineering Literacy

Many software engineers receive educations based in computer science or traditional engineering. They typically receive little or no education in systems engineering but they need to understand the concepts and processes of systems engineering to be effective participants in development of software-enabled systems. This section addresses software engineering literacy.

Lack of software engineering literacy creates issues at two levels: lack of software engineering literacy for system engineers and lack of software engineering literacy for some who are called "software engineers."

Systems engineers are usually PhSEs who have traditional engineering degrees and sufficient work experience that qualifies them to undertake systems engineering endeavors. Modern systems increasingly include more software elements, so it is necessary for systems engineers to understand the processes and methods used by software engineers and the reasons why software engineers use those processes and methods. This does not mean that systems engineers have to be software coders. Instead, they need a higher-level understanding of software engineering processes and methods and to understand why software engineers do the things they do. Section 2.4 of this text provides background information on software engineering (10 things systems engineers need to know about software and software engineering).

The second issue is software engineering literacy for software engineers. Unfortunately, the term "software engineer" is used to describe a broad range of abilities that range from those who build web sites using icon-driven software packages with no understanding of software development, to those who perform low-level testing of software modules as directed but are not qualified to write production-quality software code, to those who learned computer programming by education, training, or self-study, to those who understand and apply systems engineering principles to build complex software systems and the software elements of software-enabled systems. Software engineers in the latter category are termed "software systems engineers" in contrast to physical systems engineers.

Some organizations and individuals correctly apply the term "software system engineer" as a job title for those who have the education, experience, skills, and ability to apply the principles of systems engineering to development and sustainment of software-enabled systems that contain significant amounts of software.

3.3.2.1 Opportunities for Improving Software Engineering Literacy of Systems Engineers and Software Engineers

Systems engineers and software engineers can gain understanding of software engineering using the same kinds of opportunities listed above: directed readings, workshops, seminars, webinars, academic classes, industrial training courses, mentoring, and shadowing. In addition, the ISO/IEC/IEEE Standard 12207 for software engineering processes is a companion to ISO/IEC/IEEE Standard 15288 for systems engineering processes (ISO 2017).

The *Guide to the Software Engineering Body of Knowledge* (SWEBOK Version 3.0) and the *Software Engineering Competency Model* (SWECOM Version 1.0) are available for free downloading at the cited web sites (Bourque and Fairley 2014; Fairley et al. 2014). Another source of information is the *Software Extension to the PMBOK® Guide Fifth Edition,* which is described below.

SWEBOK Version 3.0 contains 15 knowledge areas (KAs): 10 KAs cover software engineering processes (requirements, design, construction, and so forth) and 5 are foundations for software engineering (mathematical foundations, computing foundations, and so forth).

SWECOM Version 1.0 was developed to clarify the work activities a competent software engineer is qualified to undertake, at different levels of competency for different skill areas. SWECOM can be used to assess the competencies of software engineers, at different levels of competency for selected skill areas, skill categories, and work activities. SWECOM can also be used as a guide to developing competency improvement plans for individuals and organizations. SWECOM is described in the below box (Fairley et al. 2014).

SWECOM

SWECOM Version 1.0 specifies software engineering competencies for 13 skill areas, each of which includes skill categories, each skill category is composed of activities, and each activity is specified at five competency levels. The 13 skill areas are as follows:

1. Software systems engineering
2. Software process and life cycle

(Continued)

(Continued)

3. Software requirements engineering
4. Software design
5. Software construction
6. Software testing
7. Software sustainment
8. Software quality
9. Software security
10. Software safety
11. Software configuration management
12. Software measurement
13. Human–computer interaction

Each skill area includes skill categories; for example, the requirements skill area includes the following skill categories:

- Software requirements elicitation
- Software requirements analysis
- Software requirements specification
- Software requirements verification
- Software requirement management

Each skill category in SWECOM includes activities, rather than job roles, because job roles are typically dependent on the organizational environment and context in which the work activities occur – SWECOM activities can be grouped into job roles by organizations, organizational units, and projects to satisfy their needs.

The five levels of increasing competency for SWECOM skills are the following:

1. Technician
2. Entry-level practitioner
3. Practitioner
4. Technical leader
5. Senior software engineer

SWECOM senior software engineers are referred to as software systems engineers in this text.

Another source of software engineering insight and knowledge for system engineers (both PhSEs and SwSEs) and others is provided by SWX: the *Software Extension to the PMBOK® Guide Fifth Edition*, which extends and tailors the Project Management Institute (PMI), project management body of knowledge, for management of software projects (Fairley et al. 2013).

The following quote is from the preface to SWX:

> Many of the 47 processes in the *PMBOK® Guide* are applicable to the management of software projects. This extension builds upon the *PMBOK® Guide* by describing additional knowledge and practices and by modifying some of them. It is intended to be consistent with the *PMBOK® Guide*. The primary contribution of this extension to the *PMBOK® Guide* is description of processes that are applicable for managing adaptive life cycle software projects.

Adaptive life-cycle models for software development are discussed in Section 2.4 of SWX and Section 4.5.2 of this text. An integrated-iterative-incremental model for developing software-enabled systems is presented in Chapter 5.

In summary, systems engineers and software engineers can gain increased understanding of software engineering using the same kinds of materials and opportunities listed above:

- Directed readings;
- Workshops, seminars, and webinars;
- Academic classes;
- Industrial training courses;
- Mentoring; and
- Shadowing.

Plus:

- SWEBOK;
- SWECOM; and
- SWX.

Becoming registered professional engineers (PEs) presents another opportunity for systems engineers and software engineers to increase their professional credentials, as indicated in Tables 1.1 and 1.2. Systems engineers and software engineers in the United States who have engineering degrees from ABET-accredited universities are qualified to become registered PEs by completing the required steps: passing a field exam in an engineering discipline, completing four years of supervised mentorship, and passing the PE exam after four years. Evidence of continuing engineering practice must be submitted periodically to maintain PE certification. Other countries have similar procedures for licensing systems engineers and software engineers as PEs. Some organizations and governmental agencies require PE credentials for the engineering leaders on projects, where the resulting systems might place the public at risk for safety and security (e.g. mechanical and civil engineers).

Section 3.4 describes a second inhibitor to increased synergy between systems engineers and software engineers; it includes the failure to communicate effectively because of the different ways in which the "terms of art" are used in the two disciplines. Opportunities for improvements are also presented.

3.4 Differences in Terminology

Unambiguous communication requires common terminology with commonly understood meanings of the terms used. There is a famous quote variously attributed to Oscar Wilde, George Bernard Shaw, and Winston Churchill:

> Britain and America are two nations divided by a common language.

Speakers of British English and American English use the same terms to denote different concepts. For example, British football and American football are different sports. Brits and Americans also use different terms to denote the same concept. For example, a British first floor and an American second floor are the same floor in a building.

All engineering disciplines develop terminology that has concise meanings and is used as shorthand to ensure precise communication. The differences between physical artifacts (e.g. hardware) and logical artifacts (e.g. software) and the different approaches used by system engineers and software engineers to develop systems are manifest in the different meanings of shared terms used in the two disciplines.

To paraphrase the above quote:

> Systems engineering and software engineering are two disciplines divided by a common language.

Systems engineers and software engineers, like British English and American English speakers, also use identical terms to denote different concepts and different terms to denote the same concept.

Some of the commonly used terms that have different meanings in systems engineering and software engineering are described below. Opportunities for bridging the terminology gap are presented in Section 3.4.8.

3.4.1 Hierarchical Decomposition

The term "hierarchy" denotes the decomposition of a system into subordinate elements. However, the term "hierarchy" is sometimes used with different meanings when applied to physical entities and software entities.

3.4.1.1 System Hierarchy

Hierarchy is used in systems engineering to depict functional, product, and work breakdown structures. The purpose of hierarchical decomposition is to decompose an entity or concept into successive levels of detail. Lower-level elements provide the details of higher-level elements. Conversely, higher-level elements are aggregations of lower-level elements. The template for a (partial) product breakdown structure (PBS) is depicted in Figure 3.1.

The white diamonds in Figure 3.1 denote aggregation. The aggregated elements at the tails of the lines are members of the element at the head of the line. The aggregated elements are independent parts of the aggregator, as for example a power supply or a keyboard that can be separately procured could be part of a system. The black diamond indicates composition, which is a form of aggregation in which the composed elements do not exist independent of the composer. A special purpose chip on a circuit board fabricated for sole use on the circuit board would be a composed element of the board. Software objects created at system runtime exist only as a part of an executing program. They are created at system startup or by command and are deleted by command or when the software shuts down (see below).

Guidelines for decomposing a system using hierarchical decomposition are as follows:

- Include all elements and attributes of a higher-level entity in the subordinate entities using top-down decomposition; or
- Collect all elements and attributes of subordinate entities into a superior entity using bottom-up aggregation or composition;
- Partition subordinate entities into relatively autonomous units that each has well-defined functionality and minimal but necessary behavioral interactions with the other units; and
- Decompose a system so that lowest level elements in the hierarchy can be assigned to individual engineers or interdisciplinary teams.

Figure 3.1 Template for a partial product breakdown structure.

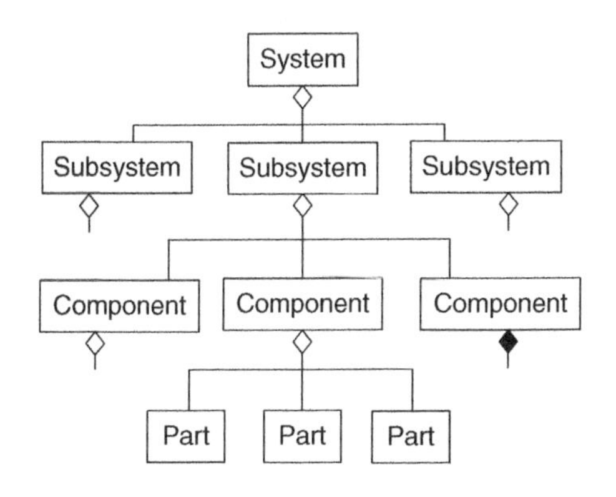

Decomposing a PBS for a software-enabled system should not be used to create artificial distinctions, such as partitioning a unit of integrated hardware/software into separate hardware and software parts.

3.4.1.2 Software Hierarchy

Packages and classes provide two ways of establishing software hierarchies; they are diagram types in the Universal Modeling Language (UML) and are used to document software specifications and designs (UML 2017). Packages are general-purpose constructs that can be used to encapsulate collections of UML diagrams of all types, including other packages. Functional hierarchies of packages can be specified when using the "divide and conquer" approach to system decomposition. However, some software engineers, not understanding the constraints imposed on development of physical elements, think that functional decomposition as used by systems engineers is old fashioned and out of date. Software engineers use packages for other forms of encapsulation.

With the advent of object-oriented software development, generalization/specialization relationships for software classes are used to depict software hierarchy, as illustrated in Figure 3.2.

A class, as illustrated in Figure 3.2, is a fundamental construct of the UML (2017). Classes have four parts: names, attributes (aka properties), operations (aka methods), and ports that provide interfaces to other UML entities. Superior classes in a class hierarchy (called superclasses) provide attributes and operations that are shared among subordinate classes (called subclasses). The subclasses are specializations of the superclass; they are said to inherit the attributes and operations of the superclass.

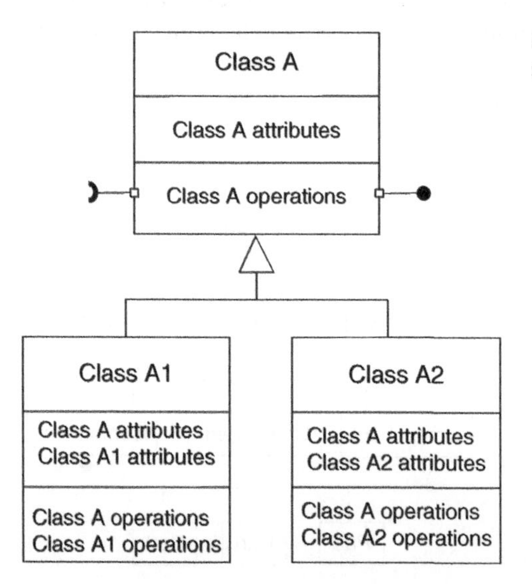

Figure 3.2 A generalization/specialization class hierarchy.

Conversely, a superclass contains attributes and operations that are common to and shared by the subclasses, thus providing generalization of the common properties of the subclasses. Subclasses that may be added to the hierarchy in the future will also inherit the attributes and operations of the superclass.

UML classes also include ports (the input and output icons in Figure 3.2). Ports are used to specify interconnection points among classes and other UML constructs. The operations in Class A might, for example, include an algorithm to receive input data through input port A from the output of another object and restructure it into a format accessible for use by Class A objects, its subclass objects, and other objects that would connect their input ports to the Class A output port B.

Generalization/specialization hierarchies for software are usually restricted to two, at most three, levels of hierarchy to avoid hierarchical complexity.

A variety of connector types are provided in UML to establish associations among classes that collaborate on the same level of hierarchical abstraction, in contrast to functional decomposition, which establishes levels of abstraction. One of the common flaws of object-oriented design is to establish a top-level hierarchical "God class" that, when instantiated, will functionally control interactions of the instantiations of other classes.

Note: Classes are design entities – they are not executable. Classes are written and stored in libraries and then used to create multiple executable instances as desired (i.e. executable objects are instantiations of classes). Object constructors are used to instantiate classes using notations such as

myObject is Class A1; yourObject is Class A1;

or

Class A1.myObject; Class A1.yourObject.

Objects can be created and destructed, as desired, during execution of a software program. An object destructor might be stated in a programming language as

Delete myObject.

3.4.2 UML Classes and SysML Blocks

The UML was developed for specification and design of software. The Systems Modeling Language (SysML) was developed for specification and design of systems. SysML is a subset and extension of UML; it includes four unaltered UML diagrams, three modified UML diagrams, and two new diagrams (UML 2017; Friedenthal 2011).

SysML blocks are extensions of UML classes. As indicated in Figure 3.2, a UML class has attributes (aka properties), operations (aka methods), and ports that provide interfaces to other UML entities. SysML blocks extend the syntax and semantics of UML classes.

Properties of SysML blocks include parts, references, and values. Ports include flow ports, full ports, and proxy ports. Flow ports provide for flows of energy and material, in addition to data. Like UML, SysML provides a variety of connector types to establish associations among blocks that collaborate.

Block diagrams include block definition diagrams (BDDs) and internal definition diagrams (IDDs). BDDs are black-box diagrams that include names and types of different block elements in different compartments of the block (e.g. operations, flow properties, parts). IDDs are white-box diagrams that specify the internal details of BDDs. They can be embedded in BDDs or shown as separate blocks connected to their BDDs.

The SysML package, unaltered from UML, is a general-purpose encapsulation mechanism that can be used to encapsulate other SysML constructs. Packages can be used to encapsulate hierarchies of blocks as desired. Aggregation and composition, as illustrated in Figure 3.1, can also be used to establish hierarchies. BDDs and IDDs can also be used to establish hierarchies.

UML classes and SysML blocks are similar constructs used for similar purposes but the differences are significant. Systems engineers who use SysML and software engineers who use UML do not need to be experts in the differences but they do need to be aware that they are speaking of similar but different entities when they use the terms "class" and "block."

3.4.3 Performance

System engineers often use "performance" to describe all of the observable functional, behavioral, and quality characteristics of a system; they may include system attributes such as real-time responsiveness, availability, reliability, and robustness. The term "performance envelope" is used both narrowly and broadly in systems engineering. In aerodynamics, for example, "performance envelope" narrowly refers to relationships among airspeed, altitude, and maneuverability. However, the terms performance and performance envelope are used more broadly in other areas of systems engineering.

In contrast, software engineers use "performance" in a much narrower sense to refer to response time and throughput of data processing in a computational system or subsystem. The term "performance envelope" is not used in software engineering.

3.4.4 Prototype

Systems engineers use "prototype" to mean an approximation of a system that includes simulated elements and is functionally executable. An early prototype may be successively elaborated until a full representation of a system is achieved to provide a basis for system implementation. The term "prototype" is also used, in the systems engineering sense, to mean a first fully functioning version of an implemented system that is evaluated in real operational environments for characteristics such as robustness, ruggedness, and fitness for use.

In contrast, software engineers use "prototype" in two other ways: first, to describe a mock-up of a user interface that is used early in a project to gain reactions and feedback from representative users as an aid to eliciting stakeholders' requirement; functionality and behavior behind the interface is simulated using other software. Second, software prototyping is used as needed during software development to investigate technical issues such as the behaviors of different algorithms under various simulated and real conditions. Software engineers use the terms "alpha version" and "beta version" to describe the first fully functioning versions of a software product.

3.4.5 Model-Based Engineering

Model-based engineering (MBE) emphasizes the use of visual models during development and sustainment of engineered systems. Model-based system engineering (MBSE) is an approach to system development that uses visual models to supplement and replace documents during system analysis, design, and development. Models are expressed in notations such as UML and SysML and are stored as digital data that is used to construct various kinds of computer-based simulations and emulations. An MBSE initiative was launched as part of the INCOSE SE 2020 Vision and continues at the present time. INCOSE and a Domain Special Interest Group (DSIG) of the Object Management Group are sponsors of the MBSE initiative (SEDSIG 2018). Modeling and simulation-based systems engineering is an extension of MBSE that includes simulation of system models.

Software engineers pursue modeling and simulation (M&SBSwE) and they also pursue model-based software engineering (MBSwE) in a different way; it is an approach that splits software development into two processes: domain engineering and application engineering. Domain engineering is concerned with developing software assets that can be used in different applications within a domain such as medical devices or automobiles. Domain engineering includes domain analysis, domain design, and domain implementation. Application engineering is used to develop individual applications within a domain using the domain assets; it includes requirements engineering, design, construction, integration, and testing of an application program.

MBSwE is widely used in development of software product lines (Donohoe 2014). Figure 3.3 illustrates packaging of a software system that includes a domain package and an application package. The application packages might provide different natural language interfaces for the domain package (i.e. English, German, Japanese).

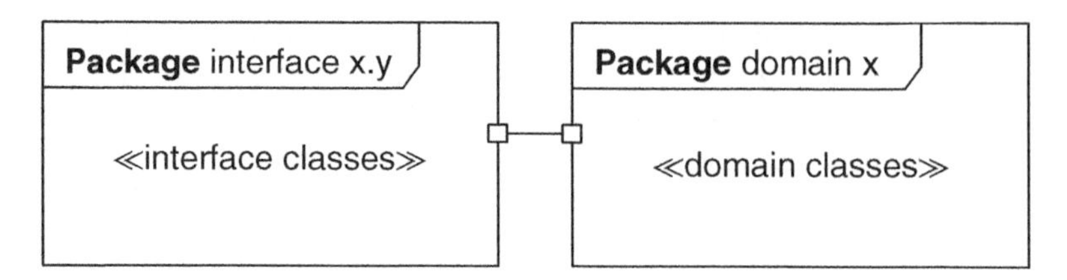

Figure 3.3 A software product line instantiation.

Different natural language ports in the domain package could be used to connect the application packages, translate natural language inputs into the language of the domain package, which performs computations, and use the appropriate output port to translate and deliver the results to the application package user interface. The association line between the interface package and the domain package indicates a two-way input/output communication between the packages.

"Model-based" thus implies some similar and some different approaches in systems engineering and software engineering.

Another difference in terminology is the contrasting use of the terms "implementation" by systems engineers, "construction" by software engineers, and "realization" by both systems engineers and software engineers.

3.4.6 Implementation, Construction, and Realization

The terms implementation, construction, and realization are sometimes used to denote the same thing and are sometimes used to denote different things.

According to Subclause 4.7.1.1 of ISO/IEC/IEEE standards 15288 and 12207, the purpose of the implementation process is "to realize a specified system element" (ISO 2015, 2017).

A "system element" is, according to the Software and Systems Engineering Vocabulary, "a discrete part of a system that can be implemented to fulfill specified requirements" (SEVOCAB 2010). System elements include physical elements, software elements, and procedures for performing manual operations.

The Merriam Webster dictionary defines "realize" as "to bring into concrete existence." Note that "realize" is used above to denote bringing individual system elements into concrete existence.

"System realization" is, according to the SEBoK Glossary, "The activities required to build a system, integrate disparate system elements, and ensure that a system both meets the needs of stakeholders and aligns with the requirements identified in the system definition stage" (Adcock 2017).

SEBoK Part 3 indicates that system realization includes the following:

- System implementation;
- System integration;

- System verification; and
- System validation.

Systems engineers thus apply "implementation" to realization of system elements and realization of total systems. In contrast, software engineers use the term "implementation" to denote integration of a software package or a software-based service into the workflow of an end-user or an end-user organization. A software engineer would refer to transitioning a database package into a live operational financial system as having implemented the database package that had previously been constructed or procured.

To avoid confusion, the term "software construction" is used in this text to denote realization of individual software code elements (i.e. designing, writing, and unit testing of code elements that are written to fulfill specified requirements). Software construction is preceded by requirements definition, architecture definition, and design definition. Detailed design is usually included as part of software construction, in contrast to hardware implementation for which detailed design precedes implementation.

3.4.7 Specialty Engineering and Supporting Processes

Specialty engineering denotes the engineering disciplines that are outside the "mainstream" engineering disciplines of analysis, design, realization (i.e. implementation of physical elements and construction of software elements), integration, verification, transition, and validation. The specialty disciplines include security, reliability, and other disciplines described in the SEBoK Part 6 KA, Systems Engineering and Specialty Engineering (Adcock 2017). Different engineers specialize in different specialty disciplines.

Chapter 10 of the *INCOSE Systems Engineering Handbook* includes 14 kinds of specialty engineering activities. These activities and the specialty engineers who perform them cover the "quality characteristics," some of which may be designated as critical quality attributes for a deliverable system (Walden et al. 2015).

Specialty engineers for various critical quality attributes may be assigned full-time to a project or their skills may be required on an as-needed basis depending on the domain, size, and complexity of a system. Large organizations may have specialty engineers on staff or specialty engineers may be contracted as consultants and subcontractors.

Software engineers often use the term "supporting processes" to denote the processes of systems engineering specialty engineering disciplines. The former version of ISO/IEC/IEEE Standard 12207 for software processes (12207:2008) included a section on "supporting processes." These processes are no longer identified as such in the current version of the software engineering process standard (ISO 12207:2017) but the term "supporting process" may continue to be encountered when software engineers refer to the systems engineering specialty disciplines. Annexes E.4 of the current versions of 15288 and 12207 provide a process view for specialty engineering (ISO 2015, 2017).

3.4.8 Opportunities for Improving Usage of Terminology

Terminology discussed above includes hierarchy; aggregation and composition; performance; prototype; model-based and model-based simulation; implementation, construction, and realization; and specialty engineering and supporting processes.

Usage of terminology among systems engineers, software engineers, disciplinary engineers, and specialty engineers can be improved by

- Consistent use of terms by respected opinion leaders;
- Focused readings, training sessions, and seminars that use consistent terminology; and
- Maintaining an organizational glossary of terms that is supplemented with additional terms for each project, as needed.

The organizational glossary can be reviewed and additions prepared during the initiation phase of each project; they can be reviewed during a kick-off meeting for a new project. New terms and special meanings of present terms may be determined by the nature of the stakeholders; an acquirer's unique terminology; the product domain; the engineering disciplines; the involved disciplinary and specialty engineers; and the product being developed, the service being provisioned, or the enterprise being enabled. Project participants can be encouraged to use the accepted terminology, which should be used in all written and verbal communication.

As mentioned above, a glossary of systems engineering and software engineering terms is provided in ISO/IEC/IEEE International Standard 24765:2010 Systems and Software Engineering – Vocabulary (SEVOCAB 2010). It can be used as a guideline for developing an organizational glossary and adding terms that have unique meanings for individual projects.

3.5 Differences in Problem-Solving Styles

Different approaches to problem solving result from the differences in education, training, and experience of systems engineers and software engineers. Most systems engineers receive a tradition engineering education based on foundations of calculus and differential equations, physics, chemistry, thermodynamics, statics, dynamics, and electromagnetism. Traditional engineers are taught to reason about physical phenomena using metrics, measures, equations, and calculations expressed in the notations of continuous mathematics. Engineering metrics include measures of well-defined physical quantities such as mass, volume, distance, temperature, energy, power, force, duration, and velocity. Measures include grams, centimeters, cubic centimeters, kelvin, joules, watts, dynes, meters, and meters per second. Precise, quantitative measurements of physical entities are based on standardized international measures for physical entities.

Software engineering education and problem solving are based on theories and applications of computer science, which is a science of the artificial (Simon 1996). Software engineers are taught to reason about logical phenomena stated as algorithms and data definitions expressed in discrete mathematical notations. Software metrics are used to measure attributes such as program size and complexity, defects by category, and measures of software quality and user satisfaction. The measures include lines of code, levels of coupling and cohesion, number of defects, and quality measures such as robustness and availability. Quantitative measurement of software attributes is less precise than traditional engineering measurements because of the logical nature of software and lack of standardized measures. Qualitative metrics (i.e. subjective measures) are used for attributes such as ease of learning, ease of use, and maintainability (Kan 2003).

Systems engineers and traditional engineers tend to approach problem solving by focusing on identifying, specifying, fabricating, and procuring physical elements that will satisfy the problem statement. The physical elements are typically specified using quantified "black-box" terms for inputs and outputs. Software engineers tend to focus on the features, quality attributes, and services to be provided to system users, operators, maintainers, and other hands-on stakeholders. System engineers tend to think "hardware first/software second" and software engineers tend to think "software first/hardware second."

A hardware first/software second approach to the automated teller machine (ATM) problem statement of "providing access to financial accounts at convenient times and locations" would first identify the hardware elements (card reader, keypad, display, etc.). An aggregation diagram is provided in Figure 3.4.

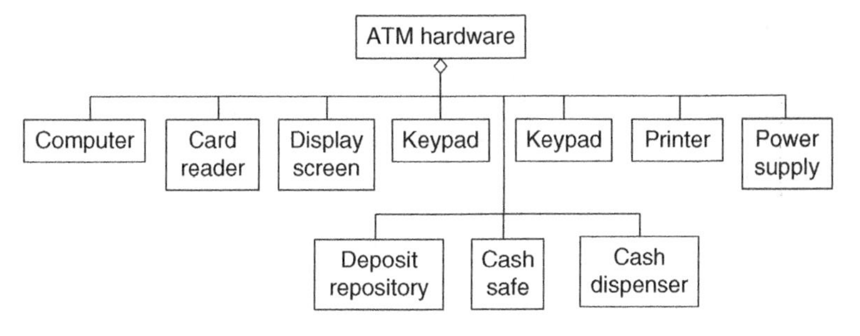

Figure 3.4 Aggregation of ATM hardware elements.

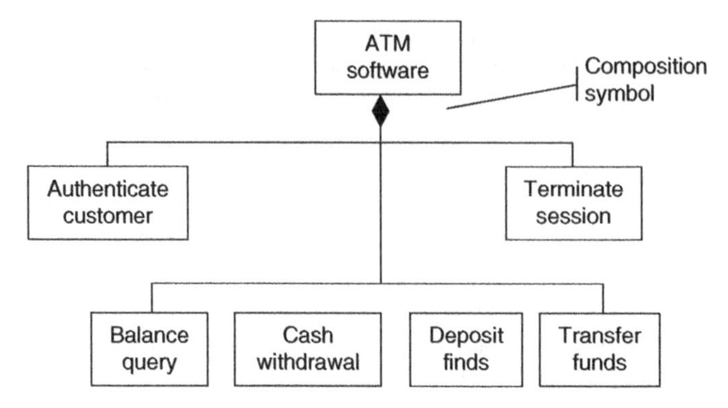

Figure 3.5 Composition of ATM software elements.

Each of the hardware elements in Figure 3.3 is part of an ATM but exist independent of the ATM, as indicated by the aggregation symbol in the figure. Some elements may be procured from available sources and some may be fabricated.

A software first/hardware second approach would identify the services to be provided (authentication, transactions to be supported, termination, etc.). A composition diagram is provided in Figure 3.5.

The software elements in Figure 3.5 are part of the ATM software but they do not exist independent of instantiations of the ATM software, as indicated by the composition symbol in the figure.

Both approaches can satisfy the problem statement but the first approach places software in a subordinate position to hardware and the second approach places hardware in a subordinate position to software.

Hardware first and software first approaches are a primary reason systems are delivered late; either the software engineers are waiting for detailed requirements from the hardware engineers or the hardware engineers are waiting for detailed specifications from the software engineers. A concurrent hardware–software approach to system development is presented in Chapter 5.

3.5.1 Opportunities for Improving Styles of Problem Solving

Different approaches to problem solving used by PhSEs and SwSEs are based on the differences between physical and logical artifacts and the resulting differences in education, training, and experience.

Frustration based on the different approaches to problem solving can be reduced by PhSEs learning to appreciate the ways in which software engineers take advantage of the different approaches to software development afforded by the malleability of software. PhSEs should also understand that while software can always be changed, the changes are not always easy or instantaneous. Ripple effects introduced by changes to software can result in undesired changes in functionality, behavior, and quality attributes in other parts of a system. Requested changes to software may not be possible within the constraints on schedule, resources, personnel, and technology.

Conversely, SwSEs need to understand the constraints and limitations imposed when engineers develop physical artifacts and that it is not unreasonable for changes to software be requested to compensate for hardware shortfalls, especially late in the system development process.

Given the close connections between system engineering and software engineering (the intimate intertwining introduced in Section 1.1), it is desirable for PhSEs (and traditional disciplinary engineers) to learn some of the approaches to problem solving in software engineering and for SwSEs (and disciplinary software engineers) to learn some of the approaches to problem solving used by PhSEs and traditional engineers.

Education and training of PhSEs (typically educated as traditional engineers) can include topics on the sciences of the artificial and the implication for solving logical, as opposed to physical, problems (Simon 1996). Education and training of software engineers can include topics on the constraints on engineering of physical systems and the implications for problem solving.

Understanding the problem-solving styles of PhSEs and SwSEs presented in this chapter do not appear to be furthered when systems engineers and traditional engineers take an introductory programming class or when software engineers take an introductory physics or engineering class. Instead, the distinctions between physical and logical entities and "process distinctions" for problem solving can be introduced into traditional engineering, systems engineering, and software engineering classes. Industry can (and should) take some responsibility by providing training, mentoring, and shadowing programs.

Both software engineering degree programs and traditional engineering degree programs tend to focus on design topics in the upper division courses with one or more required senior design projects. These courses provide opportunities to introduce approaches to problem solving in traditional engineering and in software engineering.

Most computer science programs include one or more software engineering courses and a one- or two-term senior project. Virtually all software engineering programs focus on processes, methods, and tools in the upper division courses. Opportunities thus exist for introducing basic issues in software engineering problem solving into systems engineering programs and to similarly introduce basic issues in systems engineering problem solving in software engineering programs.

Interdisciplinary courses and projects provide additional opportunities to cross-sensitize traditional engineers and software engineers. Some universities offer courses in software systems engineering for system engineers and some offer certificates in software and systems engineering. Concurrent problem solving can be experienced in interdisciplinary project courses that include software engineering, systems engineering, and traditional engineering students.

Systems engineers and software engineers apply holistic approaches and reductionist approaches to problem solving but they apply them in different ways, as described in Section 3.6.

3.6 Holistic and Reductionist Problem Solving

All systems engineers (both physical and software) engage in holistic and reductionist thinking; they initially focus on the holistic (whole) vision of a system, including a system's intended mission or business purpose; the operational needs of different kinds of system users, operators, maintainers, and other system stakeholders; environmental, resource, and technical constraints; and the social, societal, and political impact of introducing a new or modified system into one or more operational environments. Having understood, to the extent possible, the holistic view of a complex system, system engineers develop top-level system requirements and system architecture.

Reductionist thinking is then applied to recursively decompose the functional, behavioral, and quality requirements and the system architecture. Logical design is then applied to specify the relationships among the physical elements, the software elements, and the interfaces. Physical design identifies the different kinds of physical elements that will be needed, such as electrical motors, power supplies, and transformers, and mechanical structures, fixtures, and housings. Successful system engineers thus use both holistic and reductionist approaches.

PhSEs sometimes apply reductionist thinking incorrectly when software requirements are allocated to the disciplinary software engineers following system requirements definition, system architecture definition, and design definition without including SwSEs from the beginning of a project.

In these situations, the software engineers engage in the reductionist thinking needed to translate allocated requirements into software elements that will satisfy those requirements. The extreme level of concentration on details necessary to construct logically correct software further reinforces reductionist thinking because one misplaced semicolon or one incorrect but undetected "greater than" symbol rather than the intended "less than" symbol when embedded in a million lines of software code can cause software to malfunction and result in damage to or destruction of a large, complex system and perhaps result in injury to or loss of human life.

The reductionist approach of software engineers in a systems engineering environment sometimes leads PhSEs to complain that software engineers have tunnel vision and do not understand, or seek to understand, the bigger picture. Conversely, software engineers complain that their ideas of how to improve the overall system structure, behavior, and quality attributes are ignored.

When less constrained by the PhSEs and the development process, software engineers also engage in holistic and reductionist thinking, but within the environment of physical systems engineering projects, software engineers are sometimes regarded as engineering specialists who are directed to implement allocated requirements.

3.6.1 Opportunities for Improving Holistic and Reductionist Thinking

System engineers (both physical and software) apply both holistic and reductionist thinking, when possible. In some cases, software engineers are not given the opportunity to apply holistic thinking by participating in system definition, requirements engineering, and system design. Being regarded as component specialists reinforces reductionist thinking. To improve synergy and reduce frustration, PhSEs and SwSEs can jointly engage their teams of traditional engineers and software engineers in holistic and reductionist thinking to conceive, analyze, design, and develop software-enabled systems. Enhancement of both holistic and reductionist thinking, when applied to both kinds of systems engineering, can be broadened by seminars, workshops, readings, and day-to-day interactions in a system development environment that does not discourage collaboration.

Chapter 5 of this text introduces the I^3 development model that eliminates many of the problems described above by facilitating concurrent implementation of physical elements, construction of software elements, and incremental integration of physical and software elements.

3.7 Logical and Physical Design

Design of physical elements and software elements are both concerned with specifying structure and behavior. PhSEs and traditional disciplinary engineers distinguish two kinds of design: logical and physical. Logical design is concerned with specifying the characteristic of and interfaces among system elements. Physical design is concerned with the (literally) nuts-and-bolts details of physical systems needed to satisfy the logical design specifications and constraints. PhSEs tend to focus on the logical design while disciplinary engineers focus on the physical design of the system elements they implement.

The terms "preliminary design" and "detailed design" are sometimes used to differentiate architecture definition and logical design from physical design. Reviews and simulations are conducted to determine the acceptability of the architecture and the logical and physical designs. The reviews are sometimes referred to as "preliminary design reviews" (PDRs) and "critical design reviews" (CDRs). PDRs and CDRs are both important, but CDRs are critical for PhSEs and disciplinary engineers because a CDR typically precedes start of fabrication and procurement of physical system elements, which are more difficult and expensive to modify later than is software; so the goal is to "get it right" before commencing implementation.

Software designers, like designers of physical artifacts, distinguish preliminary design from detailed design. Architectural design of software, like architectural design of physical entities, is concerned with specifying structure and behavior. Detailed design of software is concerned with selecting algorithms, data representations, and the details of interfaces among software elements that will satisfy the requirements allocated to software. Software is a logical medium so software design is logical design. However, database designers sometimes use the term "logical design" to characterize the activities of defining database schema and "physical design" to characterize the details of the schema.

Detailed design of software is usually considered to be part of software construction, in contrast to design of physical elements where detailed design precedes implementation. Detailed design of software involves using logical constructs to specify the construction details for the software code that will satisfy the architectural and detailed design specifications and provide the specified functional, behavioral, and quality attributes plus the interfaces and interconnections required of the software.

It is a myth that iterative software development does not include analysis and design activities. In some cases, there may be no distinct architectural or detailed design phase but design activities are accomplished during planning and then evolved iteratively and incrementally (see Section 4.5.2).

In summary, PhSEs engage in logical design and physical design. Software engineers mostly engage in logical design. Also, detailed design of physical elements precedes implementation, while detailed design of software is typically part of construction.

3.7.1 Opportunities for Improving Logical Design and Physical Design Activities

To reduce frustration and improve synergy, software engineers need to understand and appreciate the constraints imposed on systems engineering development processes by the nature of physical entities; the importance of physical design and design reviews; and the difficulty of changing the structure, functionality, behavior, or quality attributes of physical elements after fabrication and procurement have commenced. Software engineers also need to understand why system engineers and disciplinary engineers request that software engineers address needed changes during system development.

Systems engineers and traditional engineers need to understand that software engineers may not strictly observe the development phases of analysis, architectural design, physical design, and logical design because those phases are interleaved and intermixed with software construction in various ways. And PhSEs and disciplinary engineers need to understand that while changes to software are always possible, requested changes are not always easy and may have ripple effects that will impact a system in undesirable ways. And it may not be possible to effect requested changes within the constraints of resources, schedule, and technology.

3.8 Differences in Process Models and Technical Processes

A process model is a collection of named project phases, such as requirements engineering, design, and implementation. Each phase includes inputs, activities to be performed, and outputs to be produced. Interactions among project phases may be iterated and interleaved in various ways. The process models favored by PhSEs and those favored by SwSEs are often at odds, based on the differing characteristics of physical entities and software. The technical processes also differ for the same reasons.

3.8.1 Opportunities for Improving Process Models and Technical Processes

Chapter 4 presents process models used by PhSEs, SwSEs, and their disciplinary counterparts. Chapter 5 introduces the I^3 development process that supports concurrent development of physical elements, software elements, and seamless integration. Chapters 6–9 cover the technical processes of systems engineering and software engineering when using the I^3 system development model. The processes are based on ISO/IEC/IEEE standards 15288 and 12207 (ISO 2015, 2017).

Finally, resolving the issues presented in Chapter 2 and this chapter can improve workplace respect among PhSEs, SwSEs, and their disciplinary engineers, as described Section 3.9. Failure to resolve these issues can result in loss of workplace respect.

3.9 Workplace Respect

It is human nature to discount what we don't understand, including respect for those in the workplace who differ from us in their working styles, skills, abilities, and personalities. Failure to understand the nature of others' work experiences; the roles they play at work; the processes, methods, and tools they use; and the constraints under which they do their work can result in disrespect for them and their work.

This can result in cynical questions such as

- What do they do all day?
- Why don't they do things the way I do them?

Disrespect can cause PhSEs and traditional disciplinary engineers to view software engineers in the following light:

- Software engineers are too narrowly focused; they don't show any interested in the bigger picture;
- Software engineers seem to focus on the development processes they use more than the work products they develop;
- Software engineers have a different mindset than do other engineers;
- Software engineers have different approaches to problem solving than do other engineers;
- Software engineers think we PhSEs and traditional disciplinary engineers use outdated processes; and
- Software engineers aren't "real" engineers.

Conversely, software engineers sometimes view PhSEs and traditional engineers as follows:

- They never include us in problem analysis, architecture definition, design definition, or trade studies for alternative designs;
- System requirements allocated to software are often vague and ambiguous;
- Development constraints and technical constraints are not clearly communicated;
- Interfaces to hardware elements and connections to the system environment are not adequately specified;
- Software engineers are not informed, in a timely manner, of changes to system architecture, system design, and interface specifications;
- Technical meetings with customers, other engineers, vendors, and subcontractors do not include software engineers;
- Systems engineers are not available for consultation when needed; and
- Systems engineers are "big picture" thinkers; they don't seem to care about the details.

3.9.1 Opportunities for Improving Workplace Respect

Issues and opportunities for improvements presented in Chapter 2 and this chapter can result in increased work place respect, which will in turn result in increased workplace morale and productivity and the quantity and quality of work products developed. Issues and the techniques for managing and leading systems engineering/software engineering teams are presented in Chapter 12 of this text.

3.10 Key Points

- Many who engage in systems engineering and software engineering do not have sufficient literacy in their respective disciplines or sufficient literacy in the opposite discipline.
- Systems engineers and software engineers use different process models, technical processes, and styles of problem solving based on education and experience and on the differences between physical media and the software medium.
- PhSEs distinguish between mainstream development processes and specialty engineering processes; SwSEs distinguish between mainstream development processes and supporting processes.
- Different problem-solving styles and the development processes used by PhSEs and SwSEs result, in part, from different meanings of the shared terms used in the two disciplines.
- Difference in terminology can result in ambiguous and confusing communications among PhSEs, SwSEs, and disciplinary engineers.
- Holistic and reductionist thinking is applied in both systems engineering and software engineering.
- Software engineers may not be given the opportunity to apply holistic thinking within the context of systems engineering.
- MBE has different meanings in systems engineering and software engineering and is applied in different ways.
- SwSEs and disciplinary software engineers accomplish development processes of analysis, design, implementation, verification, transition, and validation in different ways than do PhSEs and their traditional disciplinary engineers.
- Detailed design of physical elements precedes implementation; detailed design of software elements is part of software construction.
- Increased respect for counterpart engineers and their work processes can result in better workplace morale and increased productivity. Respect is based on understanding.
- Improvements can be made for systems engineering and software engineering by improving professional literacy, use of terminology, styles of problem solving, holistic and reductionist thinking, logical design and physical design, implementation and construction, process models, technical processes, and workplace respect.
- Background information for systems engineering is available in the *Guide to the Systems Engineering Body of Knowledge*, the *INCOSE Systems Engineering Handbook*, and ISO/IEC/IEEE Standard 15288 (Adcock 2017; Walden et al. 2015; ISO 2015).
- Background information for software engineering is available in the *Guide to the Software Engineering Body of Knowledge*, the *Software Engineering*

Competency Model, and ISO/IEC/IEEE Standard 12207 (Fairley et al. 2014; Bourque and Fairley 2014; ISO 2017).
- Terminology for systems engineering and software engineering is provided in ISO/IEC/IEEE 24765:2010 – Systems and Software Engineering – Vocabulary (SEVOCAB 2010).

Chapter 4 presents process models commonly used to develop physical systems and software systems. Chapter 5 presents the I^3 system development model.

This chapter concludes Part I of the text, which provides the context and background for the remainder of the text. Part II commences with Chapter 4.

Exercises

3.1. In Section 3.3, literacy is defined as the ability to read and write, but more generally, literacy is knowledge of a specific subject.
   (a) List and briefly discuss a topic other than engineering in which you are literate.
   (b) List and briefly discuss a topic in which you would like to be more literate.

3.2. Difficulties in communicating occur when different people use the same term to mean different things and when different terms are used to mean the same thing.
   (a) Briefly describe a difficulty in communicating when you and someone else used the same term to mean different things. Then briefly describe how the difficulty was overcome.
   (b) Briefly describe a difficulty in communicating when you and someone else used different terms to mean the same thing. Then briefly describe how the difficulty was overcome.
   (c) Briefly describe another difficulty in communicating you have experienced that was not based on differences in terminology. Then briefly describe how the difficulty was overcome.

3.3. Performance means different things to different people. Briefly describe what performance means to you and provide an example.

3.4. Holistic thinking involves thinking about a complex situation in its entirety. Briefly describe a complex situation to which you have applied holistic thinking. Then briefly describe how holistic thinking helped you understand the situation.

3.5. Reductionist thinking involves thinking about a complex situation by decomposing it into its elements and then focusing on the elements of the situation and how they interact. Briefly describe a complex situation to which you have applied reductionist thinking. Then briefly describe how reductionist thinking helped you understand the situation.

3.6. An analytic approach to solving a problem involves characterizing the problem using numbers, mathematical equations, and/or logical expressions and then finding a solution that satisfies the characterization. Briefly describe a problem you solved, or could have solved, using an analytic approach.

3.7. A pragmatic approach to solving a problem involves seeking a solution that meets immediate needs in the simplest way. Briefly describe a problem you solved, or could have solved, using a pragmatic approach.

3.8. A subjective approach to solving a problem involves seeking a solution that will satisfy the wants, desires, and feelings of those who will be affected by the solution. Briefly describe a problem you solved, or could have solved, using a subjective approach.

3.9. Logical design and physical design are described in Section 3.8.
(a) Briefly describe a situation where logical design alone would be an appropriate technique.
(b) Briefly describe a situation where physical design alone would be an appropriate technique.

3.10. Respect involves understanding the qualities, abilities, or achievements of a person.
(a) Briefly describe the qualities, abilities, and/or the achievements of someone you respect.
(b) Briefly describe someone you learned to respect after you understood the qualities, abilities, and/or the achievements of that person.

References

Adcock, R. (EIC) (2017). *The Guide to the Systems Engineering Body of Knowledge (SEBoK)*, v. 1.9. Hoboken, NJ: The Trustees of the Stevens Institute of Technology www.sebokwiki.org (accessed 12 June 2018).

Ardis, M., Fairley, D., Hilburn, T. et al. (2014). *Software Engineering Competency Model*, Version 1.0, (SWECOM). IEEE Computer Society. https://www.computer.org/volunteering/boards-and-committees/professional-educational-activities/software-engineering-competency-model (accessed 20 Sept 2018).

Beasley, R., D. Gelosh, M. Heisey et al. (2018), INCOSE Systems Engineering Competency Framework, INCOSE-TP-2018-002-01.0, International Council on Systems Engineering.

Bourque, P. and Fairley, R. (eds.) (2014). *Guide to the Software Engineering Body of Knowledge (Version 3.0)*. IEEE https://www.computer.org/web/swebok/v3-guide (accessed 8 June 2018).

Donohoe, P. (2014). *Introduction to Software Product Lines*. Software Engineering Institute https://resources.sei.cmu.edu/library/asset-view.cfm?assetid=423718 (accessed 12 June 2018).

Fairley, R. (2009). *Managing and Leading Software Projects*, Chapter 2. Wiley.

Fairley, R., Fewell, J., Griffiths, M. et al. (2013). *Software Extension to the PMBOK® Guide*, 5e. Project Management Institute https://www.pmi.org/pmbok-guide-standards/foundational/pmbok/software-extension-5th-edition (accessed 9 June 2018).

Friedenthal, S. (2011). Modeling System Interfaces with SysML v1.3.

INCOSE (2018a). Systems Engineering Professional Certification, International Council on Systems Engineering. https://www.incose.org/systems-engineering-certification (accessed 22 August 2018).

INCOSE (2018b). All products. The INCOSE Store URL is https://connect.incose.org/pages/store.aspx. (accessed 13 June 2018).

ISO (2015). ISO/IEC/IEEE 15288:2015, Systems and software engineering – System life cycle processes, ISO/IEEE.

ISO (2017). ISO/IEC/IEEE 12207:2017, Systems and software engineering – Software life cycle processes, ISO/IEEE, 2017.

Kan, S. (2003). *Metrics and Models for Software Quality Engineering*. Addison Wesley.

Randell, B., Naur, P., and Buxton, J. (eds.) (2001). The NATO Software Engineering Conferences. http://homepages.cs.ncl.ac.uk/brian.randell/NATO/ (accessed 22 August 2018).

SEDSIG (2018). Systems Engineering Domain Special Interest Group (SE DSIG). https://www.omg.org/syseng/ (accessed 12 June 2018).

SEVOCAB (2010). ISO/IEC/IEEE Standard 24765-2010, Systems and software engineering – Vocabulary. https://pascal.computer.org/sev_display/index.action (accessed 12 June 2018).

Simon, H. (1996). *Sciences of the Artificial*, 3e. MIT Press.

UML (2017). The Unified Modeling Language Specification Version 2.5.1. https://www.omg.org/spec/UML/About-UML/ (accessed 13 June 2018).

Walden, D., Roedler, G., Forsberg, K. et al. (2015). *Systems Engineering Handbook: A Guide for System Life Cycle Process and Activities*, 4e. Wiley.

Part II

Systems Engineering for Software-Enabled Physical Systems

Part II of this text includes Chapters 4–9, which cover the technical aspects of systems engineering for software-enabled physical systems (SEPSs). Chapter 4 presents traditional development models for SEPSs. Chapters 5–9 present an approach to development of SEPSs that facilitates concurrent development of the physical elements and software elements of SEPSs: the integrated-iterative-incremental (I^3) system development model.

Systems Engineering of Software-Enabled Systems, First Edition. Richard E. Fairley.
© 2019 John Wiley & Sons, Inc. Published 2019 by John Wiley & Sons, Inc.

4

Traditional Process Models for System Development

4.1 Introduction

In this text, a process model is defined as a set of interrelated processes for system development that are used to identify stakeholders' needs and transform them into a product or service that satisfies those needs. A development process model segments the work to be done into distinct phases because developing most software-enabled systems (SESs) is too complex to be undertaken without segmenting the work into manageable units. Top-level processes for analysis, design, implementation, and so forth can be recursively decomposed to the desired level.

Various process models for system development repeat, interleave, iterate, and recursively apply the processes and phases in various ways; the design phase, for example, may also include revision of requirements and generation of verification and validation (V&V) plans, which are accomplished using the processes of those phases in conjunction with the processes of the design phase.

The processes in each development phase transform input work products into output work products that provide tangible evidence of progress (or lack thereof). The outputs of each development phase provide inputs to other phases. The input to a design phase, for example, is a set of requirements and constraints from which design specifications are generated by applying a design process, or perhaps the input is a set of revised requirements and a previous design specification that is to be modified to accomplish needed changes.

The output of a design phase is a design specification that provides inputs for the system implementation phase and for developing integration and test plans. The output of a development process could be specification of a service that results from developing a new support system or modifying an existing system, for example, modifying the financial infrastructure of a financial institution to support operation of automated teller machines (ATMs) that will allow customers to withdraw funds at convenient times and locations.

Systems Engineering of Software-Enabled Systems, First Edition. Richard E. Fairley.
© 2019 John Wiley & Sons, Inc. Published 2019 by John Wiley & Sons, Inc.

A development phase may produce a partial work product, as in iterative and incremental development. The final development phases of a process model include system verification, transitioning the deliverable system into the operational environment, validating system performance in that environment, and obtaining system acceptance by the appropriate stakeholders.

This chapter will enable readers to understand the following:

- Some contrasting characteristics of the physical elements and software elements of an SES;
- Foundations for system development process models;
- Several well-known process models; and
- Shortcomings that provide the rationale for the integrated-iterative-incremental (I^3) development model presented in Chapter 5.

References and exercises provide the opportunity for further investigation of the topics in this chapter.

The following section provides some background needed to understand why systems engineers and software engineers use different process models.

4.2 Characteristics of Physical Elements and Software Elements

The distinctions between physical elements and software elements of an SES system provide the background for understanding the process models used by physical systems engineers and those used by software systems engineers.

Physical elements of an SES include computing hardware and other kinds of engineered elements (electrical, mechanical, civil), plus naturally occurring elements that are measured, manipulated, and controlled by engineered elements. The logical elements of an SES include software that provides system functionality, behavior, and quality attributes; software that provides interfaces among the physical system elements; software connections to entities in the system environment; and other software such as browser and router software for Internet communication, plus software that supports the manual processes performed by system operators and maintainers.

The differences between physical and logical elements of an SES present contrasting pros and cons for system development. The positive aspects of developing physical artifacts are the well-defined taxonomies and quantifications of physical elements. The positive aspect of developing logical elements, which include software and manually performed processes, is the ability to more easily modify and revise logical elements than to modify and revise physical elements because of the pliant nature of logical elements.

An example of taxonomy and quantification of physical artifacts for airborne radar units and the categories of software elements are provided in "Gulfstream V High Spectral Resolution Lidar (GV-HSRL)."

GV-HSRL

GV-HSRL is a software-enabled light detection and ranging (lidar) system that measures the properties of atmospheric aerosols and clouds. Lidar is based on radar technology but uses light emitted by lasers rather than high-frequency radio waves. GV-HSRL was developed and is operated by the Earth Observing Laboratory (GV-HSRL 2018; EOL 2018). GV-HSRL is mounted on a Gulfstream V aircraft.

An engineering taxonomy might classify GV-HSRL as airborne, electromechanical, laser, two-dimensional, signal processor, and software-enabled. Specifications for some of the HSRL technical parameters are listed in Table 4.1 (SPEC 2018).

All of the GV-HSRL performance parameters in Table 4.1, plus others, are precisely defined using standardized metrics that support black-box specification of input and output parameters for the physical elements of the system. Additional details are provided in the cited HSRL references.

Table 4.1 Physical and performance parameters of HSRL.

Physical parameters	Performance parameters	Performance specifications
Mass	Wavelength	532 nm
Volume	Pulse repetition rate	4000 Hz
Power consumption	Average power	300 mW
Heat dissipation	Range resolution – minimum	7.5 m
Ruggedness	Temporal resolution – minimum	0.5 s
Radiation hardness	Receiver channels – 4	Molecular, combined high, combined low, cross-polarization

EOL also maintains Design and Fabrication Services that provide specialized mechanical equipment and component fabrication for scientists and engineers. Products range from heavy parts for lidar pedestals to lightweight precise optical mountings for aircraft and satellite instruments.

(Continued)

(Continued)

Lidar Radar Open Software Environment (LROSE) provides the software for the HSRL system and other EOL instruments (LROSE 2018). In addition to software for acquiring real-time instrument data (as time series), LROSE provides software to format data in radial and Cartesian formats, support data visualization, and provide data in formats needed by other models.

LROSE software components include the following:

- Infrastructure software that integrates the system's hardware elements;
- Standardized data exchange formats for files and data streams;
- Algorithms and tools for data analysis, research, and generation of derived products; and
- Displays for data editing and visualization.

The list of LROSE software packages and other EOL software packages is at the cited reference; they are open source and available for download by the public (LROSE 2018).

Further details about EOL, HSRL, and LROSE can be found in the cited references.

Physical elements are not as pliable as software elements and as a result cannot, in most cases, be as easily changed to correct mistakes and accommodate changing requirements during system implementation. The difficulty and expense (and sometimes impracticality) of modifying physical elements after the implementation processes of fabrication and procurement are underway places increased emphasis on the upfront processes in an attempt to "get it right" before the implementation process is undertaken. Iterative feasibility studies, trade studies, modeling, simulation, and prototyping plus multiple stages of analysis and design may occur before committing to implementation of physical system elements.

All of the techniques used to develop physical systems can be and are applied to developing software elements but software construction differs from hardware implementation in that detailed design is usually part of software construction, whereas detailed design precedes hardware implementation in an effort to "get it right" prior to procurement or fabrication of hardware. The pliable nature of software allows iterative development of functionality, behavior, and quality attributes during software construction. In addition, software is often used to compensate for shortcomings of procured and fabricated hardware elements that would be difficult, or perhaps impossible, to modify.

Physical system elements can be specified using black-box specifications that characterize the elements in terms of quantified inputs and outputs. The system elements are "black boxes" in the sense that the internals of a black-box system element are not relevant for the purposes of functional analysis and architectural design. Nonfunctional quality attributes such as power consumption, heat dissipation, and radiation hardness may also be included in the black-box specification of a system element. Aggregation diagrams, as depicted in Figure 3.4, can be used to depict black-box systems and subsystems.

Figure 3.4 is distinguished by the aggregation symbol (open diamond). An important aspect of aggregation is that each aggregated element exists independent of the aggregator. Another aspect of aggregation is that each of the system elements can be specified as a black box and can be independently fabricated or acquired. The aggregation diagram is included in the Universal Modeling Language (UML) and Systems Modeling Language (SysML) (UML 2017; SysML 2017).

A primary issue when developing software, as compared with developing physical artifacts such as the physical elements of GV-HSRL lidar unit described above, is the comparative lack of accepted taxonomies and external "black-box" quantification of software elements. Progress has been made in this area of software engineering metrics, and continues to be made, but software does not accommodate black-box quantification of functionality, behavior, and quality attributes to the extent that characterizes physical entities. This is primarily because software is based on logical algorithms and data structures while specification of physical entities is based on physics and the traditional metrics and measures of engineering.

Software artifacts can be depicted using composition diagrams, as illustrated in Figure 3.5.

Composition is depicted in UML and SysML using a closed diamond, as illustrated in Figure 3.5. An important aspect of software composition, as compared with aggregation of physical elements, is that an instance of the composed software elements, as illustrated in Figure 3.5, exists only as part of the composer. An instance of the software elements and their interfaces is created when an instance of the composer is created (e.g. an ATM is started) and is deleted when the instance of the composer is deleted (e.g. an ATM is shut down). The startup procedure for ATM software, for example, creates an executable instance of the software code from a library of code elements; the shutdown procedure for an ATM renders inoperative the instantiation of the software elements and the memory used by the ATM instantiation is reclaimed by the operating system. Multiple instantiations can be created from the library of code elements, one for each ATM in an automated teller system.

Software engineers, like system engineers, attempt to "get it right" prior to software construction using the techniques of analysis, modeling, prototyping and architecture definition, and design definition. As mentioned above, an important distinction is that detailed design of physical elements is an aspect of design definition (as specified in the ISO/IEC/IEEE Standard 15288), whereas detailed design of software is part of the software construction process. The architecture (structure and behavior), functionality, and quality attributes of software are often modified, sometimes repeatedly, in systematic ways during software construction.

The following section provides an overview of development process foundations.

4.3 Development Process Foundations

The ISO/IEC/IEEE process standards provide foundations for the systems engineering and software engineering processes used to develop and sustain SESs:

- ISO/IEC/IEEE 15288:2015 – Systems and software engineering – System life-cycle processes (ISO 2015); and
- ISO/IEC/IEEE 12207:2017 – Systems and software engineering – Software life-cycle processes (ISO 2017).

These standards are referred to as 15288 and 12207 throughout this text. The 15288 and 12207 standards include the same 25 processes arranged into four categories:

- Agreement processes
- Organizational project-enabling processes
- Technical management processes
- Technical processes

Agreement processes are processes that govern the relationship between system acquirers and system suppliers. Organizational project-enabling processes are the processes, development process models, and associated procedures and document templates provided at the organizational level to support projects conducted within the organization. Technical processes are concerned with managing the technical work for a project. These processes are covered in Part II of the book (Chapters 4–9). Technical management processes are presented in Part III of this book (Chapters 10–12).

According to Sections 6.4 of 15288 and 12207:

> The Technical Processes are used to define the requirements for a system, to transform the requirements into an effective product, to permit consistent reproduction of the product where necessary, to use the product to provide the required services, to sustain the provision of those services and to dispose of the product when it is retired from service.

The technical processes in 15288 and 12207 are identically named with minor differences: "software system" replaces "system" in 12207 and the System Requirements Definition Process in 15288 is named the System/Software

Table 4.2 15288:2015 and 12207:2017 technical processes.

Technical processes of 15288 and 12207	Clauses
Business or mission analysis process	6.4.1
Stakeholder need and requirements definition process	6.4.2
System requirements definition process (system/software requirements definition process)	6.4.3
Architecture definition process	6.4.4
Design definition process	6.4.5
System analysis process	6.4.6
Implementation process	6.4.7
Integration process	6.4.8
Verification process	6.4.9
Transition process	6.4.10
Validation process	6.4.11
Operation process	6.4.12
Maintenance process	6.4.13
Disposal process	6.4.14

Requirements Definition Process in 12207. The changes are made to indicate that 12207 can be used in place of 15288 for systems that are predominantly composed of software. Much of the wording in 12207 is similar to the wording in 15288 with changes and notes appropriate for the technical processes of software engineering.

Table 4.2 lists the 14 technical processes in 15288 and 12207 and the clauses in the standards that cover these processes.

The first 11 processes are development processes; the last three processes (6.4.12, 6.4.13, and 6.4.14) result in a complete set of "birth-to-death" system life-cycle technical processes. The last three processes may use some or all of the first 11 processes.

The processes in Table 4.2 are listed in a linear order but different life-cycle models interleave, overlap, and iterate the processes and apply them recursively in various systematic ways.

Additional supporting information concerning the technical processes of systems engineering and software engineering can be found in

- The *Guide to the Systems Engineering Body of Knowledge* (SEBoK 2017);
- The INCOSE handbook (Walden et al. 2015);
- The NASA systems engineering competency model (NASA 2018);

- The *Guide to the Software Engineering Body of Knowledge* (Bourque and Fairley 2014); and
- The *Software Engineering Competency Model* (Fairley et al. 2014).

This section and the preceding sections have provided background information for the following sections.

4.4 Linear and Vee Development Models

System engineering and software engineering are process-oriented disciplines that include systematic ways of applying the technical processes in 15288 and 12207. Table 2.2 in this text indicates software engineering processes that have been adopted and adapted by systems engineers and systems engineering processes that have been adopted and adapted by software engineers.

However, the methods and techniques for the development processes used to develop the physical elements and software elements of an SES can be, and often are, quite different. These differences can result in confusion and frustration when developing software-enabled physical systems, as indicated in Chapter 3: Issues and Opportunities. A system development model that addresses the issues is presented in Chapter 5.

It must be emphasized that the system development models presented here provide overviews and are not complete lifecycle models for system development. As presented in this text, they do not explicitly include all 25 systems engineering and software engineering processes in 15288 and 12207. And, the processes included in 15288 and 12207 must be adapted, tailored, and instantiated for each of the various system development process models. Also, each development model must, in turn, be embedded in a project management process model for system and/or software development (PMBOK 2017; Fairley et al. 2013; Fairley 2009).

The following sections present life cycle models for developing or modifying a system. These development models include many of the attributes of better-known models such as "waterfall" and "agile" but those terms are not used here to avoid the connotations associated with them. For instance, Winston Royce is known as the father of the much-maligned waterfall model but in his seminal paper (in which he never used the term "waterfall") he clearly indicated the need to include revision processes so the term "linear-phased" is used here to remove the negative connotations of "waterfall" (Royce 1970). Iterative development models have many of the attributes of "agility" but the term "agile" is avoided because of the many variations in the meaning of "agile development" (Agile 2018).

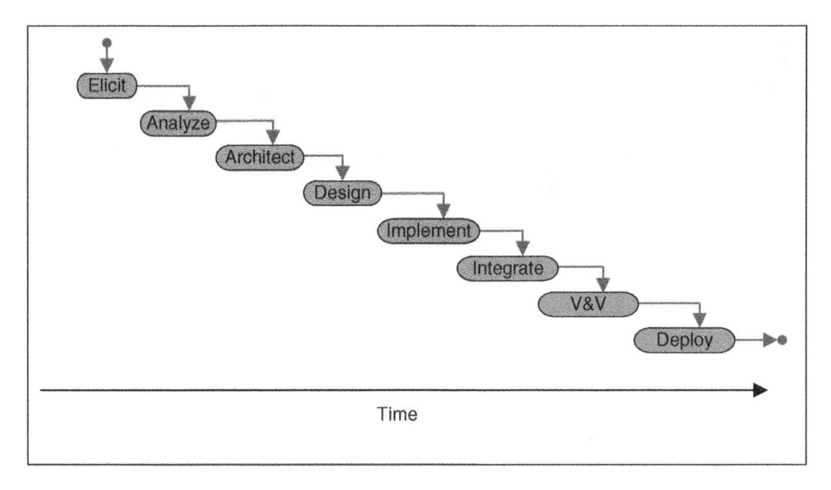

Figure 4.1 A linear one-pass system development model.

4.4.1 The Linear One-Pass Development Model

The linear one-pass development model was historically used in the evolution of systems engineering, hardware engineering, and software engineering development processes; it is illustrated in Figure 4.1. As can be observed, the linear one-pass development model includes eight development phases with "feed forward" of the work products produced in each development phase to provide inputs for the next phase.

Attempts to "get it right the first time" include milestone reviews at the end of each project phase. Reviews include "control gates," which indicate that the next phase cannot be started (the gate cannot be opened) until the present phase satisfies its completion criteria. In practice, control gates are often partially opened to allow work to begin on the next phase while remaining work for the current phase is completed. Action items to be completed before starting the next phase, or to be completed during the next phase, are (or should be) recorded and tracked to completion.

Problems are often discovered in later phases of system development because of inadequate control gate criteria, failure to track action items to completion, and the learning that occurs in later phases. This results in accumulation of technical debt, which must be paid by revisiting and revising the work products of one or more previous phases; payment can delay progress and negatively impact schedules and budgets. Treatment of technical debt is covered in Part III of this book (Fairley and Willshire 2017).

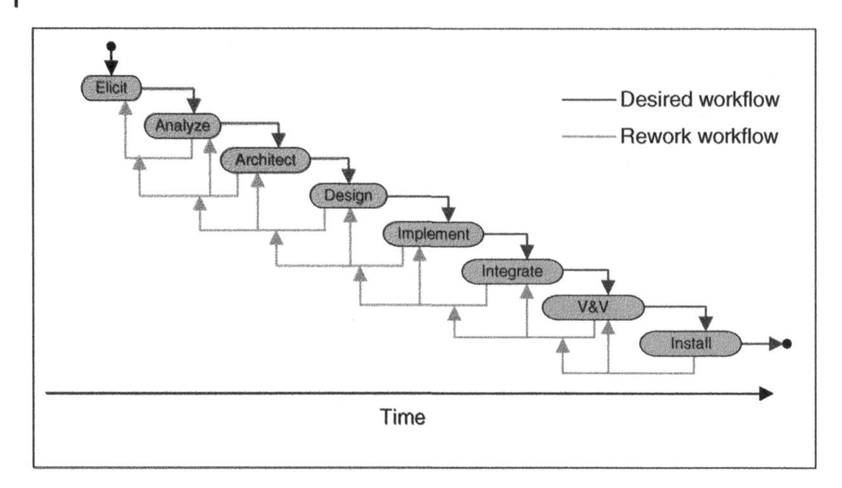

Figure 4.2 A linear-revisions system development model.

4.4.2 A Linear-Revisions Development Model

The linear-revisions development model is illustrated in Figure 4.2. The feedback arrows depict the revision links from each development phase to all previous phases. As mentioned above, the seminal paper by Winston Royce clearly indicated the need for feedback and revision when using a linear development process model.

The linear-revisions development model, like the one-pass linear model, includes milestone reviews and control gates. The primary difference between the linear-revisions development model and the linear one-pass model is that time and resources are held in reserve (or should be held) for anticipated revisions that will be needed to previously developed work products. The amount held in reserve is based on past projects, analogies, and/or rule of thumb for what has been experienced.

4.4.3 The Vee Development Model

The Vee development model is a widely used linear-revisions development model in systems engineering; it is a linear-revisions development model that has been folded into a V (Forsberg et al. 2000). A depiction of the Vee model is presented in Figure 4.3.

The feedback revision arrows between phases are an element of the Vee process model but are omitted from the figure for clarity of the presentation.

The main feature of the Vee model, as illustrated, is the emphasis placed on the development of V&V and integration plans as the corresponding work products are generated, followed by application of the plans when the integration and V&V phases occur. Early development of integration and V&V plans

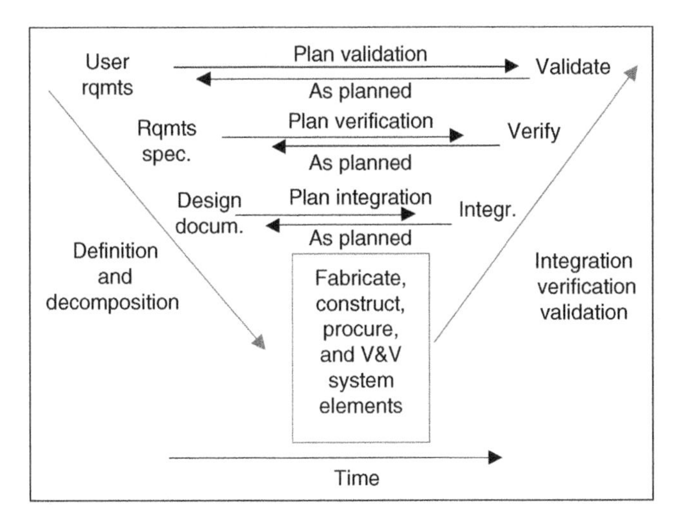

Figure 4.3 A Vee system development model.

can surface deficiencies and serve to improve the stakeholders' requirements, system requirements, and design documentation.

In the box at the bottom of Figure 4.3, requirements are often separately allocated for fabrication of hardware elements, construction of software elements, procurement of hardware and software elements, and development of manual operations. A major shortcoming of this approach is that software is treated as a discrete system component, similar to a power supply or a printer rather than as a set of elements that interact with most, if not all, of the physical components, including the power supplies and printers.

4.4.4 Incremental Vee Development Models

As systems become larger and more complex, it is desirable to develop systems in increments, as illustrated in Figure 4.4. Each increment provides a subset of deliverable system capabilities that

- Includes all capabilities developed to date;
- Can be activated and demonstrated;
- Provides the foundations for implementing the next increment; and
- Provides objective evidence of progress.

One or more increments may be developed for delivery to stakeholders to allow potential system users to experiment with a subset of final capabilities, or alternatively, the current increment of the evolving system, which incorporates all previous increments, may be inserted into the operational environment to provide a useful subset of the final system. In some cases, there may be gaps

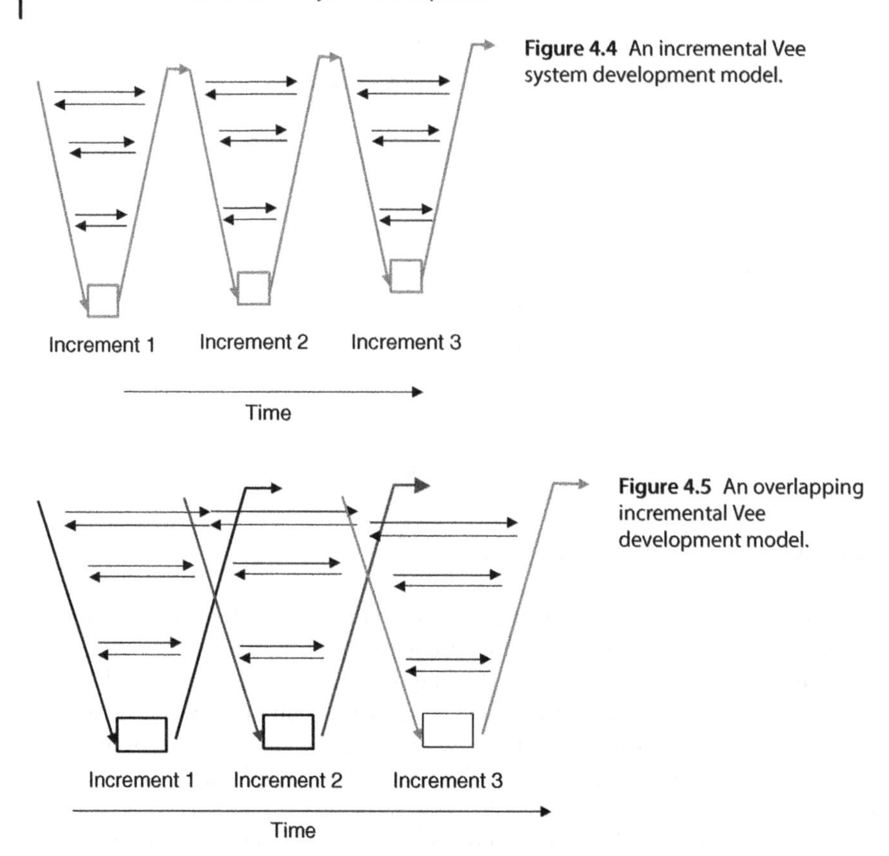

Figure 4.4 An incremental Vee system development model.

Figure 4.5 An overlapping incremental Vee development model.

between development of increments, as for example in phased procurement of a system or subsequent development of a next version of a system during operation and sustainment of the current version.

It may be possible to overlap increments, as illustrated in Figure 4.5.

Overlapping of Vee increments can compress a project schedule, at the expense of additional resources needed to engage in concurrent work activities for the overlapping Vees. Care must be taken that the next incremental Vee does not get too far ahead in making assumptions about the previous Vee that is under development. Changes to functionality, behavior, quality attributes, or interfaces of the previous Vee (or Vees) may require extensive changes to the work products of the subsequent Vee.

In summary, the primary advantage of Vee models is the emphasis placed on planning integration and V&V activities during system analysis and design, which helps to identify discrepancies early in the development process and to clarify the remaining work to be done. The primary disadvantage of Vee models and all variations of linear-development process models is

> Mistakes made first are detected last, when it is expensive and time consuming to correct them.

Iterative development is an alternative to incremental development.

4.5 Iterative Development Models

Iteration involves repeated application of one or more development processes with the goal of making progress toward a solution during each of the iterations. Iterative development can happen independent of incremental development, but most iterations result in production of one or more new or modified system elements that are integrated into the evolving system to produce a next increment of a system. There is thus no bright line between iterative and incremental development. There are, however, three characteristics that distinguish incremental implementation of physical increments from iterative development of software.

4.5.1 Iterative Fabrication of Physical Elements

Incremental implementation of physical system elements can be viewed as a kind of iterative development process because the implementation process may be repeated during each incremental phase. The three characteristics that distinguish incremental development of physical elements from iterative development of software are the following:

1. Implementing a system increment that includes physical elements often requires a longer duration than completing a software development iteration because of the time needed to fabricate or implement the needed physical elements. The duration of an incremental phase for physical elements is typically on the order of several weeks to a few months, whereas the durations of software iterations are typically in the range of two weeks or less.
2. For physical elements, the durations of different incremental phases may vary, whereas the durations of software iterations are usually fixed for most, if not all, iteration cycles.
3. The makeup of team members for incremental development of physical elements may vary from incremental phase to incremental phase depending on the skills needed, whereas the membership of iterative development software teams is usually fixed, with occasional addition of specialty engineers during some iterations (e.g. safety, security).

4.5.2 Iterative Construction of Software Elements

As mentioned above, process constraints imposed and the latitude provided when developing software are in many ways the inverse of the constraints and latitudes presented when developing physical system elements. Software development benefits from the malleability of the medium but software does not exhibit quantifiable attributes to the extent that physical elements do; conversely, implementation of physical elements benefits from quantification of attributes but physical elements are less malleable than software.

As the software elements of software-enabled physical systems became larger and more complex, the need for systematic development processes became apparent. Initially, software developers adopted the systems engineering approach, using linear-phased and incremental Vee development models. Over time, software developers have increasingly adopted iterative approaches. Figure 2.2 illustrates a capabilities-based iterative process for software development.

The iterative process illustrated in Figure 2.2 has the following attributes:

- An initial set of prioritized capabilities is prepared by the system stakeholders and a software systems engineer in consultation with the team members if they are available or in consultation with a knowledgeable software engineer who will be a team member when software development begins.
- Requirements definition and architectural design, based on desired and prioritized capabilities, precede the software construction process illustrated in Figure 2.2.
- The customer is a knowledgeable representative of the acquiring organization; he or she understands the needs to be satisfied and has the authority to guide the project.
- The customer, in consultation with the team members and other stakeholders, can add, remove, revise, and reprioritize capabilities to be constructed as the software evolves. Prioritization criteria for system capabilities are presented in Chapter 5.
- Teams are usually of fixed size and all iteration cycles are usually fixed in duration; this facilitates tracking of progress because two of the three key project variables are held constant (resources and duration); they determine the amount of work that can be accomplished during an iteration (the third variable).
- The team members, in consultation with the customer, select the next capabilities to be implemented and specify detailed requirements the selected capability or capabilities must satisfy.
- The team members then prepare V&V test scenarios and test cases prior to writing the software for the selected capability or capabilities; this approach is known as test-driven development (TDD).

- V&V includes both static analysis and dynamic tests team members use to verify that implemented capabilities satisfy the requirements, when combined with previously accepted capabilities.
- During iteration cycles after the initial one, team members refactor the existing software, add software for the new capability, and apply the verification tests.
- Refactoring involves modifying the structure of the existing software to facilitate addition of software for the new capability. Refactoring must not alter the behavior of existing software because the existing software has been validated and accepted by the customer. Object-oriented software design facilitates refactoring.
- The team members apply validation scenarios and tests to demonstrate the newly added capability in conjunction with previously accepted capabilities.
- The customer (and other stakeholders, as appropriate) validates the added capability in conjunction with the capabilities previously accepted by the customer.
- Internal iterations, not visible to the customer, typically occur on daily and perhaps hourly cycles prior to completing an "external iteration" that produces a next software increment the customer will accept, reject, or accept with requested revisions.
- Internal iterations that produce the next incremental version of an evolving system use a sandbox copy (a copy of the currently accepted version) as the environment for developing, verifying, and validating the new capabilities being added.
- Validation demonstrations may also result in the customer adding, revising, and reprioritizing the backlogged capability set.
- By prior arrangement, one or more intermediate versions of the software may be delivered for evaluation by typical users, evaluation in a "live" operational environment, or incorporation into a production environment.

Each iteration cycle ends with a private team-members' retrospective meeting to review progress and problems encountered, which may result in changes to the internal development processes being used. The next iteration begins with a private team-members' planning meeting to make work assignments and to identify potential problem areas (i.e. risk factors) for the upcoming iteration. The retrospective meeting for the completed iteration and the planning meeting for the next iteration may be combined into one meeting.

It is important to note that the customer is the development manager and is responsible for a successful outcome. The customer might be a systems engineer, a marketing representative, or an external stakeholder. He or she must balance requested capabilities with schedule, budget, and resource constraints. From this point of view, the team members are providing a service-as-requested for the customer. This approach requires continuous commitment and involvement of the customer.

The fixed number of team members and the fixed duration of iteration cycles for software iterations may result in some system capabilities being decomposed into two or more capabilities to be developed incrementally on two or more iteration cycles. It is essential that each iteration cycle produce working, demonstrable software that provides tangible evidence of progress and provides continuing feedback from the customer. Alternatively, two or more capabilities may be constructed during a single iteration cycle. Occasionally, "time out" iterations may be taken to clear a backlog of pending rework (i.e. to pay accumulated technical debt), incorporate a new procedure or software development tool, or rework the software architecture.

The duration of software iteration cycles is typically two weeks or less, and the number of team members is typically 5 or 6, but not more than 10, which facilitates detailed, ongoing communication among the small group of team members. Multiple small teams are utilized for large projects.

It is desirable that team membership be fixed at a constant number for the duration of a project. Continuity of team membership facilitates ongoing communication and coordination among the team members. Effective teamwork requires that the software team members collectively have the necessary software development skills and sufficient knowledge of the system domain; these team members are said to be cross-functional generalists. Some of the iterations may require the assistance of specialist consultants and advisors (e.g. those having safety, security, or system domain expertise).

As stated, small team size facilitates close communication and collaboration among team members but limits the amount of software that can be developed using a fixed number of iteration cycles within a stated period of time. Multiple small teams can accomplish larger projects; each team is allocated capabilities to be constructed and specifications for interfaces to be provided to other teams, plus specifications for the interfaces to be provided by other teams. Interfaces to be constructed by a team and interfaces provided by other teams are initially simulated and made available across the teams. Periodic integration and demonstrations of the outputs of multiple teams are used to verify interim versions of the evolving software. Very large systems may be segmented into subsystems.

It is important to emphasize that incremental development of physical system elements and iterative development of software elements are system development processes and are not full life-cycle development models, except perhaps for small projects. Initial and ongoing analysis, architectural design, detailed design, and management of the technical processes must be performed.

The importance of analysis and architectural design are self-evident for physical systems. In the case of software, it is a myth that the malleability of "agile development" does not need to include the full set of development processes.

More information concerning iterative processes for software development can be found in the *Software Extension to the PMBOK® Guide Fifth Edition* (Fairley et al. 2013).

4.6 The ATM Revisited

A product breakdown of ATM hardware elements as an aggregation of black boxes is presented in Figure 3.4 and a composition diagram for the ATM software elements is presented in Figure 3.5. Incremental development of the hardware in the figure is guided by the following considerations:

- Availability of a sufficient number of electrical and mechanical engineers with requisite expertise and experience, plus specialty engineers for system reliability, information security, physical security, and physical safety;
- Physical elements that can be procured to minimize the number that must be fabricated; and
- Sufficient time and resources (including money) to procure and fabricate the various physical elements.

Fewer available engineers may result in sequential incremental Vees, rather than overlapped Vees, for system development because the available engineers will all be occupied by the current incremental Vee. More available engineers would permit overlapping or perhaps concurrent parallel Vees. Lack of engineers with sufficient expertise for system elements to be fabricated may result in subcontracting with another organization, hiring additional engineers on a fulltime or consulting basis, or placing more emphasis on procurement. Time and resource constraints may result in compromising the features and quality attributes of the resulting ATM.

Iterative development of software is guided by similar considerations:

- Availability of a sufficient number of software engineers with requisite expertise and experience, plus specialty engineers for information security;
- Software elements that can be procured to minimize the number that must be constructed; and
- Sufficient time and resources (including money) to procure and construct the various software elements.

Software elements may be available in local or public software libraries or procured from a software vendor.

The remaining consideration (a big one) is how to synchronize development of physical elements with development of software elements (e.g. ATM hardware fabrication/procurement and software construction/procurement). The "hardware first" approach and the "software first" approach each has the drawbacks described in Section 3.5.

Chapter 5 presents an I^3 system development model that seamlessly integrates the iterative approach to software construction with the incremental approach to implementing physical system elements.

4.7 Key Points

- The ISO/IEC/IEEE process standards 15288 and 12207 provide foundation processes for developing and sustaining SESs.
- The 15288 and 12207 technical processes may be intermixed, interleaved, and iterated in systematic ways within a development process model.
- The differences between the physical elements and software elements of an SES present contrasting pros and cons for system development.
- Systems engineers often use the Vee and incremental Vee system development processes.
- Software engineers predominantly use iterative development processes.
- The increment of time needed to fabricate and procure the physical elements of a system increment is typically a few weeks and may be a few months in some cases.
- Procurement of some physical system elements may be initiated during the system design phase, thus shortening the duration of the incremental development phases that will incorporate those system elements.
- The increments of time and the numbers and kinds of engineers needed to fabricate and procure physical elements for a system increment may vary among increments.
- The increment of time to complete software iterations is usually fixed at two weeks or less.
- The number of team members for software development is typically fixed at 5–7, but not more than 10, team members for the duration of a project.
- Fixed durations and fixed numbers of team members for software iterations control two of the three process variables (duration and resources); these determine the amount of work that can be accomplished during construction of a software increment (the third process variable).
- Numerous internal iteration cycles, not visible to system stakeholders, may occur during iterative development of a software increment.
- Software engineers use a sandbox copy of the currently accepted incremental version of an evolving system to develop, verify, and validate the capabilities that will be provided in the next incremental version of the system.
- There is no bright line between iterative and incremental system development.
- Incremental development of physical elements and iterative development of software increments are system development processes; they are (or

should be) embedded in a full system development life cycle that includes the technical management and project management processes.

Exercises

4.1. The differences between the physical elements and software elements of a software-enabled system present contrasting pros and cons for system development.
   (a) Briefly explain the pros and cons that physical elements present for system development.
   (b) Briefly explain the pros and cons that software elements present for system development.

4.2. Physical systems engineers prefer incremental development processes and software systems engineers prefer iterative development processes.
   (a) Briefly explain why physical systems engineers prefer incremental development processes.
   (b) Briefly explain why software systems engineers prefer iterative development processes.

4.3. The following statement appears in Section 4.5: "there is no bright line between incremental development and iterative development." Briefly explain why this statement is true.

4.4. Briefly explain why the linear one-pass development model is not feasible for developing software-enabled systems.

4.5. Briefly explain the advantages of the Vee development model compared with the linear-revision development model.

4.6. Briefly explain the advantages of the incremental Vee development model compared with the Vee development model.

4.7. The increment of time needed to fabricate and procure the physical elements of a system increment is typically a few weeks and may be a few months in some cases. Briefly explain why this is true for fabrication and for procurement.

4.8. The duration of iteration cycles for software development is typically fixed at two weeks or less and the number of team members is typically fixed at 5–7, but not more than 10, team members for the duration of a project.

(a) Briefly explain why the duration of iteration cycles for software development is fixed at two weeks or less.
(b) Briefly explain why the number of team members should not exceed 10.
(c) Briefly explain how software development that requires more than 10 team members is handled.

4.9. Fixed durations and fixed numbers of team members for software iterations control two of the three process variables (duration and resources); these determine the amount of work that can be accomplished during construction of a software increment (the third process variable). Briefly explain how fixing the first two process variables determine the third one.

4.10. This exercise is concerned with the the pros and cons of buying versus building system elements.
(a) The aggregation diagram in Figure 3.4 includes the hardware elements for an ATM. Identify the hardware elements that would be candidates for buying and those that would be candidates for fabricating. Briefly explain your choices for buying and your choices for fabricating.
(b) The composition diagram in Figure 3.5 includes the software elements for an ATM. Identify the software elements that would be candidates for buying and those that would be candidates for constructing. Briefly explain your choices for buying and your choices for constructing.

References

Agile (2018). Agile Software Development. https://en.wikiversity.org/wiki/Agile_software_development (accessed 15 June 2018).

Bourque, P. and Fairley, R. (eds.) (2014). *Guide to the Software Engineering Body of Knowledge (Version 3.0)*. IEEE https://www.computer.org/web/swebok/v3-guide (accessed 8 June 2018).

Fairley, R. (2009). *Managing and Leading Software Projects*, Chapter 2. Wiley.

Fairley, R. and Willshire, M. (2017). Better now than later: managing technical debt in systems development. *IEEE Computer* May: 81–87.

Fairley, R. et al. (2013). *Software Extension to the PMBOK® Guide*, 5e. Project Management Institute https://www.pmi.org/pmbok-guide-standards/foundational/pmbok/software-extension-5th-edition (accessed 9 June 2018).

Fairley, R. et al. (2014). *Software Engineering Competency Model, Version 1.0, (SWECOM).* IEEE https://www.computer.org/web/peb/swecom (accessed 8 June 2018).

Forsberg, K., Mooz, H., and Cotterman, H. (2000). *Visualizing project management: a model for business and technical success.* Wiley.

ISO (2015). ISO/IEC/IEEE 15288:2015, Systems and software engineering – System life cycle processes, ISO/IEEE, 2015.

ISO (2017). ISO/IEC/IEEE 12207:2017, Systems and software engineering – Software life cycle processes, ISO/IEEE, 2017.

NASA (2018). NASA's System Engineering Competencies, 2018. https://www.nasa.gov/pdf/303747main_Systems_Engineering_Competencies.pdf (accessed 15 June 2018).

PMBOK (2017). *A Guide to the Project Management Body of Knowledge,* 6e. PMI https://www.pmi.org/pmbok-guide-standards/foundational/pmbok (accessed 15 June 2018).

Royce, W. (1970). Managing the Development of Large Software Systems. In: *Proceedings of IEEE WESCON 26,* 1–9. IEEE http://www-scf.usc.edu/~csci201/lectures/Lecture11/royce1970.pdf (accessed 15 June 2018).

SEBoK (2017). *The Guide to the Systems Engineering Body of Knowledge (SEBoK), v. 1.9* (ed. R.D. Adcock) (EIC). Hoboken, NJ: The Trustees of the Stevens Institute of Technology www.sebokwiki.org (accessed 12 June 2018).

SysML (2017). OMG Systems Modeling Language 1.5. http://www.omgsysml.org/ (accessed 14 June 2018).

UML (2017). The Unified Modeling Language Specification, Version 2.5.1, 2017. https://www.omg.org/spec/UML/About-UML (accessed 13 June 2018).

Walden, D. et al. (2015). *Systems Engineering Handbook: A Guide for System Life Cycle Process and Activities,* 4e, International Council on Systems Engineering. Wiley.

EOL (2018). Earth Orbiting Laboratory. https://www.eol.ucar.edu/earth-observing-laboratory (accessed 24 August 24 2018).

GV-HSRL (2018). High spectral resolution lidar. https://www.eol.ucar.edu/instruments/gv-hsrl (accessed 24 August 2018).

LROSE (2018). Lidar radar open software environment. https://www.eol.ucar.edu/lrose (accessed 9 September 2018).

SPEC (2018). GV-HSRL specifications. https://www.eol.ucar.edu/content/specifications-6 (accessed 24 August 2018).

5

The Integrated-Iterative-Incremental System Development Model

5.1 Introduction

Chapter 4 presents an incremental model for developing the physical elements of a system and an iterative model for developing the software elements. This chapter introduces the capabilities-based integrated-iterative-incremental (I-cubed or I^3) system development model that integrates incremental development of physical elements and iterative development of software elements for software-enabled physical systems.

The key features of I^3 are a capabilities-based approach to system requirements definition and an I^3 process for system implementation that includes the techniques of model-based and simulation systems engineering (M&SBSE) (Adcock 2017a). Together, system capabilities and an M&SBSE approach to system development facilitate concurrent iterative-incremental realization and integration of a system's physical and software elements, the interfaces among them, and the connections to the system environment. This development model can be adapted for all kinds of systems engineering projects but is especially useful for developing software-enabled physical systems.

The following sections of this chapter provide the opportunity for readers to gain an understanding of capabilities-based requirements definition and M&SBSE system realization. A case study is interspersed throughout the chapter. References and exercises are included to support increased understanding of the methods and techniques presented.

5.2 Capabilities-Based System Development

The term "capability" has many different meanings. In this text, a *system capability* is defined as the ability of a system to execute a course of action or exhibit a state of being. A system capability can include functionality, behavior, a quality attribute, or a combination that includes two or all three of these system characteristics.

Systems Engineering of Software-Enabled Systems, First Edition. Richard E. Fairley.
© 2019 John Wiley & Sons, Inc. Published 2019 by John Wiley & Sons, Inc.

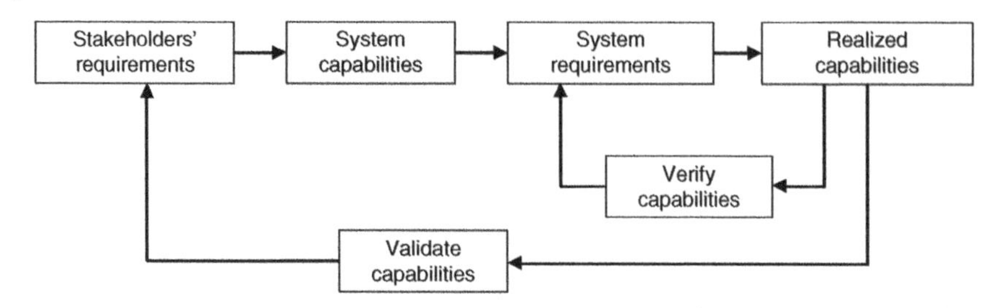

Figure 5.1 A capabilities-based approach to system development.

The process by which capabilities-based requirements are realized (i.e. made real) is illustrated in Figure 5.1.

As indicated, system capabilities are derived from stakeholders' requirements and system requirements are defined for each of the system capabilities. Physical elements and software elements are realized based on the system requirements and are verified to determine the degree to which they satisfy the corresponding system requirements. Realized capabilities are validated to determine the degree to which they satisfy the stakeholders' requirements from which they were derived.

Revisions to stakeholders' requirements, system capabilities, system requirements, and capabilities realized may also result from the illustrated process. For simplicity of presentation, system architecture definition and design definition are not shown in Figure 5.1. They are presented later in this chapter.

An example illustrates the process for a transaction processing system, which might be a point-of-sale system (e.g. an airline reservation system or an automated teller system). Suppose a stakeholders' requirement is

> Each system transaction shall have a time-stamped record.

The corresponding system capability is

> The ability of the system to provide time stamped transactions.

The system requirement to provide this capability would be

> The system shall provide a time-keeping function that has quantified accuracy, precision, and drift; parameters to be determined.

Design options for the time-keeping function include GPS connection, a URL for a master clock, or a physical clock. A GPS or URL capability will generate a requirement for an external connection. A physical clock may have additional requirements such as quantified mass, volume, power consumption, and heat dissipation.

As the system capability and the corresponding system requirement are stated, it is not clear whether time-stamped records are to be maintained internally in the transaction processing system or to be provided to one or more designated stakeholders or both; the capability and requirement should be revised to remove this ambiguity. In the latter case (provided to one or more stakeholders), it must be determined whether the time-stamped records are to be provided electronically, in printed form, or both. This decision will result in a capability to provide an electronic interface connection, a capability for a printer connection and a printer, or both. The results will be one or more quantified system requirements.

Another example illustrates additional advantages of the capabilities-based approach to system realization. Suppose a stakeholders' requirement is

> Customers of the ABC financial institution need to be able to conduct financial transactions at times and locations that are convenient, safe, and secure for the customers.

Discussions with stakeholders generated the following list of system capabilities to be provided for customers' financial accounts:

- Secure authentication
- Secure termination
- Balance query
- Cash withdrawal
- Funds deposit
- Funds transfer

A feasibility analysis indicates that traditional automated teller machines (ATMs) having the above transaction capabilities along with secure customer authentication and session termination capabilities can satisfy the stakeholders' requirement.

An alternative approach to satisfying the stakeholders' requirement would be to establish 24 × 7 walk-in offices at convenient locations staffed by financial institution personnel. Walk-in offices with appropriate physical security might have practical advantages in areas of high population density but are not deemed to be practical for the majority of ABC's customers.

System capabilities needed by other hands-on users of ATMs (e.g. ATM service personnel, IT personnel) are not presented in this example.

Table 5.1 presents the results of a feasibility study for system implementation that depicts the combined hardware and software elements that can provide the identified system capabilities. Each X in the table indicates the hardware and software elements needed to realize the capabilities listed in the leftmost column. For example, the customer authentication capability will require

Table 5.1 ATM system capabilities/feasible hardware and software.

Capabilities	Hardware and software						
Hardware Software	**Card reader**	**Display screen**	**Keypad**	**Cash safe**	**Cash dispenser**	**Funds depository**	**Printer**
Customer authentication	X	X	X				
Balance query		X	X				
Cash withdrawal		X	X	X	X		
Funds deposit		X	X			X	
Funds transfer		X	X				
Session termination		X	X				X

customer authentication software and card reader, display screen, and keypad hardware, and software.

Connections to the financial institution's data repository for customer accounts are not shown in Table 5.1.

The table reveals a key aspect of capabilities-based system development:

> For each system capability, hardware and software requirements can be defined, allocated together, and concurrently fabricated, constructed and procured to realize the system capability.

The flow diagram depicted in Figure 5.2 and the I^3 M&SBSE system development approach described in Section 5.3 include processes that can be used to pursue concurrent development of hardware and software elements, their interfaces, and their connections to the system environment.

Interfaces can be specified and implemented/constructed concurrently. For example, the first row of Table 5.1 indicates that the customer authentication capability will require hardware and software interfaces among the card reader, display screen, and keypad; the display screen column indicates that interfaces to the display screen and keypad will be required for each of the system capabilities. Other interface capabilities are also depicted in Table 5.1.

Some criteria for prioritizing implementation of system capabilities and the rationales for doing so are provided in "Capability Prioritization."

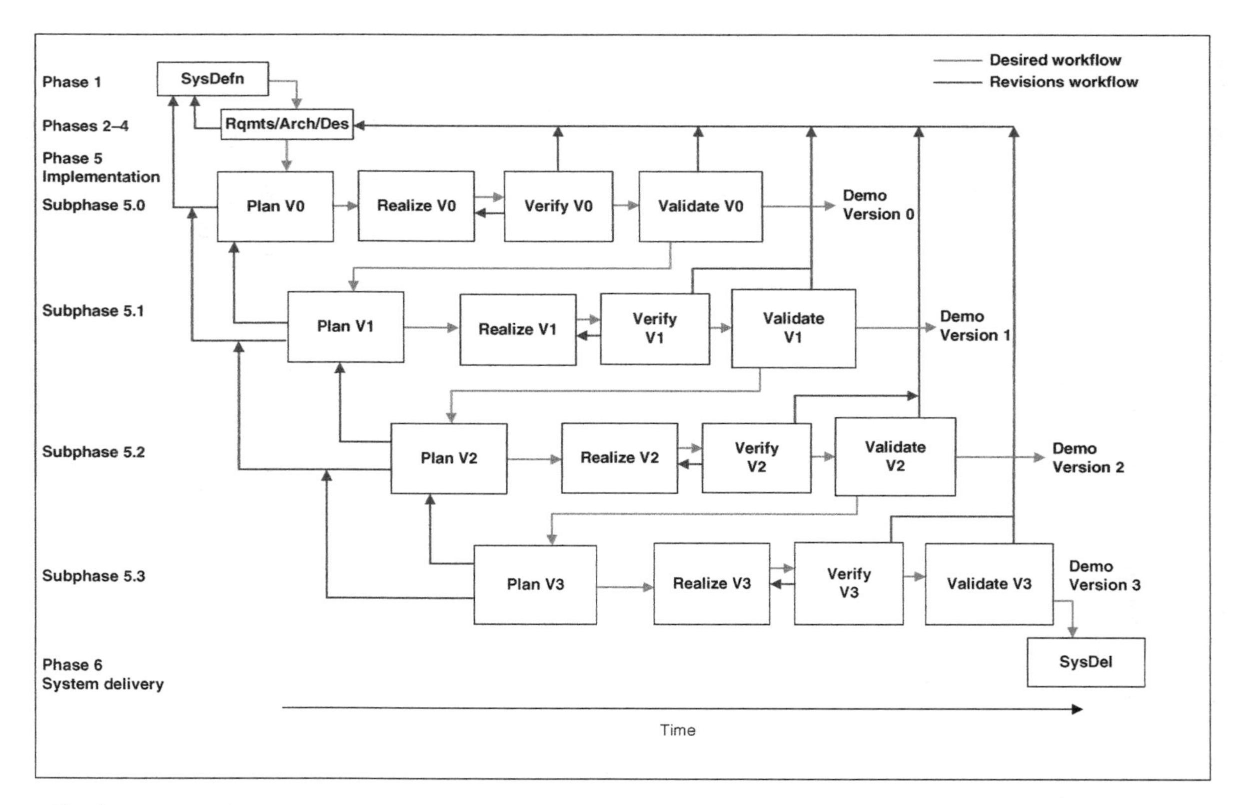

Figure 5.2 The I^3 system development model.

Capability Prioritization

Some criteria and rationale for prioritizing system capabilities are provided in Table 5.2.

Table 5.2 Capability prioritization criteria and rationale.

Criterion	Rationale
Easiest-to-realize capabilities first	To gain familiarity with the I^3 development processes and build team cohesion
Highest risk capabilities realized first	To determine that known uncertainties and complexities can be accommodated and to expose previously unknown risks and complexities
Safety and security capabilities realized first	To determine that safety and security requirements can be satisfied and to ensure that safety and security capabilities are developed first and are then tested and demonstrated most in conjunction with capabilities that are added later
External connections realized first (real and simulated)	To establish connections to the real or a simulated operational environment
Legacy-element capabilities realized first	To determine that legacy elements can be satisfactorily incorporated
Vendor-element capabilities realized first	To determine that vendor-supplied elements can be satisfactorily incorporated
Deliverable subset capabilities realized first	To be evaluated by potential users and/or placed into operational use

Various combinations of the prioritization criteria can be used, depending on particular circumstances. Project constraints (schedule, budget, resources, infrastructure, and stakeholders' needs) must be taken into consideration when prioritizing system capabilities.

Three important prioritization criteria are:

1. System capabilities realized first are verified and validated (i.e. tested and demonstrated) most frequently because previously realized capabilities are tested in conjunction with verifying and validating each subsequently realized capability. This is especially important for considerations of safety,

security, and reliability and for user, operator, and other human–machine interfaces.
2. Prioritizing the order in which system capabilities will be realized must preserve the priorities of the stakeholders' requirements, which may be renegotiated while prioritizing system capabilities.
3. It may be desirable to realize some capabilities before realizing subsequent capabilities to provide system elements, interfaces, and connections as a platform for realizing subsequent capabilities.

The capabilities to be realized for the ATM in Table 5.1 are prioritized as follows:

Priority 1: Secure customer authentication capability
Priority 2: Secure session termination capability
Priority 3: Balance query capability
Priority 4: Cash withdrawal capability
Priority 5: Funds deposit capability
Priority 6: Funds transfer capability

The customer authentication and termination capabilities are implemented first so that they can be repeatedly verified and validated in combination with subsequent capabilities when those capabilities are realized. This will provide evidence that authentication and termination will behave correctly in combination with other capabilities because, as stated, capabilities realized first are verified and validated most frequently in combination with other capabilities.

Next to be realized will be the balance query capability to obtain account balances because it will satisfy a stakeholders' requirement and it will be used by subsequently realized capabilities. Doing a balance query transaction will involve using the implemented authentication and termination capabilities, as with the other transaction capabilities.

The next capability to be realized will be cash withdrawal, which will use the balance query capability to determine whether a customer account has sufficient funds to satisfy the customer request. The funds deposit capability will be realized next; a bank teller will use the balance query capability to verify the funds in a customer's account is correct before and after the bank teller has verified the amount of the funds deposit. The funds transfer capability to transfer funds from one customer's account to another of the customer's accounts (e.g. from savings to checking) is last to be implemented; it will use the balance query, cash withdrawal, and funds deposit capabilities.

These six phases of system implementation could be additionally prioritized into essential, desired, and optional capabilities. Essential capabilities could include those to be implemented in Version 1 of an ATM (e.g. the customer authentication, balance query, cash withdrawal, and session termination capabilities). Desired capabilities to be implemented in subsequent versions could be the funds deposit and funds transfer capabilities; and optional capabilities to be included for later consideration could be to provide different currency options for withdrawals (e.g. US dollars, Canadian dollars, and Mexican pesos for North American systems; British pounds sterling, Danish krone, and euros for European versions). Other optional capabilities could include purchasing of postage stamps or payment of utility bills.

Experience has shown that the capabilities-based approach to system requirements definition and system realization results in less ambiguity in the stakeholders' requirements, increased clarity in the system requirements, more efficient system development, and better stakeholder satisfaction with the delivered system, as compared with the conventional approach of transforming stakeholders' requirements directly into system requirements without the intermediate stage of generating and analyzing system capabilities.

The conventional approach of translating stakeholders' requirement directly into system requirements can be stated as

> And then a miracle happens.

The capabilities-based approach replaces miracles with a systematic approach.

5.2.1 Issues and Ameliorations

Experience with the capabilities-based approach to system development has revealed some concerns:

1. The "overhead" of including the capabilities step in requirements definition.
2. Contradictory and inconsistent capabilities-based system requirements.
3. Synchronizing concurrent development of physical elements and software elements needed to realize the capabilities.
   1) The approach of including the capabilities step in requirements definition appears to add additional "overhead" to the development process. Experience has shown that inserting the capabilities step, as illustrated by the above examples, along with bidirectional traceability among the stakeholders' requirements, system capabilities, and system requirements improves the quality of the system requirements and the resulting system by improving the "five C's" of requirements engineering (correctness, completeness, consistency, conciseness, and clarity).
   The 5C issues are better dealt with at the requirements level, when they are comparatively easy to deal with, rather than later in system development.

2) Statements of system capabilities may be contradictory and inconsistent. For example, an exception handling capability to handle loss of communication with the financial institution's accounts database server by securely closing down the ATMs until communication is reestablished may contradict the stakeholders' requirement that an automated teller system provide 24 × 7 continuous operation that generates a capability to maintain and periodically update ATM customers' account information in each ATM. Providing 24 × 7 continuous operation by periodically updating customers' accounts in each ATM may create inconsistencies with required system performance and cyber security capabilities.
   Decisions must be made concerning these and other conflicting and contradictory capabilities, keeping in mind the impact the decisions may have on other capabilities. One or more association matrixes that show couplings and dependencies among capabilities can illustrate the impacts of each capability on other capabilities.
   A general rejoinder to concerns 1 and 2 is that increased attention to upstream activities always results in a more efficient and effective overall system development process, returns more time, effort, and cost than invested in a project, and results in a system that better satisfies the stakeholders' requirements.
3) The third concern, namely, synchronizing concurrent implementation of physical elements and software elements during realization of system capabilities, is addressed in Section 5.3.2.

5.3 The I^3 System Development Model

The I^3 system development model for software-enabled systems includes the processes needed to concurrently fabricate and procure physical elements, construct and procure software elements, and build the interfaces among them and the connections to the system environment. The I^3 system development model incorporates the capabilities-based approach described above to define and realize the system requirements for physical elements, software elements, interfaces, and connections to the system environment.

A depiction of the I^3 system development model is presented in Figure 5.2. As indicated, there are six phases in I^3:

- *Phase 1*: SysDefn (system definition);
- *Phase 2*: Rqmts (requirements definition);
- *Phase 3*: Arch (architecture definition);
- *Phase 4*: Des (design definition)
- *Phase 5*: Realize (incremental realization); and
- *Phase 6*: SysDel (system delivery: system transition and validation)

Section 5.3.1 addresses the roles of physical systems engineers and software systems engineers in each of the six I^3 system development phases.

5.3.1 Systems Engineering During the I^3 System Development Phases

During I^3 Phase 1 (SysDefn), one or more physical systems engineers and software systems engineers engage with the appropriate system stakeholders to facilitate a business or mission analysis process. If the outcome is a decision to proceed, physical systems engineers and software systems engineers (or designated others) will elicit and prioritize the stakeholders' requirements.

The physical systems engineer(s) and software systems engineer(s) also develop an initial version of the work plan for the technical activities to be accomplished during the project. Business/mission analysis and stakeholder's requirements/system capabilities definition are grouped together in I^3 Phase 1 to indicate an iterative approach to analysis and definition. Details are provided in Chapter 6.

The designated others who may accomplish Phase 1 may be members of the stakeholders' organization, consultants, business analysts, or marketing specialists of the organization that will develop the system. In these cases, physical systems engineers and software systems engineers should review and clarify business/mission analysis and the stakeholders' requirements with the stakeholders and those who prepared the stakeholders' requirements before deriving the system capabilities.

During I^3 Phases 2–4 (Rqmts, Arch, and Des), physical systems engineers and software system engineers define and prioritize system capabilities, an initial definition of system requirements, a skeletal version of the system architecture, and an initial version of the system design – these are incrementally elaborated during Phase 5 (implementation). The physical systems engineer(s) and software systems engineer(s) also elaborate the work plan for the technical activities to be accomplished during the project. System requirements, architecture, and design are grouped together in the Figure 5.2 depiction of I^3 to emphasize an iterative approach to defining system requirements, system architecture, and system design (which may also involve iterative revision of the mission/business analysis and the stakeholders' requirements).

During Phase 5 (system implementation), systems engineers (both physical and software) coordinate the work activities of the disciplinary engineers (including software engineers) and specialty engineers, who plan, realize, verify, validate, and demonstrate the system increments. The physical systems engineer(s) and software system engineer(s) may participate in some of the work activities as disciplinary or specialty engineers, but if they do, they will be playing roles distinct from their systems engineering roles.

Phase 5 includes as many incremental subphases as needed to realize a deliverable system, subject to the constraints of schedule, budget, resources, infrastructure, and technology.

As indicated in Figure 5.2, each incremental subphase of Phase 5 includes planning, realizing, verifying, validating, and demonstrating activities:

- Subphase 5.0 of Phase 5 is concerned with developing a simulation model of the skeletal system architecture (see "System Simulation").
- During subsequent subphases of Phase 5, disciplinary engineers (other than software engineers) fabricate and procure physical elements while software engineers construct and procure software elements for the selected capabilities. Disciplinary engineers who realize physical elements and software engineers who realize software elements use their preferred development processes (e.g. incremental and iterative processes).
- Disciplinary engineers (physical engineers and software engineers) work together to realize the real interfaces that replace simulated interfaces. One or more engineers coordinate development of consistent interfaces among software elements and physical elements and will be responsible for integrating the software and physical elements that provide system capabilities.
- The capability (or capabilities) to be realized during a subphase is (are) selected during the planning stage of a subphase, during which requirements, architecture, and design are reviewed, revised, and elaborated as necessary. Work assignments are discussed and agreed to. Additional details of the planning activity are presented in Chapter 10.
- The system elements being fabricated, constructed, and procured are informally verified and validated by the disciplinary engineers using a sandbox copy of the previous version of the evolving system. Completed physical and software elements that realize one or more system capabilities replace simulated elements to form a next version of the evolving system.
- The version is verified and validated during the verification and validation (V&V) activities. The realization team may accomplish V&V, or V&V may be accomplished by another functional group, depending on the structure of the system development organization and the needs of the particular project.
- During the demonstration activity, the new version of the system is demonstrated for the system stakeholders and other appropriate individuals to provide a final validation of the new version. Needed and requested changes are noted and will be addressed during the planning stage of the next subphase or prior to delivery if the new version is the deliverable system. The realization team holds a private postmortem review of the completed subphase; work activities are reviewed and plans are made for any needed changes. The postmortem review may be included as part of the planning activity for the next subphase.

As indicated above, and as illustrated in Figure 5.2, the first step in system realization is concerned with developing a simulated model of the skeletal system architecture during Subphase 5.0. The modeling and simulation-based approach to system development (M&SBSE) is described in "System Simulation."

System Simulation

A simulated model of the I^3 skeletal system architecture is developed during system realization Subphase 5.0. It includes a simulated architectural skeleton, simulated interfaces among the system elements, and simulated connections to the system environment. This approach is consistent with the modeling and simulation-based approach to systems engineering (M&SBSE) (Adcock 2017a). The simulation is verified and validated and revised as necessary until it is a true representation of the skeletal architecture (i.e. the integrity and fidelity of the simulation model is verified). Simulation and modeling tools based on SysML notations (and other notations) for system analysis and design may be used to generate the simulated skeletal system (IBM 2018). One or more specialty engineers may be needed to construct, verify, and validate the simulation model if the needed expertise is not otherwise available.

During subsequent subphases of Phase 5, newly developed system elements, interfaces, and connections for the realized capabilities replace simulated elements to produce next versions of the evolving system. An essential aspect of validating the evolving system versions is to validate the integrity (completeness) and fidelity (accuracy) of the simulation model, at the desired level of detail, as it evolves to reflect the increasingly detailed system architecture and design.

The deliverable system results when all simulated system elements are replaced with real elements, interfaces, and connections. Incremental realization of prioritized system capabilities will ensure that the most significant capabilities have been realized first if project constraints prohibit realization of all desired capabilities or if emergency deployment of a partial system is necessary.

During the subphases of I^3 Phase 5, software engineers and other disciplinary engineers use their preferred processes to concurrently fabricate/procure physical elements, construct/procure software, build real interfaces among their system elements, and implement connections to the simulated system environment. Simulated elements, interfaces, and connections are replaced for each of the capabilities realized during each incremental subphase.

Disciplinary engineers who fabricate physical elements may prefer the incremental Vee model and software engineers who construct software elements may prefer an iterative approach (see Chapter 4). One or more engineers will be designated to facilitate and document detailed design of interfaces between physical elements and software elements and to replace simulated interfaces with real interfaces as the realized interfaces become available.

Note that integration of system elements occurs continuously in I^3 because realized elements replace simulated elements when the realized elements become available. Details of integration are addressed in Chapter 9.

The final system increment (Version 3 in Figure 5.2) is the deliverable system because each system increment, including the final one, incorporates newly realized capabilities plus all capabilities previously realized, verified, and validated in all previous subphases. Verifying and validating the deliverable system in the development environment (i.e. the final increment) thus precedes I^3 Phase 6.

During I^3 Phase 6 (SysDel), one or more physical systems engineers and software systems engineers coordinate the work activities they have planned for transitioning the deliverable system from the development environment to the operational environment. The appropriate system stakeholders participate in validating the delivered system in the operational environment to determine that the system provides the needed capabilities to satisfy the stakeholders' requirements (or not). The objective of transition and validation is to install the system in the operational environment and to gain the stakeholders' acceptance of the delivered system.

The engineers who accomplish the transition and validation activities may be members of the system development team, they may be contracted, or they may be others within the development organization who will accomplish system delivery and then assume the maintenance and sustainment work for the operational system, at the discretion of the stakeholders' decision makers.

To summarize, physical system engineers and software system engineers play a leading role in I^3 system development Phases 1–4. They provide a coordinating role in Phases 5 and 6. As mentioned, physical system engineers and software system engineers may participate in system development work activities in Phases 4–6 but they are playing different roles when they do.

System development activities for each of the major phases of the I^3 development model are summarized in Table 5.3.

Three Phase 5 realization subphases are illustrated in Figure 5.2 plus Subphase 5.0. Additional subphases will be added as needed. Large systems may be decomposed into subsystems that are each developed using the I^3 development model. In this case, a top-level simulation model to integrate the subsystems is developed. The simulated elements are incrementally replaced with real integrated system elements to provide the final integration of the real subsystems. Recursive decomposition, as needed, accommodates scalability of the I^3 development model.

Table 5.3 System development activities for the I³ development phases.

I³ development phase	System development activities
SysDefn	Conduct a business/mission analysis
	Define stakeholders' requirements and generate prioritized system capabilities
	Determine project feasibility and, if approved, develop the initial work plan for technical work activities
Rqmts/Arch/Des	Generate and prioritize an initial set of system capabilities to be provided
	Establish an initial set of system requirements based on the system capabilities
	Develop a skeletal system architecture
	Develop an initial version of the system design
	Complete the work plan for technical work activities
Implementation	First develop a simulation model of the system's skeletal architecture and design (system version 0)
	Then refine the architecture and design and iteratively/incrementally plan, realize, verify, validate, and demonstrate realized system capabilities in a succession of implementation subphases
SysDel	Transition, validate, and gain acceptance of the delivered system in the operational environment

Organizational structure, organizational policies, resistance to change, job specialization, and contractual constraints may inhibit a full I³ approach to system development that includes the capabilities-based approach to defining system requirements and the M&SBSE approach to incremental system realization. Although various elements of I³ can be separately applied (e.g. capabilities without M&SBSE; M&SBSE without capabilities; model-based systems engineering [MBSE] without simulation), these inhibitors should be overcome to the extent possible; otherwise, the full benefits of using the capabilities-based I³ approach to system development that bridges the gap between fabrication/procurement of physical elements and construction/procurement of software elements will not be gained.

Details for each of the six I³ development phases are provided in Chapters 6–9.

5.3.2 Synchronizing Realization of Physical Elements and Software Elements

Concurrent realization of physical elements, software elements, the interfaces among them, and the connections to the system environment is a principal attribute of the I^3 system development model for developing software-enabled physical systems. The simulation model of system architecture and design provides an environment in which realization of physical elements (incrementally) and realization of software elements (iteratively) can proceed concurrently with some interdependence.

Depending on the nature of a capability being realized, realization of a physical element may proceed ahead of realizing the software needed for the capability. For example, one or more physical elements may be readily procured or may be otherwise available but the associated software is complicated and complex. Alternatively, fabrication or procurement of one or more physical elements may require an extended duration while realizing the software elements is straightforward or they may have been previously realized and are available for reuse.

Out-of-sync realization of physical elements and software elements for a system capability places emphasis on shared evolution of the interfaces among the elements and continuing verification of the integrity and fidelity of the evolving simulation model. Decomposing a complex system capability into smaller units may facilitate synchronization of element realizations. Additional details on synchronization of concurrent development are provided in Chapter 9.

Implementation subphases are typically scheduled at one-month intervals. Longer durations forego frequent feedback and small incremental adjustments to requirements, architecture, design, and realization processes and work products, as needed. Because system realization occurs using a simulated system model, completion of an incremental implementation subphase does not depend on specific system elements (physical or software) having been fully realized; actual system elements replace simulated elements as they become available. Emphasis is on realized capabilities as the real system elements become available. In the meantime, a capability may be provided by a combination of completed and simulated system elements.

Completing an implementation subphase results in a demonstration of the evolving system, a review work accomplished to date – as compared with work planned at the start of the subphase, problems encountered, and actions needed to resolve the problems.

5.3.3 Mapping the I^3 Development Model to the Technical Processes of 15288 and 12207

The technical processes in the ISO/IEC/IEEE 15288 and ISO/IEC/IEEE 12207 process standards provide foundations for the six phases of the I^3 system development model (ISO 2015, 2017). A mapping of the I^3 phases to the development processes in 15288 and 12207 is provided in Table 5.4. Subclause numbers in 15288 and 12207 are indicated in parentheses.

Table 5.4 I^3 development phase and 15288/12207 processes.

I^3 development phases	15288/12207 technical processes
SysDefn	Business or mission analysis process (6.4.1)
	Stakeholder needs and requirements definition process (6.4.2)
	Project planning process (6.3.1)
Rqmts/Arch/Des	Project planning process continued (6.3.1)
	System requirements definition process (6.4.3)
	Architecture definition process (6.4.4)
	Design definition process (6.4.5)
	System analysis process (6.4.6)
Implementation	Incremental project planning process continued (6.3.1)
	Implementation process (6.4.7)
	Integration process (6.4.8)
	Verification process (6.4.9)
	Validation process (6.4.11)
System delivery	Transition process (6.4.10)
	Validation process (6.4.11)

Although the I^3 processes and the corresponding 15288 and 12207 processes are listed in linear order, the work activities of the I^3 development processes are performed iteratively and incrementally.

The implementation and integration processes of 15288 and 12207 are combined in the I^3 simulation model because realized system elements replace simulated elements when they become available. The traditional separation between implementation and integration of system elements may apply when subsystems of a large system are periodically integrated.

The project planning process of 15288 and 12207 in Table 5.4 (Clause 6.3.1) is presented in Part III of this text, where technical management is presented. Technical management activities are not further addressed in Part II of the text (Chapters 5–9).

The design definition process (6.4.5) is somewhat different for the design of physical elements and software elements. For physical elements, design definition is a rigorous process used to develop specifications for the physical elements, interfaces, and connections. Design reviews are conducted to increase confidence in the design specification prior to procurement and fabrication of system elements because of the difficulty, and perhaps impossibility, of modifying the functional, behavioral, or quality attributes of a physical element or interface. As previously stated, the goal of design definition for physical elements and interfaces is to "get it right" before committing to procurement and fabrication.

Design definition for software also has the goal of "getting it right" before committing to software construction and procurement. Initial design of software precedes software construction in the same way that design of physical elements precedes implementation. However, elaboration of the initial design of software occurs iteratively as part of software construction because the malleability of software makes this possible and desirable. As stated in Chapter 3, this difference is clarified by distinguishing realization of physical system elements as "fabrication" and realization of software elements as "construction."

5.4 Key Points

- The technical processes of ISO/IEC/IEEE Standards 15288 and 12207 provide the foundations for the I^3 system development model.
- I^3 includes six development phases. Phase 5 includes incremental subphases. All phases of system development may involve revising the work products of other phases.
- A system capability is the ability of a system to execute a specified course of action or exhibit a specified state of being.
- When using I^3 to develop a software-enabled physical system, the capabilities-based approach supports concurrent and coordinated iterative construction and procurement of software elements, incremental fabrication and procurement of physical elements, and coordinated building of interfaces and connections.
- Capabilities and I^3 together eliminate the "hardware-first" and "software-first" inhibitors to effective development of software-enabled systems described in Section 3.6.
- When using the I^3 system development model, completion of an implementation subphase does not depend on complete realization of particular physical elements or software elements. Emphasis is placed on providing system capabilities.
- Synchronizing out-of-phase development of physical elements and software elements places emphasis on shared evolution of the interfaces among the elements and continuing evolution of the integrity and fidelity of the evolving simulation model.
- One or more engineers may be designated to coordinate design and realization of interfaces between physical system elements and software system elements.
- An implementation subphase of I^3 Phase 5 commences with a planning activity that may include a postmortem review of the pervious subphase, or

the postmortem review may be conducted separately prior to the planning activity.

- Ongoing V&V is essential to ensure the integrity and fidelity of an I^3 simulation model as it evolves.
- I^3 methods and techniques can be used to develop other kinds of systems in addition to development of software-enabled systems.
- Full implementation of the capabilities-based I^3 system development model may not be possible, but in those cases, the full benefits will not be obtained.

This chapter has provided an introduction to the capabilities-based M&SBSE I^3 system development model. Examples have been included. Additional details for the I^3 system development phases are provided in Chapters 6–9.

Exercises

5.1. A capability is the ability of a system to execute a course of action or exhibit a state of being.
  (a) Select and briefly describe a consumer product or a personal device (e.g. a microwave oven, a cell phone, cruise control for an automobile).
  (b) Specify three states of being exhibited by your selected product or device.
  (c) Select two states of your product or device and give them names. Then describe how the product or device transitions from one of the states to the other state.

5.2. Suppose a stakeholders' requirement is as follows: "The system should provide printed receipts for customer transactions."
  (a) Restate the stakeholders' requirement as a system capability.
  (b) State the system capability as a system requirement.
  (c) State a derived requirement needed to support the system requirement.

5.3. When using the I^3 system development model, completing an implementation subphase does not depend on complete realization of a particular physical element or software element. Briefly explain why this is true.

5.4. Capabilities and I^3 together eliminate the "hardware-first" and "software-first" inhibitors to effective development of software-enabled systems (described in Section 3.6). Briefly explain how capabilities and I^3 eliminate the "hardware-first" and "software-first" inhibitors.

5.5. During the subphases of I^3 Phase 5, one or more engineers may be designated to coordinate design and realization of interfaces between physical system elements and software elements. Briefly explain why this is an important system development activity.

5.6. An I^3 implementation subphase commences with a planning meeting that may include a postmortem review of the pervious subphase, or the postmortem review may be conducted separately prior to the planning activity.
(a) Briefly explain the purpose of a postmortem review.
(b) Briefly explain some of the possible outcomes of a postmortem review.
(c) Briefly explain the purpose of a planning meeting.
(d) Briefly explain the desired outcomes of a planning meeting.

5.7. When using I^3 to develop a software-enabled physical system, the capabilities-based approach supports concurrent and coordinated iterative construction of software elements, incremental fabrication of physical elements, and coordinated building of interfaces and connections.
(a) Briefly explain how the I^3 development model supports concurrent and coordinated iterative construction of software elements and incremental fabrication of physical elements.
(b) Briefly explain how the I^3 development model supports coordinated building of interfaces and connections.

5.8. Ongoing verification and validation is essential to ensure the integrity and fidelity of an I^3 simulation model as it evolves.
(a) Briefly describe what is meant by the integrity of an I^3 simulation model.
(b) Briefly describe what is meant by the fidelity of an I^3 simulation model.
(c) Briefly explain why it is essential to ensure the integrity and fidelity of an I^3 simulation model as it evolves.

5.9. I^3 methods and techniques can be used to develop other kinds of systems in addition to development of software-enabled systems. Briefly describe a system that is not software-enabled and then describe how the I^3 system development model could be used to implement the system.

5.10. Full implementation of the capabilities-based I^3 system development model may not be possible due to local circumstances, but in those cases, the full benefits will not be obtained. Briefly explain some local circumstances that may not permit full implementation of the capabilities-based I^3 system development.

References

ISO (2015). ISO/IEC/IEEE 15288:2015, Systems and software engineering – System life cycle processes, ISO/IEEE, 2015.

ISO (2017). ISO/IEC/IEEE 12207:2017, Systems and software engineering – Software life cycle processes, ISO/IEEE, 2017.

IBM (2018). IBM Rational Rapsody. https://evocean.com/products/ibm-rational-rhapsody (accessed 10 September 2018).

Adcock, R.D. (EIC) (2017a). *The Guide to the Systems Engineering Body of Knowledge (SEBoK)*, v. 1.9. Hoboken, NJ: The Trustees of the Stevens Institute of Technology www.sebokwiki.org (accessed 12 June 2018).

Adcock, R.D. (EIC) (2017b). *The Guide to the Systems Engineering Body of Knowledge (SEBoK)*, v. 1.9. Part Introduction to SE Transformation. Hoboken, NJ: The Trustees of the Stevens Institute of Technology www.sebokwiki.org (accessed 12 June 2018).

6

The I^3 System Definition Phase

6.1 Introduction

This chapter presents the business or mission analysis and stakeholder needs and requirements definition processes in ISO/IEC/IEEE Standards 15288 and 12207 (ISO 2015, 2017). The material in this chapter is keyed to the I^3 system development processes introduced in Chapter 5 but it can be applied independently or when using system development models other than the integrated-iterative-incremental (I^3) model.

A case study for a driving system simulator (Realistic Corporation driving system simulator, RC-DSS) is used to illustrate elements of the system definition (SysDefn) phase of I^3 system development. The case study is introduced in "The Realistic Corporation Driving Simulator System".

This chapter provides the opportunity for readers to learn about business or mission analysis, stakeholders' needs and requirements, and application of the process to a realistic case study. References and Exercises at the end of the chapter can provide additional understanding of business or mission analysis and stakeholders' needs and requirements.

6.2 Performing Business or Mission Analysis

According to Section 6.4.1 of the 15288 and 12207 standards, the purpose of the business or mission analysis process is

> to define the business or mission problem or opportunity, characterize the solution space, and determine potential solution class(es) that could address a problem or take advantage of an opportunity.

Business or mission analysis is often thought to focus exclusively on analyzing an identified problem and developing a mapping from the problem space to a characterization of the solution space. But the analysis process can also

Systems Engineering of Software-Enabled Systems, First Edition. Richard E. Fairley.
© 2019 John Wiley & Sons, Inc. Published 2019 by John Wiley & Sons, Inc.

Table 6.1 Business or mission analysis.

Input	A potential business problem, mission need, or opportunity
Process	Define a business problem, a mission, or an opportunity; then describe the solution space and determine one or more solution classes in the solution space
Output	A documented definition of a business problem, a mission, or an opportunity plus a description of the solution space and one or more solution classes

be applied to analyzing a potential opportunity to develop a new or improved product, deliver a new or improved service, or improve the functioning of an enterprise (SEBoK 2017).

Additionally, this process includes identifying potential solution spaces. For example, analyzing the solution space for renewable energy sources might identify a new renewable energy source or improvements to an existing energy source. Constraints (technical, financial, environmental, social, political, and regulatory) must be considered when conducting the business or mission analysis process.

Table 6.1 indicates the inputs, process, and outputs of the business or mission analysis process.

A business or mission analysis statement should contain the following information:

- A statement of the business problem, mission, or opportunity;
- Conformance with the organization's mission and vision statements;
- An abstract characterization of the solution space;
- Analysis of some alternative solution classes and solution strategies;
- Present and potential stakeholders;
- Primary operational concepts;
- Analysis of technical, financial, regulatory, social, environmental, political, and regulatory constraints;
- Security threats and safety hazards that need to be understood and accounted for;
- Enabling systems or services needed; and
- A recommendation to proceed or not to proceed.

An individual or group other than physical system engineers (PhSEs) and/or software systems engineers (SwSEs) may conduct the business or mission analysis process (e.g. business analysts, marketing specialists, the customer/acquirer) with the results conveyed to system engineers and/or software engineers. However, PhSEs and SwSEs should be involved to the extent possible to provide project continuity. An example of business or mission analysis for the RC-DSS driving system simulator is described in the below box.

The Realistic Corporation Driving Simulator System

This box presents an overview of the RC-DSS case study that will provide a context for the development processes of the I^3 system development model presented in this chapter and Chapters 7–9.

RC-DSS is envisioned as a realistic driving system simulator to be developed by the fictitious Realistic Corporation (RC). The RC-DSS project is small enough to be presented in considerable detail and large enough in scope and complexity to provide a realistic example of developing a software-enabled system using the I^3 system development model.

The RC-DSS is envisioned to have five major hardware/software subsystems: a vehicle cabin, a hydraulic platform, an instructor's workstation, an administrator's workstation, and a server (i.e. a computer with data repositories and connections to the other subsystems). The server, microprocessors, other digital devices, and the associated software will provide interconnections among the subsystems and coordinate internal interactions within the vehicle cabin and among the vehicle cabin, the hydraulic platform, and the instructor workstation.

The vehicle cabin of RC-DSS Version 1 will have realistic hardware, including the usual features of an automobile: a realistic cabin enclosure, door, seats, steering wheel, windshield, side window, rear window, and mirror displays, dashboard indicators controls, and so forth. Software will coordinate interactions within the vehicle cabin and among the subsystems. In addition, software will provide much of the system functionality, behavior, and quality attributes. The vehicle cabin will be mounted on a hydraulic platform that is controlled by software to simulate realistic road conditions and emergency situations during driving scenarios.

The instructor's workstation will allow an instructor to create and store scripts for driving scenarios, monitor and control driving sessions, and insert simulated emergency situations during student sessions. The instructor's workstation will also support the needs of service technicians.

The administrator's workstation will be used to enroll students, produce student progress reports, generate completion certificates, and support financial accounting processes.

The server will provide the data repositories and system interfaces, coordinate interacts among the other system elements, and support the instructor's and administrator's workstations.

(Continued)

(Continued)

The package diagram in Figure 6.1 illustrates the five DSS subsystems and the interconnections among them as a package of packages (see Appendix 8.A for more information).

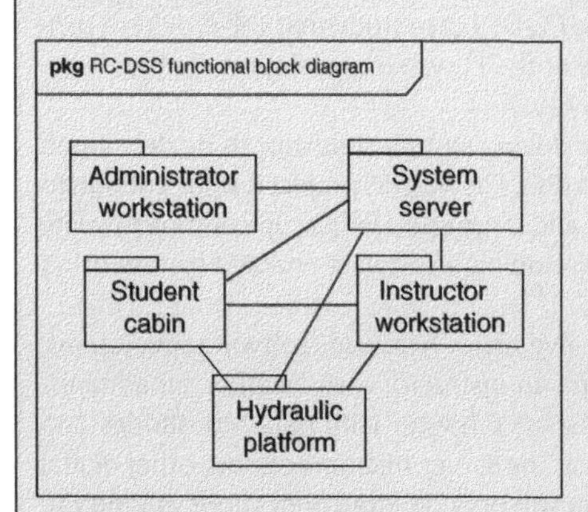

Figure 6.1 Package diagram of the RC-DSS driving simulator.

Additional details include:

1. Software sensors, displays, and a projection system will be provided in the vehicle cabin. The sensors and displays will sense and control the steering wheel, accelerator and brake pedals, shift control (manual shift optional?), windshield wiper control, cruise control system, headlights on/off, high beam/low beam, horn, speedometer, tachometer, temperature indicator, oil pressure indicator, gasoline indicator, GPS, radio/tape player/CD stereo system, air conditioning and heating controls, and any other features stakeholders may want. The controls, sensors, and displays will be driven and sensed by hardware/software elements.
2. The projection display in the vehicle cabin will provide realistic views of the windshield, side windows, rear window, and mirror displays. The physical projection system will be driven by software.
3. The vehicle cabin will be mounted on a hydraulic platform controlled by equations of motion implemented in software to provide a realistic feel of driving. The hydraulic system will be controlled by prerecorded scripts, and by student and instructor inputs to the hydraulic platform software.
4. The instructor's workstation will allow instructors to prepare and store driving scenarios, monitor student sessions, and create various driving conditions in real time (such as obstacles in the road, slick roads, rain, snow, fog, sharp curves, changing speed limits). The instructors' workstation will be a hardware/software workstation connected to the DSS server. This workstation will also provide the system interface used by the service technician.

> 5. The RC-DSS server will coordinate the interactions of the other RC-DSS components. Driving scenarios, student scores, hydraulic platform scripts, and other information will be retained in an RC-DSS data repository that resides in the server. Administrative information will also be stored in a separate data repository of RC-DSS server.
>
> The initial version of the RC-DSS will support training of beginning drivers. It is envisioned that future versions will have various modes of operation for various classes of users: beginning driver, safety refresher, emergency vehicles, police, military, race course, off-line diagnostics, maintenance, and so forth.

6.2.1 Business Analysis for the RC-DSS Project

Business analysis has identified a business opportunity for the RC to build and sell or lease realistic driving system simulators. The results of business analysis for the RC-DSS project includes the following information:

- Market analysis reveals a business opportunity to develop and sell or lease driving system simulators that will simulate driving of land-based vehicles having four or more wheels under various driving conditions.
- Characterization of the solution space ranges from simple laptop-based joystick applications to a simple steering wheel display and screen arrangement to full-scale vehicles instrumented with sensors, actuators, controls, and displays.
- Consideration of alternative solution classes and solution strategies have resulted in selection of the flight training simulators used by airlines to train and certify commercial pilots as a solution class for a variety of land-based vehicles having four or more wheels.
- Potential RC-DSS customers include the following:
  - Secondary school districts;
  - Trucking companies;
  - Departments of motor vehicles;
  - Police and fire departments;
  - Hospitals; and
  - Military installations.
- Driving system simulators that provide realistic simulations for various user classes will be installed in realistic facilities to be used by individuals who are learning to drive or are enhancing their driving skills. Driving system simulators will also be leased to and installed in the facilities of other organizations and agencies.
- The initial version of an RC-DSS will be configured as a modern automobile for beginning student drivers; it will be designed to be configurable for other kinds of realistic driving simulators.

- A statement of intent to license one or more RC-DSS Version 1 systems has been signed by a large secondary school district in a major metropolitan area (the Metro school district). In return, a major price incentive has been offered to the school district.

6.2.2 RFPs, SOWs, and MOUs

In some cases, one or both of the business or mission analysis and stakeholder needs and requirements definition processes may have been accomplished by the acquiring organization, which may be an external organization or one internal to the organization that will develop the system.

An external acquiring organization will typically generate a request for proposals (RFPs) for potential bidders unless the work is to be accomplished by a designated system developer (i.e. sole source procurement). The RFP will typically contain a statement of work (SOW) that includes a summary of business or mission analysis and stakeholders' requirement and specifies the contractual conditions under which the work is to be accomplished. An SOW becomes a contractually binding document after of the RFP has been negotiated with the selected supplier. The RFP may have been developed by a contractor for or consultant to the acquirer.

Projects internal to an organization are, or should be, governed by a memo of understanding (MOU), which is similar to, but less formal than, an SOW. Internal projects include those for which one department develops a system for use by another department, a project in which a department develops a system for use within that department, or a project that develops a system to take advantage of a marketing opportunity, as in the case of the RC-DSS.

Elements of SOWs and MOUs include the following:

- Technical goals to be achieved;
- Scope of work to be accomplished;
- Identification of acquirer, customers, and key stakeholders;
- Key development personnel needed and responsibilities;
- Cost, schedule, resource, and technology constraints;
- Policies, guidelines, regulations, and standards to be observed;
- Shared responsibilities and dependencies between the acquirer and supplier; and
- Other goals, objectives, and constraints that will affect system development, installation, and sustainment.

Internal projects are often conducted in an informal manner. Failure to develop and negotiate an informal MOU can result in misunderstandings and miscommunication.

6.2.2.1 An RC-DSS MOU

The acquirer of RC-DSS Version 1 will be the marketing department of the RC, which will be acquiring the DSS on behalf of the Metro school district; the school district will be the initial customer for an RC-DSS. Other acquirers and customers are anticipated. The supplier will be the RC engineering department. An MOU is developed to document the relationship between the marketing department (acquirer) and the engineering department (supplier).

An MOU for the RC-DSS to be negotiated and mutually accepted by the marketing department lead, the RC-DSS project manager, his or her manager, the lead PhSE, and lead SwSE includes the following information:

- The RC-DSS is a software-enabled driving system that will be developed by RC.
- The RC marketing department will be the acquirer of the RC-DSS Version 1 developed by RC engineers plus contractors as needed.
- The above-mentioned Metro school district will be the initial customer for Version 1 of the RC-DSS.
- The senior physical systems engineer of RC will be the project manager who will plan and supervise the project and will participate in the technical aspects of developing Version 1 of the RC-DSS. There will be a designated lead software system engineer who will provide collaborative technical leadership. System developers will include electrical, mechanical, civil, and software engineers plus safety, security, and other engineering specialists as needed.

Additional elements of the RC-DSS MOU include the following:

- Technical goals to be achieved:
  Development of a fully operational RC-DSS Version 1
  Installation and continued sustainment of the RC-DSS Version 1
- Scope of work to be accomplished:
  A full I^3 development cycle will be executed, starting with identification of stakeholders' needs and concluding with delivery and installation of an RC-DSS at the customer site.
  The RC engineering department will install and provide continuing sustainment services to the Metro school district.
- Identification of acquirer, customers, and key stakeholders:
  The acquirer is the RC marketing department.
  The initial customer is the Metro school district.
  Other potential customers and other stakeholders will be determined during the I^3 SysDefn process.
- Key development personnel and responsibilities:
  A member of the RC marketing department will be the product manager and will facilitate interactions with the customer. A designated PhSE

and a designated SwSE will provide systems engineering leadership. Senior mechanical, electrical, civil, and software engineers will provide disciplinary engineering leadership. The physical systems engineer and the SwSE will coordinate their work.

- Cost, schedule, resource, and technology constraints:
  Development of a product-line architecture suitable for implementing various RC-DSSs is a major design goal for RC-DSS Version 1.
  Cost, within reasonable limits, is not a major constraint. Delivery of a fully operational RC-DSS Version 1 is planned for 12 months from project initiation.
  Key personnel resources are adequate; some contracted specialists may be needed.
  Some additional software development tools may be needed. Existing, mechanical, electrical, and software expertise, technologies, and infrastructure are adequate.
- Policies, guidelines, regulations, and standards to be observed:
  RC policies, guidelines, and procedures will be observed, with tailoring as needed and approved by the RC quality assurance department.
  National, state, and local safety regulations will be observed. Current versions of ISO/IEC/IEEE Standards 15288 and 12207 will provide guidance for software-enabled system development.
  There are no external acquirer conditions to be satisfied other than the stakeholders' requirements expressed by the acquirer and the initial customer (i.e. the RC-DSS marketing department and the Metro high school district).
- Shared responsibilities and dependencies between the acquirer and the supplier:
  The RC marketing department, as system acquirer, will represent potential customers, including the Metro high school district.
  The Marketing lead, the RC-DSS project manager, the lead PhSE, the lead SwSE, and the leaders of the development team will share responsibility for timely development of RC-DSS Version 1.
  One or more representatives of the initial customer will participate in evaluating demonstrable RC-DSS increments as they evolve.
- Other goals, objectives, and constraints that will affect system development, installation, and sustainment:
  - RC-DSS systems will not be developed for land-based vehicles having fewer than four wheels or for airborne or water-borne vehicles.
  - Physical safety is the primary design goal for RC-DSS systems; they must never create unsafe situations for student drivers, technicians, operators, instructors, or others during startup, operation, and shutdown, in both normal operation and system failure.

- The system architecture for RC-DSS Version 1 will be structured to accommodate a product line of DSS systems that will accommodate various kinds and forms of land-based vehicles having four or more wheels.
- RC-DSS systems must be configured to permit remote monitoring of performance and remote diagnostics.
- RC-DSS systems must be configured so that two trained technicians can accomplish installation and full operation of an RC-DSS in a pre-prepared facility in two calendar days.
- RC-DSS systems must be configured so that a trained technician can accomplish diagnostics and routine maintenance of an RC-DSS with not more than four hours of downtime per week.

6.3 Identifying Stakeholders' Needs and Defining Their Requirements

Guidance for the stakeholder needs and requirements definition process is provided by Section 6.4.2 of 15288 and 12207. According to Section 6.4.2, the purpose of the stakeholder needs and requirements definition process is

> to define the stakeholder requirements for a system that can provide the capabilities needed by users and other stakeholders in a defined environment.

Stakeholders who will directly interact with a DSS (i.e. the users) and other stakeholders who will influence or be influenced by the system during its developmental and operational life cycle are identified and their needs are elicited. Their needs will be analyzed and a set of stakeholders' requirements, including intended interactions with the operational environment, will be determined. An initial set of system capabilities will be developed (see Section 5.2).

Table 6.2 indicates the inputs, process, and outputs of the stakeholder needs and requirements definition process.

The stakeholder needs and requirements definition process is usually applied in an iterative manner, both internally within the process and in conjunction with the business or mission analysis process, as indicated in Figure 5.1 and Table 5.2.

The concept of operations (ConOps) listed as an output of the stakeholder needs and requirements definition process is discussed in Section 6.3.4.

Individuals or organizations other than PhSESs or SwSEs may conduct the stakeholder needs and requirements definition process (e.g. requirements analysts or marketing specialists) with the results conveyed to PhSEs and SwSEs.

Table 6.2 Stakeholder needs and requirements definition.

Input	A documented statement of business need, mission statement, or opportunity
Process	Identify stakeholders and characterize the operational environment
	Elicit, categorize, and prioritize operational requirements
	Identify needed system capabilities, constraints, and risk factors
	Conduct a feasibility study
	Develop a documented agreement between the acquiring organization and the supplying organization
Outputs	Stakeholders' requirements, system capabilities, constraints, and identified risk factors
	A statement of work or memo of understanding
	A concept of operations that includes operational scenarios

However, PhSEs and SwSEs should review with the appropriate parties all information provided to them and request revisions, as needed; PhSEs and SwSEs should be involved to the extent possible in stakeholder needs and requirements definition to provide project continuity.

6.3.1 Identifying System Stakeholders

Identifying the system stakeholders who will use, operate, and maintain the envisioned system in a hands-on manner is an initial step in the stakeholder needs and requirements definition process. Hands-on users of a driving system simulator, for example, will be student drivers, instructors, and system technicians.

The needs of other stakeholders who will not directly interact with a system but who will affect or be affected by development and operation of the system should also be identified and their needs determined. Examples of other stakeholders include those who might provide inputs to a system, use system outputs, or potential system acquirers who might procure or lease a system.

Other stakeholders who will not directly interact with the operational system in a hands-on manner include, in the case of an RC-DSS enterprise, automobile manufacturers, retail sellers, insurance agencies, drivers' educators, and agencies that issue automobile licenses. Individuals who work for those organizations may operate automobiles, but when they do so, they become members of a different stakeholder class (i.e. automobile drivers).

In some cases, system operators and hands-on users will be the same individuals, as in the case of a microwave oven, and in other cases, they will be distinct, as in the case of an enterprise information system for which IT personnel support operation of the system for the system's users. And in still other cases, there will be different classes of users (hands on and other) who will interact with a system in different ways, for example, the users of an automated teller system who include the user classes of bank customers, bank tellers, auditors, personnel who service automated teller machines (ATMs), and IT personnel who maintain the automated teller system infrastructure.

6.3.1.1 Some Clarifying Terminology

The system supplier is the organization or organizational group that develops a system for an acquirer. An acquirer is the designated individual or organizational representatives who interact directly with the supplier. A customer is a designated individual or organizational representatives who provide the stakeholders' requirements to the acquirer, on behalf of the stakeholders. In many cases, the acquirer and the customer will be the same individual or group of organizational representatives. In other cases, the acquirer who interacts with the supplier may represent multiple customers (e.g. a consortium of financial institutions).

The acquirer and customer representatives are primary system stakeholders. An acquirer might, for example, engage a knowledgeable customer to provide expertise by interacting with the system developer (i.e. the supplier), or an acquirer might represent multiple customers as, for example, an airline, a hotel chain, and a car rental company that collaboratively sponsor development of a coordinated reservation system.

The role of the customer, or customers, is to represent the stakeholders, provide the stakeholders' requirements, interact in an ongoing manner with the supplier and the acquirer (if there is a separate one), and accept or reject the deliverable system on behalf of the system stakeholders. In some cases, there may be subcontractors and vendors who provide system elements to the system supplier (i.e. the system developer). In other cases, there may affiliated projects, as when multiple suppliers develop a large system; there is usually a designated lead organization (i.e. a prime contractor) in these cases.

The general situation is illustrated in Figure 6.2.

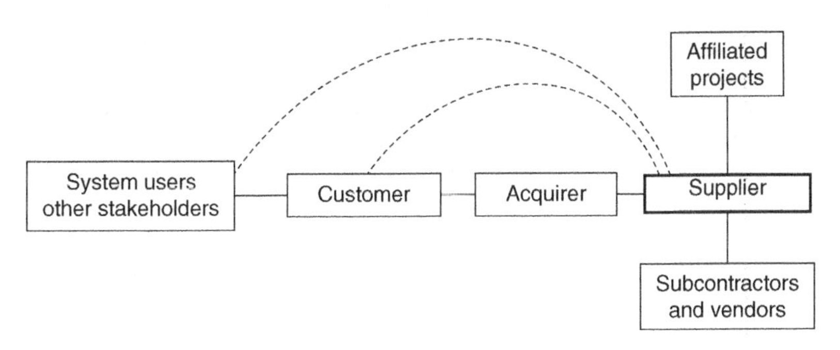

Figure 6.2 Some project relationships.

The accompanying "Project Relationships" describes some potential problems with the relationships in the figure.

Project Relationships

The dotted line connections between system users and acquirer and between supplier and system users/customer(s) are highly desirable but may not exist. For example, an agency in the U.S. Department of Defense may form a contract with a supplier, as the acquirer for a military defense system. The customer may be the U.S. Army. The system users may be the soldiers. Without the dotted line connections, the acquirer and customer stand between the supplier and the soldiers who will use the delivered system. The supplier will have no opportunity to determine the true needs of the system-user soldiers.

This situation also occurs in commercial organizations, as in the RC-DSS case study. The potential for miscommunication is presented if the dotted line connections don't exist when the supplier is a system development department, the acquirer is the marketing department, the customer is the Metro school district, and the system users to-be are high school students.

System developers (i.e. the supplier in Figure 6.2) are sometimes discouraged from communicating with system users, especially in contracted situations, because the system developers will want to include features requested by the system users when inclusion of those features would exceed the cost and schedule constraints of the SOW, MOU, or contract.

Affiliated projects are those whose outcomes are to be combined with the outcomes produced by a supplier (i.e. a system development organization) to satisfy a business need or a mission, as in the case of a program to develop a satellite system that involves three different contractors to develop the ground-based launch system, the telemetry and communication system, and the space-borne satellite.

Failing to identify all stakeholders and their requirements, plus other involved organizations, such as affiliated projects, subcontractors, and vendors during the stakeholder needs and requirements definition process can be a costly mistake, as happened in the case of a project that failed to identify the user class of smoke-jumping fire fighters and their needs until late in a project to develop a satellite surveillance system for forest management, including management of forest fires (author's personal experience).

A simpler relationship is illustrated in Figure 6.3. An even simpler version of Figure 6.3 would illustrate the situation when the customer and stakeholders are the same entities.

Figures 6.2 and 6.3 reveal an important aspect of systems development projects:

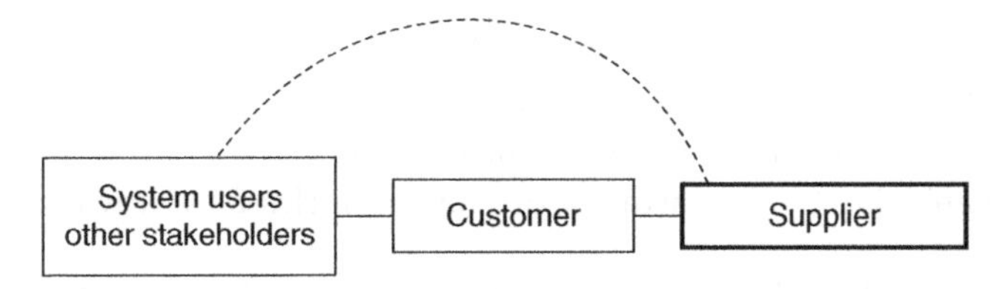

Figure 6.3 Simpler project relationships.

All systems development endeavors are two or more projects.

A Venn diagram would reveal separate activities to be conducted by each party in Figures 6.2 and 6.3 plus shared duties among the various parties.

The next activity of the stakeholder identification and requirements definition phase is to identify system stakeholders and elicit their needs, wants, and desires.

6.3.1.2 Identifying RC-DSS Stakeholders

As indicated above, stakeholders include users, operators, maintainers, and other hands-on stakeholders plus others who will not directly interact with the RC-DSS system in a hands-on manner but will affect or be affected by development and/or deployment and operation of a system.

Hands-on users of systems are grouped into user classes; the users in a user class are individuals who will use a system in a similar manner. Different user classes for the RC-DSS, for example, use the DSS in different ways. User classes for the RC-DSS include the following:

- Student drivers who may be subdividing into separate subclasses such beginning drivers and emergency vehicle drivers.
- Instructors who will provide orientations for new RC-DSS drivers, prepare scripts for driving scenarios, monitor student performance, and inject departures from prepared scripts into real-time simulations.
- Service technicians who will maintain the RC-DSS hardware and software.
- Potential customers who will evaluate the RC-DSS for purchase or lease.
- Mechanical, electrical, civil, and software engineers who develop and demonstrate incremental capabilities during system development, maintain the RC-DSS system elements, evaluate new technologies, and generate new versions of driving system simulators. They will also develop the physical housing, power connections, interlocks and warning devices, and other site preparations for installing an RC-DSS.
- Specialty engineers who will advise on how best to include hardware, software and procedures for information security, physical security, and physical safety.
- Automobile manufacturers and researchers who may use RC-DSS installations to evaluate new technologies and driver responses.

Student drivers can be further categorized as follows:

- Secondary school students and others who are learning to drive;
- Drivers who are learning to drive detached-trailer trucks;
- Drivers who are applying to a department of motor vehicles for a driving license;
- Students of commercial driving schools who include novice drivers, older people who are learning to drive or are taking a refresher class, and drivers taking court-ordered remedial classes;
- Policemen, fire fighters, emergency medical technicians, and other emergency vehicle drivers; and
- Military personnel.

Other RC-DSS stakeholders who will affect or be affected by development and operation of a DSS but will not interact with an RC-DSS in a hands-on manner include the following:

- RC decision makers;
- RC-DSS administrators who enroll students, generate reports and certificates, and manage the financial aspects of DSS installations;
- RC-DSS customers and acquirers who may include some or all of school districts, departments of motor vehicles, police and fire departments, courts and other legal entities, and military and other governmental agencies;
- Insurance companies that may provide discounts for RC-DSS-certified drivers; and
- Departments of motor vehicles that may grant various categories of driving licenses to RC-DSS students who exhibit acceptable driving skills on an RC-DSS.

6.3.2 Eliciting, Categorizing, and Prioritizing Stakeholder Requirements

Eliciting, categorizing, and prioritizing stakeholders' requirements are essential activities in the stakeholders' requirements definition process. Each is considered in turn.

6.3.2.1 Eliciting Stakeholders' Requirement

Elicitation is the process of engaging with stakeholders to determine their needs, wants, and desires for a proposed system. Techniques for eliciting stakeholders' requirements include (BABOK 2015) the following:

- Brainstorming;
- Questionnaires;
- Surveys;
- Observations;

- Structured interviews;
- Prototypes;
- Focus groups; and
- Structured workshops.

These techniques are often used in combination; for example, questionnaires and surveys may be used as background information for preparing interview scripts and conducting interviews.

It is important to record all initial stakeholders' statements of needs and expectations. Some initial statements may be vague and ambiguous, such as "highly reliable," "realistic," and "easy to use" but these statements are meaningful to those who provide them. Later, these initially vague statements must be understood in objective terms, to the extent possible, and consensus on the objective statements must be obtained from the affected stakeholders. Statements that cannot be restated objectively are classified as design goals that will influence design decisions; for example, "easy to use" will influence the emphasis placed on design of user interfaces and other modes of human–machine interactions.

6.3.2.2 Categorizing Stakeholders' Requirements

Stakeholders' requirements can be categorized as follows:

- *User features*: Functional and behavioral characteristics a system will exhibit to system users;
- *Quality attributes*: Externally visible quality attributes a system will exhibit to system users; and
- *Design constraints*: Design decisions stated in the stakeholders' requirements.

Features are the functions and behaviors a system will provide for system users. It is important to note that a system user is any entity in the system environment that interacts with a system. These users may include humans, computing hardware, other kinds of physical elements, software data repositories, and naturally occurring elements.

A function is an input/output relationship; for example, an RC-DSS instructor or a student driver might provide an "ignition requested" input and the system would respond with a notification output of "vehicle enabled" or an exception notification of "vehicle failed to enable."

Behavior is exhibited to a system user as a sequence of functional interactions over time. For example, turning the RC-DSS steering wheel would create the simulated behavior of a vehicle changing directions by adjustments made to the hydraulic platform and the visual displays. Behavior can be specified in operational scenarios that document the various ways users will interact with a system. Use cases and use case diagrams are widely used to document operational scenarios. Examples of a use case and a use case diagram are provided in Section 6.3.3.1.

Quality attributes of interest to system stakeholders are those that are externally visible; they include attributes such as safety, security, availability, reliability, robustness, dependability, ease of use, and response time. Other quality attributes that are important to system developers but will not be apparent to system stakeholders include testability, ease of installation and modification, and level of supporting documentation (see Table 7.2).

Design constraints are design decisions stakeholders have placed in their requirements; they will be reflected in the system design. A design constraint might dictate that a specified radar unit or a specific computer operating system must be used for compatibility with other system elements or other systems. Stakeholders' design constraints for the RC-DSS might include a secure connection to be used by authorized customers and students' parents or guardians, or perhaps stakeholders will require that RC-DSSs provide access to the vehicle cabin for physically challenged student drivers.

Design constraints limit design options and may affect the project's schedule, budget, and needed resources. Design constraints such as unnecessarily rapid response times or generation of unneeded reports may require expensive and unneeded design solutions.

Stakeholders' constraints may be well founded and necessary, but in many cases, stakeholders are unaware of more desirable alternatives and may unknowingly state inappropriate or overly restrictive constraints.

In a recent example, a questionable design constraint for response time resulted in a cost and time overrun for system development because the stakeholder(s) who specified the design constraint could not be identified and the system's customer/acquirer was unwilling to change the design stated by an unidentified (and perhaps important) stakeholder. Traceability should be maintained between stakeholders' requirements and the stakeholders who included them; this may cause stakeholders to think more carefully about the features, quality attributes, and design constraints they specify.

Different stakeholders may state conflicting requirements; negotiation and compromise among the collective stakeholders is usually necessary, provided the stakeholders can be identified. For example, an RC-DSS stakeholders' requirement might state that providing realistic look and feel of driving is of top priority and another stakeholders' requirement might state that safety and security are of top priority. These requirements may present conflicting options for the system architecture and design definition.

A useful rule of thumb is the following:

> Always question stakeholders' design constraints.

Stakeholders who state design constraints should be identified and their contact information recorded so that options can be discussed with them, if desired. For example, a stakeholders' design constraint to use an unproven technique for ATM customer authentication may be questionable and should be discussed with those who stated the constraint.

Requirements triage is conducted to determine the feasibility and desirability of user requirements in light of constraints on schedule, budget, available resources, technologies, and infrastructure (Davis 2003). The appropriate stakeholders are consulted.

6.3.2.3 Prioritizing Stakeholders' Requirements

Stakeholders' requirements can be prioritized in a scheme such as

- *Essential requirements*: Must-have stakeholder requirements;
- *Desirable requirements*: Should-have stakeholder requirements;
- *Optional requirements*: Might-have stakeholder requirements;
- *Will not have*: To control stakeholders' expectation; and
- *Must not have*: For safety, security, reliability, or other considerations.

Essential requirements are those that must be satisfied by the first fully functioning version of a deliverable system. An essential requirement for an RC-DSS might be that the initial deliverable version must provide realistic driving scenarios for beginning drivers under conditions of fair weather and good roads.

Desirable requirements are also important but are candidates for deferred delivery in a later version of the system, if necessary. Constraints on schedule, budget, resources, or the urgency of immediate operational needs are reasons to defer some stakeholder requirements to the desirable category.

A desirable requirement for an RC-DSS might be that it should provide driving scenarios for rain, fog, snow, and slick roads, each of varying intensity. Another desirable requirement would allow instructors to insert real-time situations into the recorded scripts for driving scenarios during student driving sessions. Deferral of user requirements to the desirable category should be accomplished with concurrence of the affected stakeholders. Some desirable stakeholders' requirements may be scheduled for a Version 2 delivery on an agreed-to schedule.

Optional requirements describe capabilities that would be nice to have if time and resources permit their inclusion. Optional requirements may also include capabilities that are currently not needed but may be needed in the future or that are not feasible given the current state of technology. Many infeasible requirements became feasible at a future time, based on advances in technology such as the invention of the computer chip and continuing advances in computing technology. Optional requirements can also provide a placeholder for recording issues that might be otherwise forgotten. An optional requirement for a DSS could be to provide an autonomous driving option that will allow humans to experience monitoring and human intervention for driverless vehicles.

The order in which system capabilities will be implemented may also influence prioritization of stakeholders' requirements. It may be desirable to implement some capabilities to provide context and interfaces as a platform for implementing subsequent capabilities.

It is also important to include the following statements:

- Capabilities that the system *will not* provide in the foreseeable future; and
- Capabilities the system *must not* provide based on considerations of safety, security, reliability, regulations, or corporate policy.

Capabilities a system will not provide serve to control stakeholders' expectations for a system to be developed and to frame system development for the system developers. For example, a capability an RC-DSS will not provide in the foreseeable future is simulation of hands-off autonomous driving. And, an automobile driving simulator must not provide any capability that would place a simulator vehicle, a service technician, or others who directly interact with a DSS in a hazardous situation, in normal operation and during system failure (i.e. must provide fail-safe operation). If safety is of top priority, it may inhibit some desired features such as simulating a vehicle crash, which could endanger or traumatize a student driver.

6.3.3 Defining RC-DSS Stakeholders' Requirements

RC-DSS stakeholders' requirements can be categorized as user features, quality requirements, and design constraints.

6.3.3.1 Specifying RC-DSS User Features

RC-DSS user features can be categorized for different user classes. Examples include the following:

- Student drivers must experience the realistic look and feel of driving a real automobile under various driving conditions;
- Instructors must be able to prepare driving scenario scripts, inject unanticipated driving events into driving scenarios during student driving sessions, and review and prepare reports for student drivers' performance;
- Service technicians must have features that support routine maintenance, diagnostics, repair, and updates to RC-DSS hardware and software; and
- Administrators will need computer support to enroll students, track their progress, prepare reports and certificates, and manage finances.

Use cases and use case diagrams are commonly used to document operational scenarios as interactions between system users and a system. A use case includes sequences of interaction between a system and a hands-on user class; both normal and exception interactions should be documented in a use case. It is not unusual that a hundred or more use cases may be needed to document the operational scenarios for a system of moderate size and complexity such as an ATM or a driving system simulator. Use cases for large systems are developed for each of the subsystems.

A template for and an example of a use case are provided in Table 6.3 for a student driver initiating an RC-DSS driving session.

Observe that the normal or expected scenario is documented in the use case; exception condition scenarios are noted. Also note that Comment 2 mentions a state diagram and an activity diagram; they are displayed in Figures 6.4 and 6.5.

The state diagram in Figure 6.4 illustrates the states an RC-DSS enters and exits during driver authentication and session termination. Note that three tries are allowed to enter the correct PIN.

The state machine diagram indicates the states the RC-DSS software transitions through during driver authentication and termination of a driving session but it does not indicate that the student driver inserts his or her identity (ID) card into the card reader and that the instructor enters the PIN. The activity diagram in Figure 6.5 illustrates this. Together, the two figures provide an example of the need for multiple views when documenting a system.

A use case diagram, illustrated in Figure 6.6, depicts a set of user cases, the interactions among them, and interactions with one or more external actors.

The use case diagram illustrates a student (primary actor) enrolling in an RC-DSS driving class, with the assistance of the administrator acting as registrar (secondary actor). The three use cases are indicated in the ovals. The ≪extend≫ stereotype indicates that the "Check status" use case is invoked, if needed, to handle any problems with the registration request. In general, ≪extend≫ is used to denote a supplemental behavior that is used in conjunction with the extended use case under stated conditions. The extending use case can be thought of as analogous to a subroutine that is used in a computer program under stated conditions.

The ≪include≫ stereotype in Figure 6.6 is used to denote a use case that notifies the student's sponsor that the student was successfully enrolled or provides an explanation stating why the enrollment was blocked. In general, ≪include≫ indicates that the included use case inserts behavior into the including use case; it provides a mechanism for decomposing a complex use case into smaller elements and indicated that the included use case may be useful in more than one including use case. In the example, "Notify sponsor" is a separate use case that can be used to notify sponsors in conjunction with other use cases.

Table 6.3 An RC-DSS use case.

Use case #7 for RC-DSS
Use case ID: RC-DSS 7 *Use case name*: Driving session initiation *Created by*: D. Fairley *Date created*: 06/23/20xx *Date last updated*: na *Last updated by*: na *Primary actor that initiates the use case*: Student driver. *Secondary actors, if any*: Instructor. *Statement of purpose*: Allows an RC-DSS driver and an instructor to initiate a driving session. *Preconditions that must be true before this use case can be "executed"*: 1. The RC-DSS is in the idle state; 2. The driver has a valid RC-DSS identity card (ID); and 3. The instructor has a valid password (PIN). *Primary scenario that describes the normal course of this use case*: 1. Driver inserts ID card into a card reader. 2. RC-DSS reads the driver ID information, verifies it in the RC-DSS data repository, displays it on the instructor workstation, and requests entry of the instructor's PIN. 3. Instructor enters his or her PIN. 4. System opens a driving session for the student in the RC-DSS data repository and enters the "initiate driving session" state. *Exceptions to be handled*: 1. Bad ID card; 2. Incorrect PIN; and 3. Timeout during driving session initiation. *Features incorporated in this use case*: Feature #6 from the RC-DSS stakeholders' requirements. *Business rules for this use case*: 1. RC-DSS drivers must be registered in the RC-DSS data repository. 2. RC-DSS authorization PINs are issued to certified instructors. *Post conditions that must be true after this use case is "executed"*: System is in the "initiate driving session" state OR the sign-on failed, the ID card has been ejected, and a "sorry" message has been displayed OR the initiate period timed out and the ID card was ejected. *Comments*: 1. This use case demonstrates a feasible sign-on mechanism using ID cards and instructor's personal identification numbers. It does not mandate that this authentication mechanism must be used. 2. The attached state diagram and activity diagram document the initiation scenario for driving sessions.

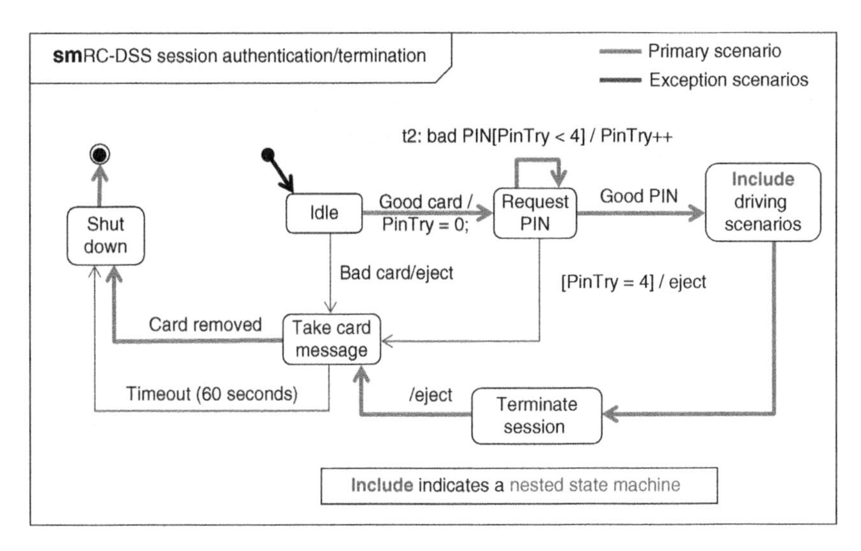

Figure 6.4 A state machine diagram for RC-DSS authentication and termination.

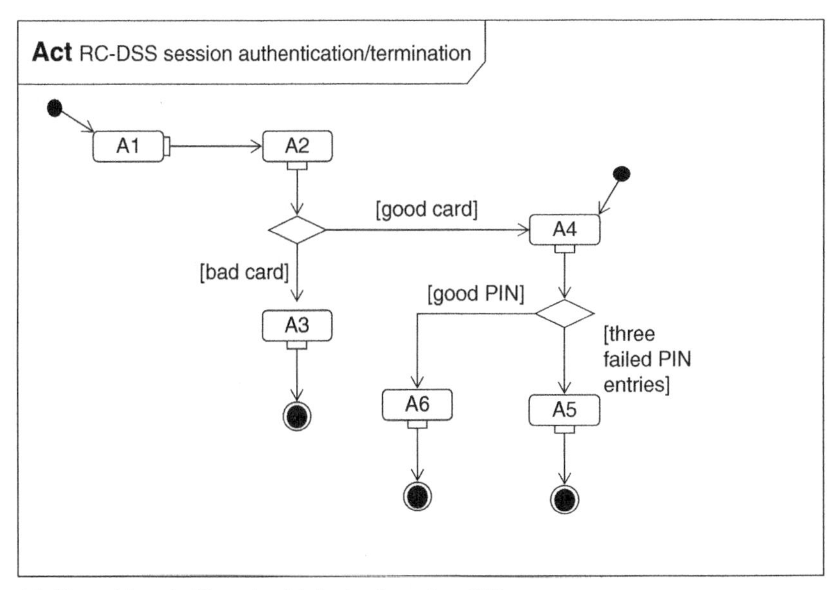

A1: Sturent inserts ID card
A2: System checks card
A3: Bad card ejected and session terminates
A4: Instructor enters PIN
A5: System shuts down after failed PIN entries
A6: System enters include nested state then terminates after the driving session

Figure 6.5 An activity diagram for RC-DSS authentication and termination.

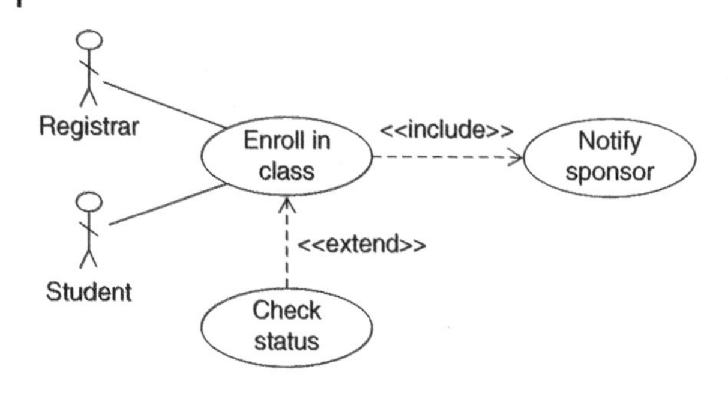

Figure 6.6 A use case diagram.

As illustrated, a use case diagram can be regarded as a "visual table of contents" for a set of use cases that involve one or more actors with stereotypes to indicate circumstances when the use cases would be invoked.

Use cases, use case diagrams, state diagrams, and activity diagrams are notations provided by UML (2017) and SysML (2017). Additional details for use case diagrams, state machine diagrams, and activity diagrams are provided in Appendix 8.A.

6.3.3.2 Specifying RC-DSS Quality Attributes

The primary quality attributes for the RC-DSS are physical safety of student drivers and service technicians plus security of student enrollment data and driving performance ratings. Other important quality attributes for the RC-DSS are realistic response times and the haptic sensations of driving a real vehicle. Some quality attributes, such as safety, security, and reliability, may be developed and certified by specialty engineers who are members of the RC-DSS system development team, either as RC employees or as consultants.

6.3.3.3 Specifying RC-DSS Design Constraints

The primary stakeholders' requirement that an RC-DSS "provide realistic look and feel of driving a modern vehicle under various driving conditions" could be regarded as a design constraint that an RC-DSS must satisfy; however, design constraints are focused statements for specific system attributes rather than general statements.

An RC-DSS design constraint stated by the initial customer for an RC-DSS (the Metro high school district mentioned above) might be that a data interface, in a specified format, be made available for the high school's IT system to be used for receiving student driving records and reports. This design constraint would probably result in the design of a SysML package with an abstract RC-DSS IT interface that could be instantiated for various kinds and formats of data that might be required by other RC-DSS customers.

6.3.4 The Concept of Operations

The outcomes of the I^3 SysDefn process are based on business or mission analysis and stakeholder needs and requirements definition. These outcomes are typically documented in a concept of operations (i.e. a ConOps) that also includes the results of a risk analysis that identifies risk factors for the technical aspects of the proposed project.

Key elements of a ConOps include descriptions of the following:

- The current system or situation;
- The challenge or opportunity;
- The envisioned system;
- Key stakeholders;
- Desired features, quality attributes, and design constraints;
- Design constraints to be observed;
- Priorities among desired features and quality attributes;
- Envisioned modes of operation;
- Operational scenarios;
- Operational policies and constraints;
- Functions and behaviors the envisioned system will not provide;
- Limitations and restrictions on what the envisioned system must not do;
- Managerial and technical risk factors; and
- Impacts during development and operation.

IEEE Standard 1362-1998 documents the format and contents of a ConOps. It is available for purchase from IEEE but the content of 1362 has been absorbed in the ISO/EAC/IEEE Standard 29148:2011 for requirements engineering (ISO 2011). The following example uses the format of the 1362 standard.

6.3.4.1 The RC-DSS ConOps

A ConOps for the RC-DSS includes the following elements:

1. *The current system or situation*:
   There are no full-scale, realistic, and commercially available driving system simulators that can be purchased or on which time can be rented; there are no organizations that sell or lease realistic DSS systems.
2. *The challenge or opportunity*:
   The challenge and opportunity are to develop a product line of driving system simulators.
3. *The envisioned system*:
   A line of configurable driving system simulators for various user classes and kinds of land-based vehicles having four or more wheels.
4. *Key stakeholders*:
   RC decision makers, potential customers, system engineers, software engineers, representative students, instructors, and maintainers.

5. *Desired features and quality attributes*:
   Features that provide realistic driving experiences for students and provide support for instructors, maintainers, and administrators.
   Primary quality attributes:
   - Safety of student-drivers and system maintainers is the primary quality attribute during normal operation and during system failure.
   - Realistic response times for driving under various conditions must be provided.
   - Monitoring of student performance during driving simulations must be provided.
6. *Envisioned modes of operation*:
   - Instructor supervised driving;
   - Preparation and validation of driving scripts;
   - Diagnostics and maintenance; and
   - Administration.
7. *Operational scenarios*:
   User stories, use cases, activity diagrams, and state diagrams are used to document operational scenarios for student drivers.
8. *Operational policies*:
   DSS systems must satisfy:
   - Federal, state, and local regulations for safe operation of vehicles;
   - Regulations of the American Disabilities Act; and
   - Parental consent for vehicle drivers under age 18.
9. *Technical constraints*:
   DSS systems must be self-contained and deployable with installation in a prepared site in at most two days for two trained technicians.
10. *Technical risk factors*:
    - Lack of engineering experience in developing driving system simulators; and
    - Development tools and infrastructure may be inadequate.
11. *Limitations of the envisioned system*:
    Limited to simulated driving of land-based vehicles having four or more wheels.
12. *Impacts during development and operation*:
    The primary impact for RC-DSS developers is that system developers will not be 100% available on a full-time basis; they must maintain existing systems and engage in other work while developing Version 1 of the RC-DSS. The primary development impact for organizations that buy or lease an RC-DSS for on-site installations is that they must prepare the necessary infrastructure while the system is being developed and/or prepared for installation.

The primary operational impact for the RC is instructors, maintainers, and administrators must be recruited, trained, and supervised to support operation of RC-DSS systems.

The primary operational impact for system acquirers is changes to operational policies and procedures that will occur with the use of DSS training facilities. For example, changes in a school that replaces human driving instructors with DSS facilities.

In addition, the framework for the system engineering management plan (SEMP) must be developed to guide the I^3 SysDefn process. Structure and contents of an SEMP and the relationship of the SEMP to the system project plan (SPP) are presented in Section 10.2 of this text.

6.4 Identifying and Prioritizing System Capabilities

System capabilities are specified courses of action or states of being a system must provide to satisfy the stakeholders' requirements (user features, quality attributes, and design constraints). System capabilities are derived from stakeholders' requirements and provide the basis for defining the system requirements. Further details are provided in Section 5.2.

Chapter 7 (system requirements definition) presents development of system requirements based on system capabilities.

6.4.1 Identifying RC-DSS System Capabilities

User features, quality attributes, and design constraints provide the basis for identifying system capabilities to be provided by an RC-DSS.

Suppose a stakeholders' requirement for the RC-DSS states the following:

> RC-DSS vehicle cabins will provide the physical look and feel of driving a real land-based vehicle having four or more wheels under various driving conditions.

Some capabilities the RC-DSS Version 1 vehicle cabin must have to simulate driving a realistic modern automobile include the following:

1. *Interior environment*:
   Realistic driver and passenger seats, arm rests, seat belts, center console, steering wheel, dashboard with gauges and displays, etc.
2. *Controls*:
   Gear shift, steering wheel, accelerator, brakes, turn signals, etc.
3. *Dashboard displays*:
   Temperature gauge, fuel level indicator, and so forth.

4. *Visual displays*:
   Windshield, side windows, and mirrors.
5. *Haptic adjustments to the driver's feedback for a moving automobile*:
   Feel of steering wheel, accelerator, brakes; sound of engine and road noise; sound and feel of wind, falling rain, and snow.
6. *Realistic feel of vehicle movement with six degrees of freedom*:
   For acceleration, deceleration, turning, road vibrations, and so forth.

Note that some capabilities of the vehicle cabin have implications for other system elements. For example, capability five has implications for the capabilities that must be provided by the hydraulic platform on which the vehicle cabin is mounted.

6.4.2 Prioritizing RC-DSS Capabilities

The first priority for an I^3 capability-based M&SBSE development project is implementation of a simulation model for the system architecture that includes the major subsystems (vehicle cabin, hydraulic platform, server, instructor and technician workstation, administrator workstation) and the interfaces among them as portrayed in Figure 6.1. Each of the subsystems can be simulated with simulated interfaces to the other subsystems, as appropriate.

A simulation model for the RC-DSS server will provide interfaces to and coordinate interactions among the simulated subsystems. Each subsystem will be decomposed into system elements that will provide the system capabilities. The simulated elements within each subsystem will be replaced as they become available; for example, interfaces to the real server and interfaces to the simulated steering wheel in the vehicle cabin can be modified, if necessary, when the real server and the real steering wheel interfaces replace the simulated ones. The RC-DSS will be fully realized when all simulated system elements are replaced with real elements.

For each of the subsystems, including the RC-DSS vehicle cabin, the simulation approach provides flexibility in the priorities for implementing the real system elements and permits concurrent, coordinated development of the hardware elements and software elements needed to realize system capabilities (see Chapters 5 and 9 for additional details on concurrent development of hardware and software).

Some priority considerations presented in Chapter 5 are repeated here for ease of reference (Table 6.4).

In addition, capabilities-based system requirements should be established initially for the physical components that are anticipated to have long procurement and fabrication schedules.

Each of the subsystems will interact with the other subsystems to provide the overall RC-DSS system capabilities. With appropriate architecture and design,

Table 6.4 Capability prioritization and rationale.

Criterion	Rationale
Easiest-to-implement capabilities first	To gain familiarity with the I^3 development processes and build team cohesion
Highest risk capabilities first	To determine that known uncertainties and complexities can be accommodated and to identify previously unknown risks and complexities
Safety and security capabilities first	To determine that safety and security requirements can be satisfied and to ensure safety and security – capabilities developed first are tested most in combination with added capabilities
External interfaces first (real and simulated)	To establish connections to the real or a simulated operational environment
Legacy-element capabilities first	To determine that legacy elements can be satisfactorily incorporated
Vendor-element capabilities first	To determine that vendor-supplied elements can be satisfactorily incorporated
Deliverable subset capabilities first	To be evaluated by potential users and/or placed into operational use

system requirements for the behaviors, functions, and interfaces of each system element can be defined for and allocated to each of the subsystems.

6.5 Determining Technical Feasibility

The basis for an initial assessment of the technical and managerial feasibility of a proposed project is provided by the following:

- The characterization of the abstract solution space identified during the business or mission analysis process;
- The stakeholders' requirements for user features, quality attributes, and design constraints;
- The system capabilities needed to satisfy the stakeholders' requirements; and
- Constraints on schedule, budget, infrastructure, and available expertise and resources.

It is important to note that assessment of feasibility may result in a decision to not pursue the project, defer the project to a later date, or significantly redefine the business opportunity or mission before proceeding.

Technical feasibility can be determined by examining the following:

- Available expertise;
- Local experience with similar projects;
- Analogies from other organizations and systems;
- Historical precedents for similar systems; and
- Maturity of the needed technologies.

Project feasibility is concerned with determining whether schedule, budget, technical and managerial infrastructure, and available resources and skills (kinds and numbers) are (or are not) adequate for conducting the technical work activities of the proposed project. The project planning processes of 15288 and 12207 (Subclause 6.3.1) provide guidance for determining the managerial feasibility of a proposed project.

In general, planning the technical work activities is concerned with the following:

- Developing plans for overall schedule, schedule milestones, budget, and resources (including numbers of personnel by engineering discipline);
- Specifying supporting technologies and quantitative measures to be used;
- Identifying risk factors that could impede progress in performing the technical work to be accomplished; and
- Assessing likely customer relations that could facilitate or impede progress.

Plans for the technical work to be done must adhere to the overall project constraints on schedule, budget, available resources, and technologies. Work plans are assessed for feasibility during the I³ SysDefn phase of a project and completed during the I³ Rqmts&Arch phase because the limitation on schedule, budget, resource, and technology will determine the system requirements that can be implemented within those constraints, thus imposing limits on the stakeholders' requirements that can be satisfied.

Technical work plans are documented in an SEMP. Planning the technical work activities and the format and content of SEMPs are presented in Part III of this text.

6.5.1 Determining RC-DSS Technical Feasibility

The technical feasibility of implementing a realistic driving system simulator can be demonstrated by identifying hardware and software elements that could be used to implement these capabilities. The following analysis of technical feasibility for an RC-DSS is based on the abstract characterization of a solution space provided by flight training simulators identified during the business analysis phase of the I³ system development model.

- *Capability*: The look and feel of a realistic vehicle cabin.
  Feasible hardware: A real vehicle retrofitted with physical interlock sensors for doors, seatbelts, ignition key, and other physical elements of the vehicle.

Sensors and actuators to be provided for the physical interlocks; signals to be provided by A/D and D/A converters*.
*A/D: Analog-to-digital; D/A: Digital-to-analog.
Software to: Receive interlock signals and send response signals.

- *Capability*: Realistic feel of vehicle movement with six restricted degrees of freedom.
  Feasible hardware: A hydraulic platform.
  Software to: Receive digital signals from other system elements' A/D converters; send platform control signals to the hydraulic platform's D/A converters and receive response signals from them; generate signals to notify other system elements of platform status.
- *Capability*: Visual effect of driving a real vehicle.
  Feasible hardware: LCD screens for windshield, side windows, and mirrors.
  Software to: Receive sensor signals for vehicle status; generate real-time signals for display screens.
- *Capability*: Displays of vehicle status and performance on front panel gauges.
  Feasible hardware: Vehicle speedometer; fuel level, oil pressure, and engine temperature gauges; rpm indicator; and other displays of vehicle performance; all performance indicators retrofitted with A/D and D/A devices.
  Software to: Sense vehicle status parameters; send signals to the performance indicators; receive and send digital signals from and to A/D and D/A devices.
- *Capability*: Tactile feel of driving a real vehicle.
  Feasible hardware: Sensors and actuators for the steering wheel, accelerator pedal, brake pedal, shift indicator, turn signals, cruise control, and other vehicle control devices with *pressure* control and realistic response times that depend on the current state of the vehicle.
  Software to: Receive digital signals from vehicle A/D control sensors; generate and send digital signals to vehicle D/A actuator devices.

Note that the RC-DSS hardware, software, and interfaces identified to demonstrate technical feasibility incorporate several design decisions, namely, a real vehicle mounted on a hydraulic platform could be used, the display screens could be implemented using software-driven LCDs, and realistic equations of motion could be developed for the hydraulic platform.

It must be clearly understood that identification of system capabilities and feasibility of implementing those capabilities do not represent the final implementation decisions. The purpose of a feasibility analysis is to determine whether a project should proceed, be delayed until a later time, or cancelled while is easy to do so without excessive loss of invested time and effort.

The hardware, software, and interface requirements stated here demonstrate the technical feasibility of developing a realistic driving system simulator. Architectural alternatives for hardware, software, and interface elements of a

DSS will be examined and specifications are generated during the I^3 architect phase (see Chapter 8).

Techniques for determining the feasibility of conducting the RC-DSS project (schedule, budget, resources, infrastructure, cost) are presented in Part III of this text.

6.6 Establishing and Maintaining Traceability

Traceability links among the various work products of a system development project portray the relationships and continuity of workflow among the work products, which can be used to determine that nothing is omitted from the work products and that nothing is included in a later work product that is not based on a previous work product. For example, the architecture and design specification cover the system requirements and no design elements are introduced that are not traced to the system architecture, either directly or by derivation. Traceability also supports impact analysis to determine which work products need to be changed when related work products are changed.

When conducting the SysDefn phase of I^3 system development, bidirectional traceability should be maintained between the opportunities and problems identified during the business or mission analysis process and the stakeholders' requirements and system capabilities identified during the stakeholder needs and requirements definition process.

An example of traceability analysis is provided in Section 7.4.1.2.

6.7 Key Points

- The I^3 SysDefn phase is based on the business or mission analysis and the stakeholder needs and requirements definition processes of ISO/IEC/IEEE Standards 15288 and 12207.
- The RC-DSS driving simulator case study is used in this chapter and Chapters 7–9 to provide examples of the technical processes for development of software-enabled systems.
- The business or mission analysis process defines a business problem, a mission, or an opportunity; it then describes the solution space and determines one or more solution classes in the solution space.
- The outcome of business or mission analysis may result in a decision to not proceed with a mission or business opportunity.
- RFPs, SOWs, and MOUs are used to facilitate agreements between system acquirers and system suppliers.

- The stakeholder needs and requirements definition involves identifying system stakeholders; characterizing the operational environment for the envisioned system; and eliciting, categorizing, and prioritizing stakeholders' requirements.
- Stakeholders' requirements:
  - Include desired features, observable quality attributes, and design constraints;
  - Can be categorized as essential, desirable, and optional; and
  - Should include statements of what the envisioned system will not do and what the system must not do.
- The ConOps documents the outcomes of stakeholders' requirements definition.

In addition, stakeholders' requirements are used as the basis for generating the capabilities the envisioned system must provide to satisfy the defined requirements. Two-way traceability is essential to trace the correspondences among the outcomes of business analysis, stakeholders' requirements, and system capabilities.

The phases of I^3 system development, based on the ISO/IEC/IEEE 15288 and 12207 process standards, are continued in the following chapters.

Exercises

6.1. Projects selected for development by business analysis should conform to the organization's mission and vision statements.

(a) Select a business of interest to you and read the mission and vision statements for that business. Briefly describe how the products and services provided by that business are, and are not, consistent with the mission and vision statements.

(b) Select a product or service provided by the organization. Briefly describe the intended stakeholders for the product or service, in your opinion.

(c) Briefly describe, in your opinion, how the selected product or service does, or does not, satisfy the intended stakeholder wants and needs.

6.2. A capability is the ability of a system to exhibit a course of action or a state of being.

(a) Describe a course of action exhibited by your selected product or service from Exercise 6.1.

(b) Describe a state of being exhibited by your selected product or service.

6.3. Stakeholders' requirements can be categorized as essential, desirable, and optional.

(a) Select and briefly describe two features, quality attributes, or design constraints (any combination of two) for the product or service selected in Exercise 6.1 that you would categorize as essential.

(b) Select and briefly describe two features, quality attributes, or design constraints for the product or service selected in Exercise 6.1 that you would categorize as desirable.

(c) Briefly describe two features, quality attributes, or design constraints not present in the product or service you selected in Exercise 6.1 that you would categorize as optional.

6.4. Search and find a template for, or an example of, a ConOps. Compare the template or example to the items listed in Section 6.3.4. Briefly describe how your template or example is, and is not, like the description of a ConOps in Sections 6.3.4 and 6.3.4.1.

6.5. Search and find additional material that describes a statement of work (SOW) and additional material that describes a memorandum of understanding (MOU).

(a) Briefly describe how SOWs and MOUs are alike and how they are not alike.

(b) Compare the SOW and MOU material you have found to the items in Section 6.2.2. Briefly describe how the SOW and MOU material are like, and are not like, the items in Section 6.2.2.

References

BABOK (2015). *A Guide to the Business Analysis Body of Knowledge (BABOK Guide)*, 3e. International Institute of Business Analysis.

Davis, A. (2003). The art of requirements triage. *IEEE Computer Society* 36: 42–49.

ISO (2011). ISO/IEC/IEEE Standard 29148:2011 Systems and software engineering – Life Cycle Processes – Requirements Engineering.

ISO (2015). ISO/IEC/IEEE 15288:2015, Systems and software engineering – System life cycle processes, ISO/IEEE, 2015.

ISO (2017). ISO/IEC/IEEE 12207:2017, Systems and software engineering – Software life cycle processes, ISO/IEEE, 2017.

SEBoK (2017). *The Guide to the Systems Engineering Body of Knowledge (SEBoK)*, v. 1.9. R.D. Adcock (EIC). Hoboken, NJ: The Trustees of the Stevens Institute of Technology www.sebokwiki.org (accessed 12 June 2018).

SysML (2017). OMG Systems Modeling Language 1.5. http://www.omgsysml.org/ (accessed 14 June 2018).
UML (2017). The Unified Modeling Language Specification Version 2.5.1. https://www.omg.org/spec/UML/About-UML/ (accessed 13 June 2018).

7

System Requirements Definition

7.1 Introduction

Defining system requirements is the initial technical phase of system development. As illustrated in Figure 5.2, system requirements definition is Phase 2 of the integrated-iterative-incremental (I^3) system development model (preceded by the Phase 1 system definition phase). System requirements definition, system architecture definition, and design definition are grouped together in Figure 5.2 as Phases 2–4 of I^3 (Rqmts/Arch/Des) to emphasize the close iterative coupling of these three system development phases. Figure 5.2 is repeated here as Figure 7.1 for ease of reference, with Rqmts highlighted.

For purposes of exposition, requirements definition is presented in this chapter and architecture definition and design definition are presented in Chapter 8.

ISO/IEC/IEEE Standard 29158:2011 includes the processes and products for requirements engineering of systems and software throughout the life cycle. ISO/IEC/IEEE Standards 15288 and 12207 (herein referred to as 15288 and 12207) provide guidance for the various phases of the I^3 system development model (ISO 2015, 2017). *The Guide to the Systems Engineering Body of Knowledge* (SEBoK) and *Guide to the Software Engineering Body of Knowledge* (SWEBOK) provide background information.

Background information for this chapter includes the overview of the I^3 system development model and the example of applying I^3 presented in Chapter 5 plus the material on mission or business analysis and the stakeholder identification and requirements definition process presented in Chapter 6. The RC-DSS case study introduced in Chapter 6 is continued in this chapter.

The I^3 system development model provides the context for the methods and techniques of system requirements definition presented in this chapter but the material in this chapter can be applied when using system development models other than I^3 or when performing requirements engineering separate from system development.

Systems Engineering of Software-Enabled Systems, First Edition. Richard E. Fairley.
© 2019 John Wiley & Sons, Inc. Published 2019 by John Wiley & Sons, Inc.

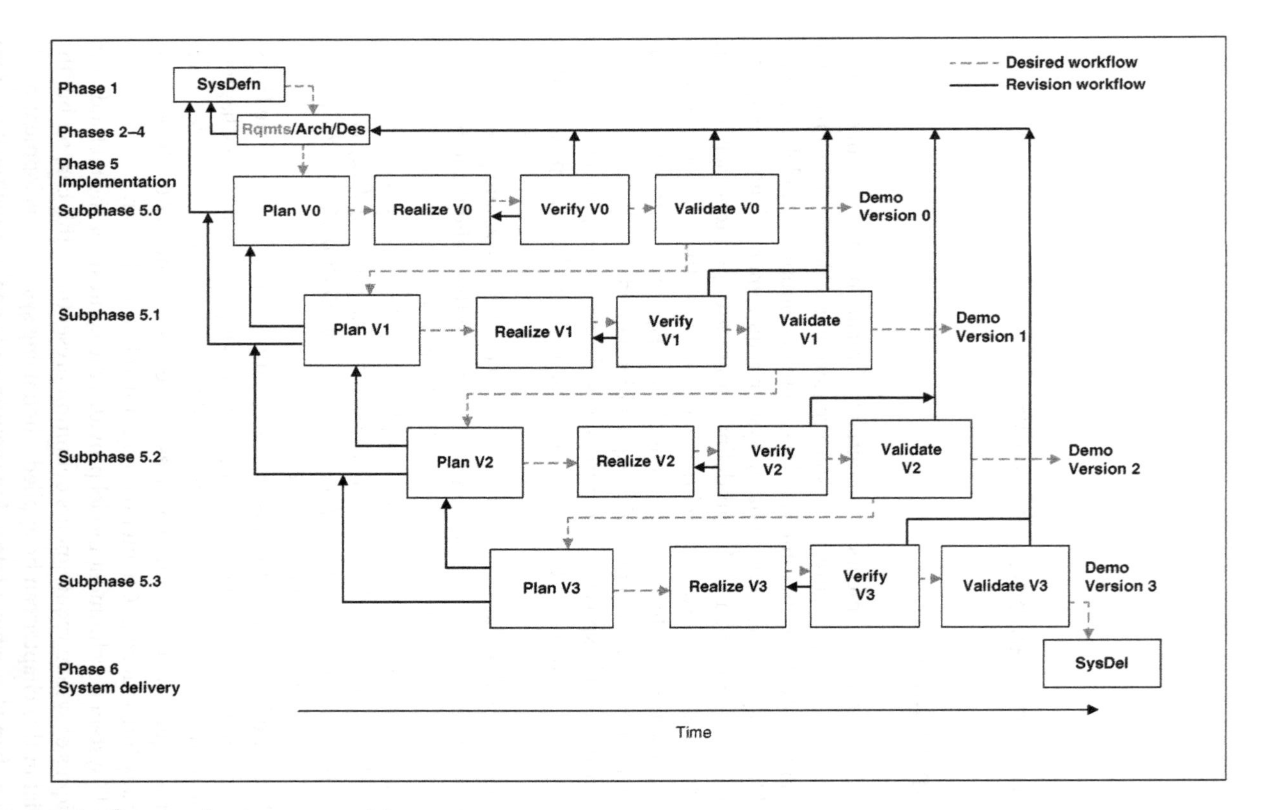

Figure 7.1 The I^3 system development model.

This chapter provides the opportunity for readers to learn about the following:

- The system requirements definition process;
- A taxonomy of requirements;
- The capabilities-based approach to defining system requirements;
- Defining, verifying, and validating system requirements; and
- System requirements definition for the RC-DSS case study.

References and exercises at the end of this chapter provide the opportunity to gain increased understanding of system requirements definition and the context in which it occurs.

7.2 The System Requirements Definition Process

According to the 29148 Standard, requirements engineering is

> An interdisciplinary function that mediates between the domains of the acquirer and supplier to establish and maintain the requirements to be met by the system, software or service of interest. Requirements engineering is concerned with discovering, eliciting, developing, analyzing, determining verification methods, validating, communicating, documenting, and managing requirements.
>
> (ISO 2011)

Note that this definition includes the stakeholders needs and requirements definition process and the system requirements definition process of the 15288 and 12207 standards. Stakeholder needs and requirements definition is presented in Chapter 6.

According to Sections 6.4.3.1 of 15288 and 12207, the purpose of system requirements definition is

> to transform the stakeholder, user-oriented view of desired capabilities into a technical view of a solution that meets the operational needs of the user.

Table 7.1 indicates the inputs, processes, and outputs of system requirements definition.

Table 7.1 indicates that system capabilities are an input to the requirements definition process. As defined in this text, a system capability is the ability of a system to execute a specified course of action or the ability to exhibit a state of being. In a capabilities-based approach to system development, system capabilities needed to satisfy stakeholders' requirements are identified and provide the basis for defining the system requirements.

Table 7.1 Inputs, processes, and outputs of system requirements definition.

Inputs	Stakeholders' requirements (operational features, quality attributes, and design constraints)
	System capabilities needed to satisfy the stakeholders' requirements
	A preliminary work plan
Processes	Develop, verify, and validate the system requirements, initially and incrementally during system development
	Complete the preliminary work plan for the technical work activities
Outputs	System requirements definition
	Completed work plan
	Revised stakeholders' requirements and system capabilities (as necessary)

System capabilities thus provide the bridge from stakeholders' requirements to system requirements, as illustrated in Figure 5.1 and discussed in Section 5.2. The capabilities-based approach is used to bridge the gap of transforming stakeholders' requirements into system requirements and can be applied independent of I^3 using the I^3 system development model.

System requirements needed to provide the system capabilities are defined during initial system requirements definition and elaborated during incremental phases of system implementation. Revisions to the system requirements are made as needed if stakeholders' requirements and/or system capabilities are revised during system development.

A preliminary plan for the technical work to be accomplished is prepared by during the I^3 SysDefn phase of systems projects and finalized during the requirements definition phase. Techniques for developing the technical work plan are presented in Part III of this text.

A taxonomy of requirements is presented in the following section.

7.3 A Requirements Taxonomy

The requirements taxonomy illustrated in Figure 7.2 includes stakeholders' requirements, system capabilities derived from the stakeholders' requirements, and the system requirements needed to implement the system capabilities. The elements in the taxonomy are discussed in the following sections.

Each category in Figure 7.2 plus quality attributes, interface requirements, and the will-not and must-not design constraints are discussed in turn.

7.3.1 Stakeholders' Requirements and System Capabilities

Stakeholders' requirements (discussed in Chapter 6) include operational features, quality attributes, and design constraints. Operational features are

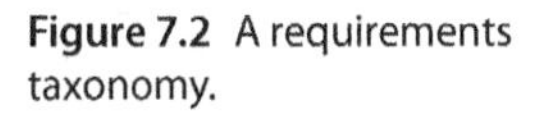

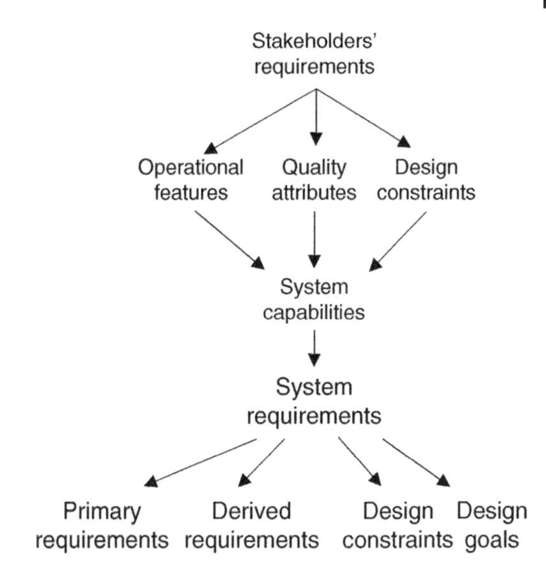

Figure 7.2 A requirements taxonomy.

Table 7.2 Some examples of quality attributes.

Visible to system users	Not visible to system users
Safety	Testability
Security	Modifiability
Reliability	Reusability
Availability	Configurability
Performance	Serviceability
Ease of use	Installability

externally observable functions and behaviors the stakeholders desire in the delivered system. Stakeholders' quality attributes, like operational features, are those that are observable during system operation; they include system attributes such as safety, security, and reliability (see Table 7.2). Stakeholders' design constraints are "must have" elements that must be included in the system design and system implementation, for example, incorporation of legacy elements or providing specific interfaces to elements in the system environment.

System capabilities are derived from stakeholders' requirements and provide the basis for defining the system requirements. Section 5.2 provides a discussion and an example of deriving system capabilities from stakeholders' requirements. Figure 5.1, repeated here as Figure 7.3 for ease of reference, indicates

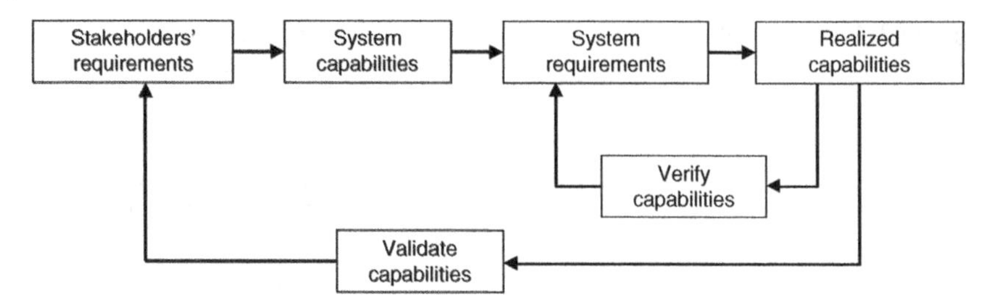

Figure 7.3 A capabilities-based approach to requirements realization.

the relationships among stakeholders' requirements, system capabilities, and system requirements.

As indicated, system capabilities are generated from stakeholders' requirements and system requirements are defined for the system capabilities. Implemented system requirements provide the capabilities, which are verified with respect to the system requirements. Implemented capabilities are validated with respect to the stakeholders' requirements on which they are based. For simplicity of presentation, system architecture definition and design definition are not shown in Figures 5.1 and 7.3.

Iterative revisions to stakeholders' requirements, system capabilities, system requirements, and requirements realization may result from the illustrated process.

7.3.2 System Requirements

As indicated in Figure 7.2, system requirements can be categorized as primary requirements, derived requirements, design goals, and design constraints. Each is discussed in turn.

7.3.2.1 Primary System Requirements

Primary system requirements are based on system capabilities that are directly transformed into system requirements stated in a manner that permits objective verification with supporting evidence of verification.

For example, suppose a system capability for an RC-DSS driving simulator is

> Instantaneous steering wheel responses will be provided.

To quantify, "instantaneous response" experiments with experienced drivers could be conducted using a prototype mockup of a steering wheel interface. It might be determined that response times in the range of 0.3–0.5 seconds becomes perceptible and thus a verifiable upper limit of 0.25 seconds is defined to satisfy the stakeholders "instantaneous response time" requirement.

For some driving scenarios, such as driving on slick roads, instantaneous response may not be desirable in order to provide the feel of driving a modern automobile under various conditions. Experiments to determine desired lags in response time for different kinds of interactions can be determined and quantified.

It can be challenging to state primary system requirements in a manner that will permit development of objective verification criteria used to determine the degree to which a requirement has been satisfied. "Objective Verification Criteria" provides an example of quantifying a capability based on the "easy to learn and easy to use" stakeholders' requirement.

Objective Verification Criteria

A stakeholders' requirement resulted in a system capability to be provided by point-of-sale terminals: they should be "easy to learn and easy to use."

The capability was transformed into an easy-to-learn and easy-to-use system requirement that was objectively verified and validated as follows:

- A three-hour training session and a one-hour exercise were developed for the organization's sales clerks who will operate the point-of-sales terminals.
- Thirty sales clerks were selected at random for training on the new system (30 being a statistically significant number).
- With stakeholders' agreement, a two-step process was developed to objectively assess "easy to learn and easy to use."
  (1) *Verification*: Twenty-seven successful completions of the training exercise in the afternoon following the morning training class (~95% confidence level).
  (2) *Validation*: One week later, the 27 participants received a one-hour refresher session. Successful "easy to learn and easy to use" was determined if 25 of the 27 participants successfully completed a different training exercise following the refresher training (~90% overall confidence level).

A successful outcome resulted in customer acceptance of the system and the organization adopted the two-step training program as the certification process for point-of-sale clerks.

7.3.2.2 Derived System Requirements

A derived system requirement is a requirement that is not defined directly from a system capability but that must be implemented to supplement a primary requirement. Derived system requirements, like primary system requirements, have objective verification criteria.

For example, there may be an RC-DSS requirement to include date and time stamps for all hands-on interactions with a Realistic Corporation driving system simulator (RC-DSS) but there is no stakeholders' requirement to display date and time (RC-DSS hands-on users include administrators, students, instructors, and IT support and maintenance personnel). A derived requirement would be added to the system requirements to provide a clock function within the system's internal design. The clock function would be quantified by specifying parameters such as resolution, precision, drift, and interface specifications for power and functionality to be provided.

7.3.2.3 System Design Constraints

Some system design constraints are derived from stakeholders' design constraints. For example, an RC-DSS stakeholders' requirement to provide compatibility with a particular stakeholders' database may result in one or more design constraints for the external interfaces to be provided by the RC-DSS being developed.

System design constraints may also be based on organizational policies, contractual requirements, feasibility studies, use of available (off-the-shelf) system elements, or other constraints imposed by product-line considerations.

Other design constraints might be based on the overriding constraint that the first version of an RC-DSS must incorporate design attributes that will facilitate expansion of the RC-DSS Version 1 architecture into a product-line architecture. Still other design constraints might come from the will-not and must-not requirements stated in the RC-DSS stakeholders' requirements.

System design constraints must have objective verification criteria.

7.3.2.4 System Design Goals

Primary requirements, derived requirements, and design constraints are quantified or otherwise stated in a manner that will permit objective evaluation to determine the degree to which a system realization satisfies those requirements. System requirements that are not (perhaps not yet) stated objectively are categorized as design goals.

For a variety of reasons, it may not be possible to initially transform some of the system capabilities into objectively stated primary requirements, derived requirements, or design constraints. For example, it may be that the technical meaning and feasibility of quantifying stakeholders' requirements such as "instantaneous response" and "highly reliable" cannot be determined without further study and discussion.

Over time, some design goals will be better understood and recategorized as quantified system requirements when feasibility issues are resolved and stakeholder consensus is obtained. Clarification is typically based on research, experiments, and prototypes.

But some stakeholders' requirements may remain as unquantified design goals. For example, a stakeholders' requirement to build the "best driving simulator in the world" should not be ignored even though the meaning of "best" is ambiguous. And even if the meaning of "best" is understood, it would not be feasible to survey all driving simulator systems in the world and demonstrate that the delivered system is best one, by the understood meaning of best.

These unquantified requirements may be important to stakeholders and should not be ignored because they will influence design and implementation decisions. If "best" means "safest DSS in the world," and if safety is a top priority for stakeholders, design and implementation decisions that reinforce safety will take precedence over conflicting design goals such as "most features" or "ease of modification."

Each unresolved design goal should be flagged with a to be determined (TBD) or to be resolved (TBR) indicator. A TBD flag indicates that the issue will be examined at some time in the future. A TBR flag is used to indicate that the topic is under active consideration but consensus has not been obtained on how to quantify the quality requirement. Each TBD and TBR should be accompanied with an action plan that includes timelines for periodic status reviews. These action plans should be tracked to closure, which results in recategorization of the design goal or a determination that the design goal cannot be stated in an objective manner.

A previously stated, it is nevertheless important to not ignore design goals in a requirements specification because "best," by whatever meaning it may have, is important to the system's stakeholders. The meaning of "best" (Most reliable? Shortest response times?) must be considered when making architecture, design, and implementation decisions.

7.3.2.5 System Quality Requirements

System quality requirements do not appear in the taxonomy of system requirements in Figure 7.3. Stakeholders' quality attributes and internal quality guidelines developed by physical systems engineers, software systems engineers, and quality control personnel are transformed into objectively stated system design constraints, when possible. Otherwise, they are categorized as design goals (for now).

Categorizing system quality attributes as system design constraints when possible and as design goals when they cannot (yet) be categorized as objectively stated design constraints provides important inputs that must be satisfied by the system architecture definition because

> Quality attributes are the primary determinants of system architecture.

There may be several ways to satisfy system requirements for functionality and behavior but fewer ways to satisfy system quality requirements.

A system quality requirement might be based on a system capability derived from a stakeholders' requirement. For example, a stakeholders' requirement for instantaneous response for Type B status indicators may become a capability to provide "a maximum response time of 2 seconds for continuous updates of a Type B status indicator," which then becomes a design constraint in the system requirements.

A stakeholders' subjective requirement for "rapid response time" might be restated as a system capability to be provided, which becomes an unquantified design goal in the requirements specification (for now). Subjectively stated system quality requirements are not uncommon during the initial definition of system requirements because the impact of quantification must be analyzed before a commitment can be made.

Unquantified design goals for system quality will be recategorized as design constraints when "rapid response time" and similar design goals are sufficiently understood so they can be quantified. For example, a design goal for "instantaneous steering response" in a driving system simulator may become a design constraint of "a maximum response time of 0.25 seconds for each driver-DSS steering wheel interaction when driving on dry, straight roads is simulated."

Another way to categorize system quality attributes is to distinguish between those quality attributes that are visible to system users and those desired by the system engineers, software engineers, system developers, and system maintainers but that are not observable by system users. Some examples are provided in Table 7.2.

Some quality attributes may not be directly measurable, so surrogate measures must be used. For example, ease of modifying a system will be indirectly determined by the kinds of modifications anticipated and a system architecture and system design that will facilitate those modifications. Modification scenarios can be used to evaluate the ease or difficulty of making specific modifications to the system.

Quality attributes, those externally visible and those internally desired, must be prioritized because there typically will be tradeoffs among the quality attributes. For example, emphasis placed on information security and physical security may negatively impact system performance and vice versa. In addition, each quality attribute, along with the other system requirements, should be bidirectionally traced to and from the sources of those requirements and to and from the associated verification procedures that will be used to determine the degree to which the system satisfies the quantified quality attributes.

7.3.2.6 System Interface Requirements

Interface requirements, like quality attributes, are important system design constraints; they specify the system requirements that must be satisfied to provide connections to entities in the system environment. A context diagram can be used to illustrate a system's boundary, its external interfaces, and entities in the system's environment with which the system will interact. As illustrated in Figure 7.4, context diagrams are deceptively simple. The name

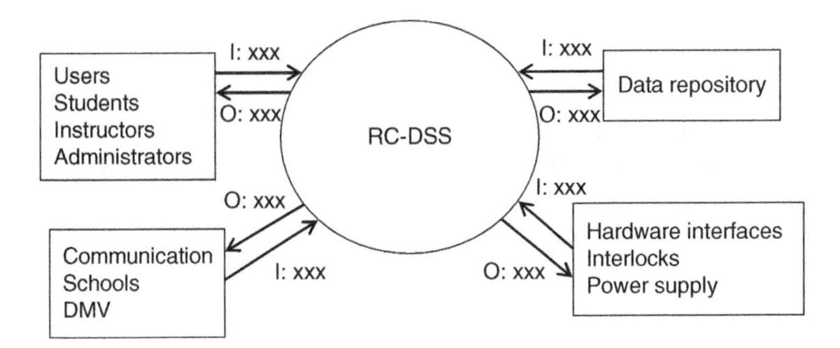

Figure 7.4 Context diagram for the RC-DSS driving system simulator.

of the system is placed inside the context circle. The perimeter of the circle is the boundary between the system and its environment. Emphasis is placed on the entities in the environment and the interfaces between the system and the external entities.

The external entities are denoted as "terminators." Each interface is given a meaningful name and accompanying text to include whatever details are available during system requirements definition. Inputs and outputs are documented in an interface requirements specification (IRS) at the requirements level, which will evolve into an interface control document (ICD) during system design definition. Some interfaces in the IRS may be flagged as «reconfigurable». The corresponding interfaces in the ICD would be designed to permit customization for different system environments.

Lack of sufficient information may result in an external interface being temporarily categorized as a design goal that is flagged as a TBR.

Some interface requirements may be generated from stakeholders' design constraints. For example, a stakeholders' design constraint may result in a system capability that will provide a connection to the acquiring organization's financial transactions repository. This may result in a detailed system requirement to provide an interface connection to an Oracle database using a specified protocol.

External interface specifications will become more precise during subsequent system development phases. Internal interfaces among system elements are defined during architecture and design definition and external interfaces are revised and elaborated.

Some interfaces may be connections to other systems, as for example a connection between an RC-DSS installation and a customer's data repository. Some interfaces may be connections to display devices to provide different kinds of displays for different kinds of system users, for example, different RC-DSS interfaces for administrators, instructors, and maintenance technicians.

Requirements for a hardware interface typically include the following:

- The types of devices to be supported by the interface;
- Error codes to be provided;
- Data and control interactions between the hardware and the associated software; and
- The communication protocols to be used.

Details of a hardware interface will be added during the design phase of a project; they will include the configurations of the connectors, the pin assignments, and the voltage/signal levels.

Requirements for a software interface will typically include the following:

- Precondition to be satisfied before using the interface;
- Postconditions that will be satisfied after using the interface;
- Specifications for concurrent or sequential operation;
- Method of synchronization;
- Communication protocol to be used;
- The priority level of the interface; and
- The error codes.

Data requirements for a software interface include the names, descriptions, units of measure, limits, ranges, accuracy, and precision of the data transmitted across the interface.

For simple, straightforward systems, the interface requirements can be specified as annotations attached to the context diagram. For large, complex systems, interface requirements are typically specified in an appendix to the requirements specification as an IRS.

The context diagram, as illustrated above, serves as a "visual table of contents" for the IRS. A shared IRS is especially important when multiple organizational entities or multiple organizations are involved in system development or modification. In these cases, the IRS provides an informal contract among different system developers and development groups for the external interfaces they will provide and the interfaces they will use. The IRS must be placed under version control so that all involved parties are aware of the current status of interfaces as they undergo changes.

As stated above, an IRS may (should) evolve to become an ICD or interface design document (IDD) during the architecture and design definition phases of system development. For large systems, a separate ICD or IDD may be maintained for the interfaces among major subsystems, with individual ICDs or IDDs for each subsystem and a master ICD or IDD for the interfaces among the subsystems.

7.4 Verifying and Validating System Requirements

Verifying system requirements is concerned with determining the following:

- The degree to which the system requirements cover the system capabilities for which they were defined; and
- The degree to which implementing the system requirements is feasible.

Validating system requirements is concerned with determining the following:

- The degree to which the system requirements, when implemented, will provide the capabilities for which they were defined; and
- That the system requirements, as documented, will provide information needed by the intended users of the requirements.

Verification and validation (V&V) methods and techniques are described in the following text.

Different users of the system requirements will use them for different purposes. Those who will use the documented system requirements as a basis for performing their job functions include customers, project managers, physical system engineers, software systems engineers, system architects, system designers, system implementers, test planners, deliverable system verifiers and validators, system installers, operators, and system maintainers, and other stakeholders who affect or are/will be affected by the system requirements.

The different ways some of the users of system requirements will use the requirements are indicated in Table 7.3.

Note that system stakeholders, other than a knowledgeable customer, are not listed as users of the system requirements. Stakeholders' interests are addressed in the stakeholders' requirements and system capabilities. The quantified and objectively stated system requirements are often too detailed and too technical to be of interest to most system stakeholders. However, they should assess the traceability links to ensure the stakeholders' requirements and system capabilities are adequately covered by the system requirements.

Determining "the degree to which" systems requirements satisfy the V&V criteria stated above is not an overall yes/no determination. Some system requirements may adequately cover the corresponding system capabilities and adequately satisfy the stakeholders' requirements. Some additional system requirements may need to be added and some system requirements may need to be revised in order to provide the system capabilities and satisfy the stakeholders' requirements.

Also note that verifying and validating the system requirements is not the same as verifying and validating an implemented system increment or a deliverable system. Verifying and validating system requirements is concerned with determining the degree to which the requirements provide a suitable basis for the next phases of defining of the system architecture and system design

Table 7.3 Some users and uses of system requirements.

Requirements users	Requirements used to
Customer	Assess the validity of the requirements and prepare a facilities plan
Project manager	Prepare estimates and plans; conduct risk analysis
Physical system engineers and software engineers	Aid project manager by preparing estimates and plans for needed technologies and technical disciplines (kinds and numbers of engineers)
	Prepare plans for the technical work to be done
	Conduct risk analysis for the technical work
System architects, designers, and implementers	Assess the primary requirements, derived requirements, design constraints, and design goals to determine the architecture, design, and implementation requirements that must be satisfied
System verifiers and validators	Determine the needed V&V facilities
	Develop initial test plans, test scenarios, and test cases
	Establish traceability links
System operators and maintainers	Assess the degree to which operational and maintenance requirements are adequately addressed
Safety, security, reliability, availability, and other engineering specialists	Analyze the degree to which the necessary and desired system attributes are included in the system requirements

and preparing for system realization. In contrast, verifying and validating an implemented system increment or a deliverable system is concerned with determining the degree to which a system realization satisfies the system requirements (verifying) and the degree to which the system realization satisfies the system capabilities and stakeholders' requirements (validating).

7.4.1 Verifying System Requirements

Verifying the feasibility of system implementation and analyzing bidirectional traceability are the primary techniques used to verify that the system requirements can be implemented within the constraints of schedule, cost, resources, and technology and that the system requirements adequately cover the system capabilities derived from the stakeholders' requirements.

7.4.1.1 Verifying Implementation Feasibility

The feasibility of implementing system requirements is determined by examining the constraints that may inhibit system realization. Constraints on the time and/or cost needed to procure and fabricate physical system elements, modify existing software elements, and construct and procure other software elements may be prohibitive.

An additional feasibility consideration is availability of resources needed to fabricate and procure physical elements and to modify, construct, and procure software elements. Resource constraints may include insufficient kinds and numbers of engineers and restraints on acquiring additional engineers; lack of sufficient infrastructure, including lack of software development tools or lack of machines and tools needed to fabricate physical elements; and lack of organizational resources such as configuration management, quality assurance, procurement assistance, and needed engineer training.

Another feasibility consideration is practicality of the technologies needed to implement the stakeholders' requirements. Lack of expertise in algorithms for self-driving vehicles, for example, may inhibit the ability to satisfy the system requirements that correspond to a stakeholders' requirement that driving system simulators provide a self-driving capability.

And another feasibility inhibitor is the possibility of incompatibilities of behaviors, functionality, and interfaces among the legacy elements to be incorporated; the physical elements and software elements to be fabricated, constructed, and procured; and/or the external interfaces needed to satisfy the system requirements for a software-enabled system.

In summary, system feasibility may be inhibited by the following:

- Constraints on time and cost;
- Availability of resources;
- Practicality of technologies; and
- Incompatibilities among legacy elements and interfaces.

Constraints on schedule, cost, resources, and/or technologies plus system incompatibilities may require negotiation with a system's stakeholders to make revisions to the stakeholders' requirement and the corresponding system capabilities. In some cases, procurement of available physical elements and/or reuse of available software elements that satisfy most, but not all, of the system requirements for one or more capabilities may be the preferable choice.

7.4.1.2 Verifying Coverage of System Capabilities

Verifying system requirements coverage of system capabilities involves determining the degree to which the system requirements cover the system capabilities derived from the stakeholders' requirements. As indicated above, "the degree to which" is not typically a yes/no determination; some system requirements may fully cover the corresponding system capabilities and others may need additional work.

Traceability can be used to determine the degree to which all system capabilities are covered in the specification of system requirements and that no extraneous system requirements are included, such as those that might violate will-not and must-not stakeholders' requirements or system capabilities. Traceability can also be used to determine that the system requirements, when implemented, might provide additional features the stakeholders did not request (but that the system developers are sure the stakeholders will want – without asking them).

Traceability correspondences between system capabilities and system requirements are typically documented in tables (i.e. traceability matrices), as indicated in Table 7.4.

As illustrated, SR-1 is in a desirable one-to-one correspondence with SC-1 but there are several issues to be addressed for the other table entries: system requirement SR-2 does not trace to any system capabilities; SR-3 traces back to two system capabilities, namely SC-2 and SC-3 (i.e. SC-2 and SC-3 both trace to SR-3); SR-4 is in one-to-one correspondence with SC-3 but SR-3 and SR-4 both trace to SC-3; and capability SC-4 does not trace to any system requirements.

A system requirement that does not trace to any system capabilities (e.g. SR-2) presents the possibility that a system requirement will be implemented for which there are no stakeholder requests. Ominously, SR-2 could be a system requirement inserted by someone for a malevolent purpose.

As shown in Table 7.4, system requirement SR-3 traces to both of the SC-2 and SC-3 capabilities. SR-3 might be a requirement that is legitimately allocated to multiple capabilities, as indicated. For example, a cybersecurity requirement may apply to all or most of a system's elements. However, a single system requirement that provides part or all of two or more capabilities (e.g. mapping SR-3 to SC-2 and SC-3) may be too broadly stated and result in a complex implementation that will be hard to verify and validate.

Also, maintaining a one-to-one relationship between capabilities and associated system requirements may be important for safety, security,

Table 7.4 A traceability example.

	SC-1	SC-2	SC-3	SC-4
SR-1	X			
SR-2				
SR-3		X	X	
SR-4			X	

SR, system requirement; SC, system capability.

reliability, and operational requirements. A system requirement that traces to two or more system capabilities thus indicates the need to examine and perhaps revise the capabilities and/or the corresponding system requirements.

And, a requirement that traces to two or more system capabilities may be implemented twice (or more) if two or more different engineers or engineering teams work on implementing the two or more capabilities. This happened, for example, when a traceability analysis revealed that two development groups in two different geographic locations (Phoenix, Arizona, and the Netherlands) were implementing the same system requirements that were defined to provide the two different capabilities the two teams were implementing (author's personal experience).

In some cases, a system capability may generate more than one system requirement. Note that capability SC-3 in Table 7.4 traces to system requirements SR-3 and SR-4. In these cases, the mapping from the system capability to multiple system requirements may be appropriate but it may indicate that the capability is too broad in scope and should be decomposed into two or more derived capabilities that are subsumed by the original capability.

Note that capability SC-4 in Table 7.4 does not link to any system requirements. Without traceability analysis, a missing capability-to-requirements link may go undetected. This could result in failure to detect the omitted implementation of a system capability until system validation just prior to system delivery.

It is thus desirable, though not always possible, to maintain a one-to-one or, more typically, a one-to-multiple correspondence between system capabilities and system requirements, where a one-to-many correspondence for a capability may result in several nonoverlapping system requirements that fully cover the capability. However, a one-to-many correspondence may indicate the need to decompose the capability into a set of derived capabilities. Issues such as those indicated in Table 7.4 should result in revisiting the statements of system capabilities and system requirements and perhaps revising some of them and updating the corresponding traceability links.

Traceability of quality requirements can be problematic in this regard. Some quality requirements may apply to all of the capabilities of an entire software-enabled system (e.g. security), some quality requirements may apply to some of the capabilities (e.g. safety), and some to a single capability (e.g. ease of use). Nevertheless, it is important to verify the traceability correspondences between quality requirements and the corresponding capabilities.

It is desirable, and sometimes mandatory, that test plans and test scenarios for verification of a deliverable system be developed during the system requirements phase of system development. Bidirectional traceability between the system requirements and these work products can be used to document and analyze the correspondences between them.

An important note: Traceability analysis cannot verify the sufficiency of system requirements generated from system capabilities, but only that one or more system requirements is associated with each system capability. Analyzing preliminary test plans when conducting the requirements analysis process can expose inadequacies (e.g. one or more system requirements that has no traceability links to test planning) but it cannot assess the adequacy of test planning content. Traceability is thus a necessary but not sufficient mechanism for verifying the system requirements defined for the system capabilities.

Mechanisms for assessing the technical content of system requirements include the following:

- Analysis;
- Reviews;
- Inspections; and
- Usage scenarios.

Some additional observations about requirements traceability: Traceability is sometimes insufficiently accomplished because it is a labor-intensive activity. It is, however, an essential on-going activity of system development. Tools for requirements management provide traceability capabilities that can be used to enter traceability links and generate analysis reports but the tools do not eliminate manual entry of the links.

Traceability tools usually provide predefined tags and the ability to define additional tags that can be placed on system capabilities and system requirements that are entered into the tool. Tags can be used for a variety of purposes, including

- Categorizing the essential, desirable, and optional capabilities;
- Two-way linking of system capabilities and system requirements;
- Linkage of requirements to test plans, test scenarios, and test cases;
- Indicating the priorities within essential and desirable capabilities that will guide incremental system implementation;
- Indicating whether a requirement is primary, derived, design constraint, or a design goal; and
- Indicating the level of difficulty or risk anticipated for implementing a capability or a requirement.

Requirements tools can also be used to generate reports that document the kinds of issues presented in Table 7.4. Those issues are apparent in the small table but may not be apparent in a large traceability matrix.

And, it is essential that traceability be accurately maintained as linkages are changed – outdated traceability links are worse than useless because they are misleading and can result in costly mistakes. Also, traceability matrices must be maintained under version control so that changes and the time series log of changes are made visible to all involved parties.

Finally, it should be noted that the principles and techniques of traceability can be, and should be, used to link the various work products of system development and to provide traceability among them, in addition to tracing requirements. Traceability analysis can provide assurance that all necessary elements of all work products are included and that extraneous elements are not included.

Comprehensive traceability also provides a mechanism for tracing the impact on other work products of changing a work product, for example, determining the impact on system architecture of changing some requirements or capabilities or the impact on system integration plans of changing the system architecture or the planned sequence of implementing capability sets.

7.4.2 Validating System Requirements

Validating system requirements is concerned with the following:

- Determining the degree to which the system requirements, when implemented, will provide the specified system capabilities and thus satisfy the stakeholders' requirements; and
- Determining the degree to which the system requirements, as documented, will adequately provide for the needs of the intended users of the requirements.

As indicted above, validating system requirements should not be confused with validating implemented system increments or with validating a deliverable system.

Techniques for validating system requirements include the following:

- Traceability analysis;
- Technical analysis;
- Reviews;
- Scenarios; and
- Prototyping.

System users, operators, and maintainers are those who will be most directly affected by the operational features and quality attributes of a delivered system. Determining the degree to which the system requirements, when implemented, will satisfy their operational needs can be determined by examining the relationships between the system requirements and the system capabilities generated from the stakeholders' requirements.

Two techniques for validating system requirements are (i) analysis of use cases to determine if the different users of the system requirements (user classes) are adequately represented and (ii) using the traceability links between system capabilities to system requirements discussed above to analyze the suitability of the system requirements for the intended users of the requirements.

Use cases document the external entities (actors) that initiate and participate in desired operational scenarios. Exception scenarios are also identified, along with other issues such as preconditions for correct instantiation of an operational scenario and the postconditions that should be true after the use case is "executed." Related issues that may be documented in a use case include requirements addressed by the use case, associated use cases, quality attributes, and relationship of the use case to a business case or a mission concept. Bidirectional traceability of use cases to and from system requirements may reveal the need to revise a use case, decompose it into smaller use cases, create additional use cases, or revise the associated system requirements. An example of a use case is provided in Section 6.3.3.1.

Traceability can also be used to link system capabilities to validation plans for acceptance tests and demonstrations.

The distinction between using traceability analysis to verify the system requirements in contrast to using traceability analysis to validate the system requirements should be noted. Verifying system requirements is concerned with determining the degree to which the system capabilities are covered by the system requirements and not with the adequacy of the system requirements to satisfy stakeholders' needs; the latter is the purpose of system requirements validation.

The following conditions will render system requirements invalid:

- System requirements that will not, when implemented, satisfy the stakeholders' requirements and system capabilities;
- Inadequately documented system requirements;
- System requirements expressed in notations that are not understood by the intended users of the requirements;
- Requirements that are not updated to reflect ongoing revisions;
- Requirements that are not kept current using version control;
- Traceability links that are not updated;
- Traceability matrices that are not kept current using version control; and
- Updates to system requirements and traceability matrices that are not distributed to all affected individuals.

Note that system requirements can be invalid because they do not adequately serve the needs of the intended requirements of the users, even though the requirements may be verified for adequate coverage of the system capabilities.

System requirements, if verified and validated as specified, and revised as necessary, will provide the system capabilities needed to satisfy the stakeholders' requirements. System requirements that are not complete, correct, consistent, concise, and clear (the five C's of requirements engineering) will probably not provide the system capabilities that will satisfy the stakeholders' requirements.

Table 7.5 Examples of RC-DSS system capabilities and system requirements.

Capability	Requirement category	System requirement
Braking	Primary requirement	Simulated stopping distances for 20, 40, 60, and 80 mph speeds on dry pavement, wet pavement, and ice shall be as determined using the Forensic Dynamics calculator (Forensic Dynamics Inc. 2018)
Braking	Derived requirement	Intermediate stopping distances shall be determined by linear interpolation of stopping distances determined at 20, 40, 60, and 80 mph speeds on dry concrete, wet concrete, and ice
System interface	Design constraint	RC-DSS systems shall provide an Oracle database interface to allow customer access to student-driver records
Usability	Design goal	RC-DSS systems shall provide the look and feel of realistic driving experiences

7.5 System Requirements for the RC-DSS Case Study

System requirements are defined for the system capabilities derived from the stakeholders' requirements. Some examples of system capabilities and the corresponding system requirements for the RC-DSS driving system simulator are presented in Table 7.5.

7.5.1 RC-DSS Operational Requirements

System requirements provide the specifications for designing, implementing, verifying, and validating a deliverable system. Operational requirements are requirements that must be satisfied for successful installation and operation of a system. Operational requirements are documented in a concept of operations, operational scenarios in use cases, and other sources as requirements expressed by RC-DSS managers, legal experts, customers, users, operators, and maintainers.

Operational requirements for an RC-DSS might include the following:

1. Installation of an RC-DSS shall require 400 ft^2 of floor space (20 × 20), overhead clearance of 20 ft, a 110 or 220-V power supply capable of delivering 100 kW of power, and Internet access.
2. RC-DSS facilities shall satisfy the American Disability Act (ADA) accessibility requirements in the United States. Local accessibility requirements shall apply for RC-DSS installations in other countries.

3. Each RC-DSS shall be housed in a secure facility that provides door interlocks and warning lights to indicate that a driving session is active.
4. Two RC-certified technicians shall be able to install an RC-DSS system in a customer's facility that satisfies the installation requirements in two normal workdays.
5. RC-DSS trained and certified instructors shall
   - Initiate, monitor, and terminate student driving sessions;
   - Prepare scripts for various driving scenarios that will be stored, retrieved, and activated as desired;
   - Inject unanticipated situations into driving scenarios, as appropriate; and
   - Maintain records of student performance.
6. RC-DSS trained and certified system administrators shall
   - Enroll students;
   - Maintain and provide status reports periodically and as requested;
   - Certify driving course completions and issue certifications; and
   - Maintain financial records for students and for RC-DSS acquirer accounts.
7. Realistic Corporation trained and certified technicians shall have four work-hours of dedicated access to each installed RC-DSS, each week, to perform routine maintenance. Emergency access will be provided as needed.

Note that these requirements will influence analysis, design, implementation, delivery, operation, and stakeholders' satisfaction with RC-DSS systems.

Unfortunately, there are systems for which operational requirements were not specified, or if specified, were not realized:

- Operators and technicians not adequately trained or certified;
- Infrastructure, interfaces, and tools for maintenance not arranged;
- Routine access to the system for maintenance not arranged; and
- Facilities not adequately prepared, such as a facility where the airplanes do not fit through the hanger doors for the specified installation site.

7.6 Key Points

- The requirements taxonomy includes stakeholders' requirements (operational features, quality attributes, and design constraints), system capabilities, and system requirements (primary requirements, derived requirements, design constraints, and design goals).
- A system capability provides the ability of a system to execute a specified course of action or to exhibit a state of being.
- Stakeholders' requirements, system capabilities, and the corresponding system requirements should be sufficiently defined so that

- ○ Complexities can be identified;
- ○ Technical risk factors and mitigation strategies can be identified; and
- ○ Initial estimates of effort, schedule, and resources needed to implement the requirements can be made.
- System capabilities are derived from stakeholders' requirements.
- System requirements are defined to provide the system capabilities.
- System requirements that are not complete, correct, consistent, concise and clear (the five C's) will probably not result in acceptable delivered systems.
- Some system design goals may be recategorized as objectively stated primary requirements, derived requirements, or design constraints; some design goals may continue to be subjectively stated.
- Subjectively stated design goals should not be ignored; they are meaningful to the system stakeholders.
- Some design goals may be recategorized as objectively stated requirements after discussions and prototyping; other may remain as design goals.
- System quality requirements and system connections to the system environment are categorized as system design constraints.
- System quality requirements are the primary determinants of system architecture and system design.
- System connections to the system environment are depicted in context diagrams and documented in IRSs.
- Verifying system requirements determines the degree to which the requirements, when implemented, will provide the system capabilities.
- Validating system requirements determines the degree to which the system requirements, when implemented, will satisfy the stakeholders' requirements.
- Techniques for verifying and validating system requirements include traceability analysis, technical analysis, reviews, inspections, and prototyping.
- Traceability is a necessary but insufficient technique for V&V of system requirements.
- Verifying and validating system requirements should not be confused with verifying and validating the deliverable system that results from implementing the requirements.
- System requirement should be documented in sufficient detail to serve the needs of the intended users of the system requirements definition.
- The RC-DSS driving simulator case study provides examples of systems requirements definition.

Exercises

7.1. System requirements should be defined in sufficient detail to be useful to the intended users of the requirements.

(a) Identify a system of interest to you.
(b) List, by job title, some individuals and groups who will use the system requirements and then briefly explain how they will use the requirements.

7.2. The systems requirements taxonomy in Figure 7.2 includes the categories of primary requirements, derived requirements, design constraints, and design goals. Briefly explain each category and how it is different from the other categories.

7.3. Traceability is a necessary but an insufficient technique for verifying and validating system requirements, when used alone.
(a) Briefly explain the ways in which traceability, when used alone, can be used to verify and validate some aspects of system requirements.
(b) Briefly explain why traceability, when used alone, is not a sufficient mechanism for verifying and validating system requirements.

7.4. Briefly explain the following:
(a) The purpose of verifying system requirements;
(b) The purpose of validating system requirements; and
(c) How the purposes of verifying system requirements and validating system requirements are different.

7.5. Table 7.5 provides an example of a primary requirement, a derived requirement, a design constraint, and a design goal for the driving system simulator. Briefly describe an additional primary requirement, a derived requirement, a design constraint, and a design constraint for the RC-DSS. State the primary requirement, derived requirement, and design constraint using objective, quantified terms. *Note*: The "RC-DSS" is introduced in Section 6.2.

7.6. Section 7.5.1 describes some operational requirements for the RC-DSS. Briefly describe three additional operational requirements for the RC-DSS that you think would apply.

7.7. System quality attributes are the primary determinants of system architecture and system design. Briefly explain why this is a true statement.

7.8. Subjectively stated design goals should not be ignored. Briefly explain why they are important.

7.9. System quality requirements and system interface connections to the system environment are categorized as system design constraints in the system requirements taxonomy. Briefly explain why they are design constraints.

7.10. Verifying and validating system requirements should not be confused with verifying and validating the deliverable system that results from implementing the requirements. Briefly explain the difference between verifying and validating system requirements and verifying and validating a deliverable system.

7.11. System requirements that are not complete, correct, consistent, concise, and clear (the five C's) will probably not result in acceptable delivered systems. Briefly explain each of the following:
(a) System requirements that are not complete will result in the following problems:
(b) System requirements that are not correct will result in the following problems:
(c) System requirements that are not consistent will result in the following problems:
(d) System requirements that are not concise will result in the following problems:
(e) System requirements that are not clear will result in the following problems:

References

Forensic Dynamics Inc. (2018). Stopping (braking) distance calculator. http://forensicdynamics.com/stopping-braking-distance-calculator (accessed 25 June 2018).

ISO (2011). ISO/IEC/IEEE 29148:2011, Systems and software engineering – Life cycle processes – Requirements engineering, ISO/IEEE, 2011.

ISO (2015). ISO/IEC/IEEE 15288:2015, Systems and software engineering – System life cycle processes, ISO/IEEE, 2015.

ISO (2017). ISO/IEC/IEEE 12207:2017, Systems and software engineering – Software life cycle processes, ISO/IEEE, 2017.

8

Architecture Definition and Design Definition

8.1 Introduction

Defining the architecture of a system is concerned with defining the structure and behavior of a system that, when implemented, will satisfy the system requirements, which will in turn provide the system capabilities needed to satisfy the stakeholders' requirements. Design definition is concerned with decomposing the architecture into design elements, specifying the interfaces among design elements and connections to entities in the system environment, identifying the technologies to be used to implement the design, and determining the feasibility of system implementation and deployment using those technologies.

Sections 6.4.4 and 6.4.5 of ISO/IEC/IEEE Standards 15288 and 12207 provide guidance for architecture definition and design definition (ISO 2015, 2017). Current versions of these standards separate architecture definition and design definition into distinct but closely coupled system development phases in recognition of the importance of directing attention to each of them. Former versions of 15288 and 12207 did not separate architecture definition and design definition into distinct development phases. Other, earlier versions of process standards denoted the two design phases as "preliminary design" and "detailed design."

The distinction between architecture and design is more important for developing physical elements than for developing software elements because physical system elements are less malleable than software and thus more difficult to revise during system implementation if the architecture or design is incorrect. It may be prohibitively expensive in time, resources, and/or finances to later modify or redo a physical system element, so it is important to "get it right" (as nearly as possible) before fabricating or procuring physical system elements. This is not to deny the importance of software design but to emphasize the importance of carefully architecting and designing the physical elements.

Architecture definition and design definition are depicted in Figure 5.2 as activities that occur during Phases 3 and 4 of the I^3 system development model.

Systems Engineering of Software-Enabled Systems, First Edition. Richard E. Fairley.
© 2019 John Wiley & Sons, Inc. Published 2019 by John Wiley & Sons, Inc.

Figure 5.2 is repeated here as Figure 8.1 for ease of reference – architecture and design (Arch/Des) are highlighted in the figure.

Background material for this chapter includes the overview of the I^3 system development model presented in Chapters 5 and 6 on mission or business analysis, and stakeholder identification and requirements definition, and Chapter 7 on system requirements definition. The Realistic Corporation driving system simulator (RC-DSS) case study introduced Chapter 6 is continued in this chapter.

This chapter provides readers with the opportunity to learn principles and practices of architecture definition and design definition and the techniques used to control the complexity of system architecture and design.

Architecture definition is covered in Sections 8.2–8.5. Design definition, the RC-DSS case study, and controlling the complexity of architecture and design are covered in subsequent sections. Key Points of the chapter are summarized in Section 8.10. References and Exercises provide the opportunity to learn more about system architecture and system design.

Appendix 8.A provides an overview of the System Modeling Language (SysML) diagrams that can be used to define system architecture and system design.

8.2 Principles of Architecture Definition

According to subclauses 6.4.4.1 of ISO/IEC/IEEE Standards 15288 and 12207, the purpose of architecture definition is (ISO 2015, 2017)

> to generate system architecture alternatives, to select one or more alternative(s) that frame stakeholder concerns and meet system requirements, and to express this in a set of consistent views.

Architecture definition is concerned with defining the structure and behavior of a system to be implemented. System structure includes the major system elements, the interfaces among them, and the connections of system elements to external entities in the system environment. A large system may be decomposed into subsystems that each has an architecture and design specifications.

Behavior is determined by the observable results of interactions among system elements over time in response to external stimuli and in response to stimuli generated internally. System performance is the external manifestation of system behavior.

As noted in 15288 and 12207, architecture is expressed in a set of consistent views. Multiple views of system architecture are usually needed to understand the structure and behavior of a complex software-enabled system.

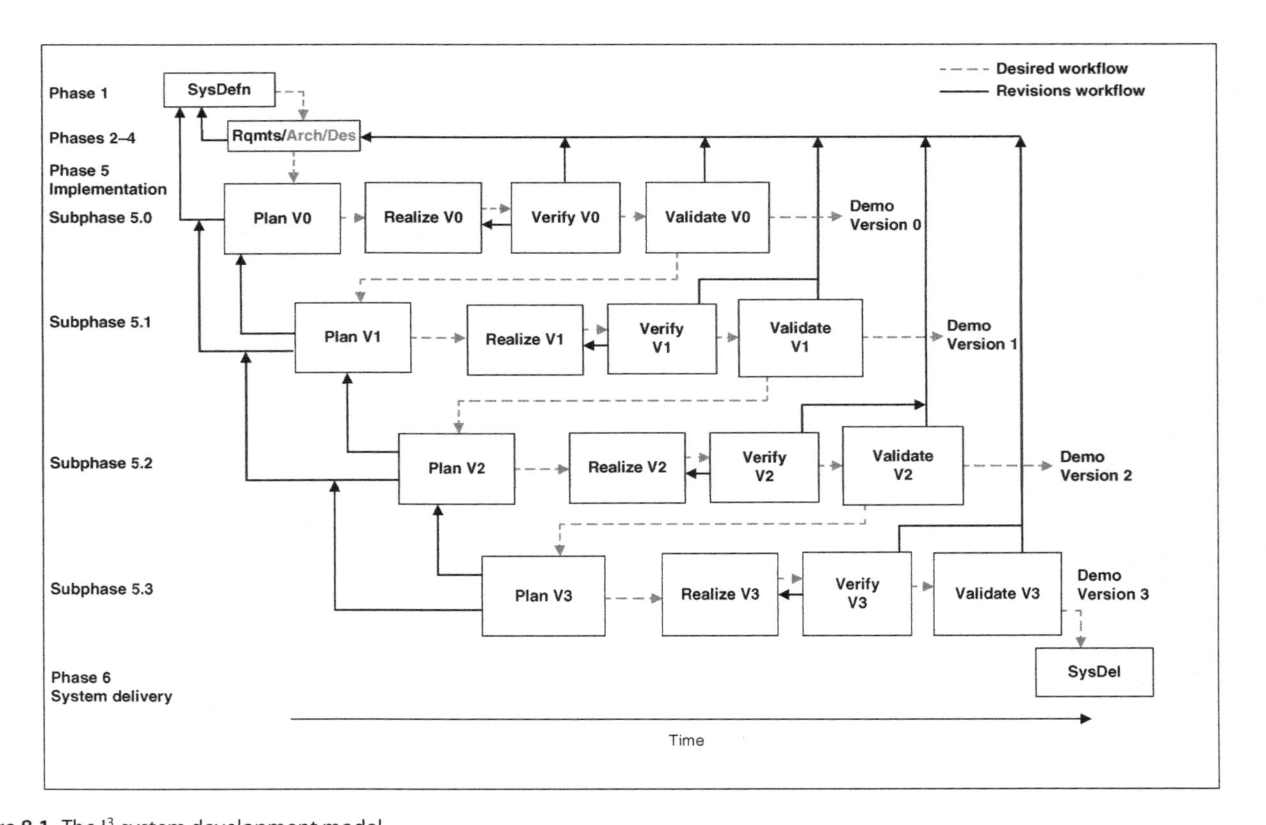

Figure 8.1 The I^3 system development model.

Defining system architecture is a creative process based on the system designer's (or designers') background, knowledge, and experiences; research, brainstorming, and analogies are used. Candidate system architectures may be routine, innovative, or revolutionary.

Routine architectures are based on well-known principles learned from past experiences with similar systems, for example, in architecting the structure and behavior of a network of electronic control units (ECUs) for the engine-transmission drive train of a traditional gasoline-powered automobile (see Appendix B).

Innovative architectures are the result of applying past experience and well-known design principles in new and different ways, for example, in architecting a next-generation ECU network for the drive train of a hybrid automobile.

Revolutionary architectures are few in number; they incorporate new design paradigms that include new principles of architecture and new design patterns – for example, in designing the ECU architecture for a highly autonomous (i.e. Level 5) self-driving system for automobiles (SAE 2018).

Two or more candidate architectures should be developed and evaluated for suitability using predefined evaluation criteria (see Section 8.4). Candidate architectures are often based on an abstract characterization of potential solutions developed during the business or mission analysis process covered in Chapter 6 of this text.

Common failings of architecture definition are

- Failure to generate alternative candidate architectures;
- Failure to prespecify objective criteria for evaluating candidate architectures;
- Failure to rigorously apply the evaluation criteria;
- Failure to document the design rationale for the chosen architecture; and
- Failure to document the rationales for rejecting the alternative architectures.

Another common failing is defining architectural alternatives to be rejected in favor of a designer's biased choice; generating alternative architectures requires a great deal of open-minded mental discipline to avoid bias, either conscious or unconscious. A similar failing is defining three system architectures, two of which are unrealistic and will be rejected to satisfy a policy statement that requires at least three alternative architectures be developed and evaluated.

The selected architecture must be verified to determine that the architecture adequately covers the system capabilities and system requirements that when implemented will satisfy the stakeholders requirements – in particular, the architecture must cover the quality requirements, which are the primary drivers of system architecture. There may be several architectures that will satisfy the behavioral requirements but fewer that will satisfy the quality requirements *and* the behavioral requirements.

The selected architecture must be validated to determine the extent to which the architecture, as documented, will satisfy the needs of the intended users of the architecture documentation. Intended users of system architecture include physical systems engineers; software systems engineers; system designers; and implementation engineers who will plan, realize, verify, validate, demonstration, and (perhaps) maintain the system.

The key elements of the architecture definition process are thus

- Reviewing and revising stakeholders' requirements, system capabilities, system requirements, and interface requirements for connections to the system's environment;
- Developing objective evaluation criteria;
- Generating two or more proposed architectures;
- Providing structural and behavioral views of the architectures;
- Evaluating the proposed architectures by applying the evaluation criteria and conducting trade studies;
- Selecting an architecture in consultation with the appropriate decision makers;
- Documenting the design rationale for the chosen architecture along with rationales for rejecting the other candidates;
- Developing comprehensive models and documentation that will serve the needs of the intended users of the architecture definition; and
- Revising, reverifying, and revalidating the selected architecture during system development to satisfy stakeholders' requirements, system capabilities, system requirements, and operational requirements that may be revised.

These activities may be iterated and intermixed in various systematic ways, along with the activities of system design.

8.2.1 Activities of Architecture Definition

The activities of architecture definition are accomplished to address the key elements listed above. Inputs, processes, and outputs of architecture definition are presented in Table 8.1.

8.3 Defining System Architectures

As indicated in Table 8.1, the first step in defining a system architecture is to review, and revise as necessary, the stakeholder's requirements, the system capabilities needed to support the stakeholder's requirements, and the system requirements defined to satisfy the system capabilities. Constraints on schedule, budget, resources, infrastructure, and technology must also be considered and revised as needed and permitted. Revisions, as appropriate, are made to

Table 8.1 Architecture definition inputs, process, and outputs.

Inputs	Stakeholders' requirements, system capabilities, system requirements, and constraints on resources, schedule, and budget, infrastructure, and technologies
Processes	Review and revise the inputs, as needed
	Develop structural and behavioral views of two or more candidate architectures
	Select an architecture that best satisfies predefined evaluation criteria and trade studies
	Verify and validate the selected architecture definition
	Document the design rationale for the chosen architecture along with rationales for rejecting the other candidates
	Provide documentation for the selected architecture to serve the needs of those who will use the design documentation
	Develop an initial system integration plan for the deliverable system
Outputs	A verified and validated architecture specification that covers the system requirements and is sufficiently documented to serve the needs of users of the architecture definition
	A preliminary plan for system realization, verification, validation, delivery, installation, and acceptance

these work products, which may be further revised during the architecture, design, and implementation phases of system development.

The context diagram generated during requirements definition is reviewed and revised as necessary to specify the entities in the system's environment that will require interfaces (see Figure 7.4). External interface specifications are further developed during system design.

Two or more candidate architectures should be defined. Multiple views of system structure and system behavior are typically developed for each of the candidate architectures because system architecture is usually too complex to be depicted in a single structural or behavioral view. The SysML provides notations that can be used to define system requirements, system architectures, and system designs. Appendix 8.A provides an overview of SysML.

8.3.1 Defining System Structure

As indicated in Appendix 8.A, the three SysML diagrams for defining system structure are

- Package diagrams;
- Block definition diagrams; and
- Internal block diagrams.

Package diagrams are the basic mechanism used to place SysML diagrams in various containment groupings. A SysML package can contain any arbitrary grouping of SysML diagrams, including other packages. An example of a package diagram that contains packages is provided in Section 8.7 for the RC-DSS case study.

Block definition diagrams (bdds) are extended types of SysML packages that can be used to compose the functional-hierarchical structure of a system or subsystem. Internal block diagrams (ibds) are also typed packages; they depict the composition of bdd internals. Bdds are black box representations of system structure; ibds are white box representations. Bdds and ibds can be used in combination or independent of one another. An example of a bdd that has an associated ibd is provided in Section 8.7 for the RC-DSS case study.

Examples of using packages and blocks to specify system structure are provided in Appendix 8.A. Two additional kinds of SysML diagrams, parametric diagrams and requirements diagrams, are also described in Appendix 8.A.

8.3.1.1 Allocating System Requirements to System Architecture

System requirements are allocated to the functional elements of system architecture. Requirements may be individually allocated or allocated to multiple system elements, as illustrated in Figure 8.2.

System requirements are categorized as primary requirements, derived requirements, design constraints, and design goals in the I^3 system development model (see Chapter 7). One or more requirements of each kind may be

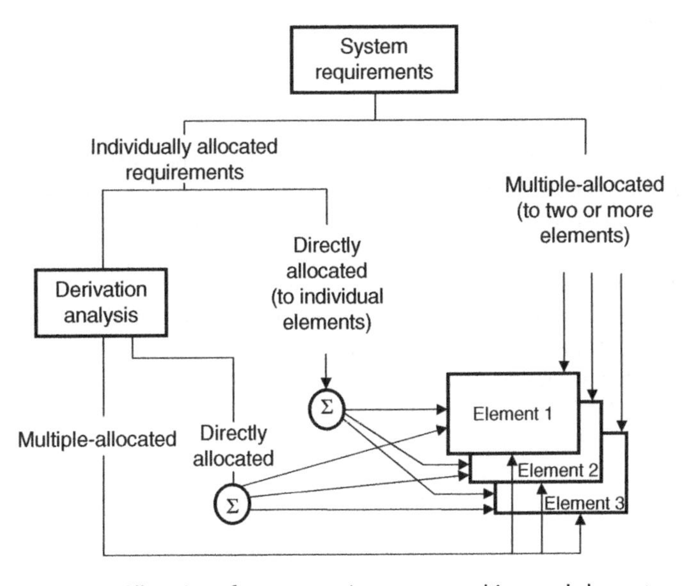

Figure 8.2 Allocation of system requirements to architectural elements.

allocated to an architectural element. The ≪allocate≫ stereotype can be used to attach SysML requirements diagrams to the packages, bdds, and ibds of system architecture.

As indicated in Figure 8.2, a system requirement may be individually allocated or allocated to multiple system elements. The time-stamping requirement will be directly allocated to the system element that provides the clock function. The response time requirement may be allocated to multiple system elements that determine response time for various system behaviors. Quality requirements (e.g. safety and security) are often allocated to multiple system elements. The summation symbol ($\sum$) in Figure 8.2 indicates that the sum of all directly allocated requirements and multiple allocated requirements should cover all requirements allocated to all of the design elements.

Requirements for internal interface connections among system elements and external connections to the system environment that were defined during system requirements definition are revised as necessary and extended during architecture definition. The interface requirement specification (IRS) may morph into an interface definition document (IDD) or interface control document (ICD).

8.3.2 Defining System Behavior

The four SysML diagrams for defining system behavior are described in Appendix 8.A and illustrated in Section 8.7 for the driving system simulator case study (RC-DSS). They are the following:

- Sequence diagrams;
- Activity diagrams;
- State machine diagrams; and
- Use case diagrams.

Sequence diagrams can be used to show interactions of external entities with system elements and receive responses. Sequence diagrams can also be used to depict interactions among system elements, initiated by one of the elements (physical or software).

Activity diagrams are flow diagrams. They are composed of nodes connected by lines with arrowheads that indicate the direction of flow. The nodes in an activity diagram are actions that transform inputs into outputs, which in turn provide inputs to other nodes or deliver the outputs. Decision points can be used to direct flows. A branch construct indicates start of concurrent subdiagrams and a join construct indicates synchronization points for concurrent elements. SysML activity diagrams can include flows of energy and material, in addition to data flows that are specified in the Universal Modeling Language (UML).

A state machine diagram includes a set of states that are interconnected by transition arrows. The behavior of a system is determined by transitions through connected states. Input stimuli can drive a system through a sequence of behavioral states to define an operational scenario. State machine diagrams can include hierarchically nested state machines that can be recursively decomposed to any desired level. They can also be partitioned to depict concurrent behavior.

Use case diagrams can be used to indicate the ways in which one or more actors interact with a set of use cases. The ≪include≫ and ≪extend≫ stereotypes can be used to indicate relationships among the use cases. Use case diagrams provide a high-level view of interactions between actors and use cases and are easily understood.

More information about SysML structural and behavioral diagrams are provided in Appendix 8.A. Two additional kinds of SysML diagrams, parametric diagrams and requirements diagrams, are also described in Appendix 8.A.

8.3.2.1 Analyzing System Behavior

Folmer et al. (2004) have described three ways of analyzing the behavior of system architecture to determine the extent to which the architecture satisfies behavioral requirements:

- Scenarios;
- Simulation; and
- Mathematical modeling.

Simulation and mathematical modeling are closely related in that a simulation may be based on mathematical models of system properties. For example, simulated Petri nets (on which the semantics of SysML activity diagrams are based) can be used to simulate flow through a network to identify potential bottlenecks and system deadlocks (Reisig 2011).

A scenario may be in the form of a narrative sequence of interactions between an actor and the system, as in a use case scenario, but a scenario also can be driven by a mathematical construct, such the statistical distribution of an input stimulus provided by the system environment.

Behavioral scenarios can be traced through each of the four kinds of SysML diagrams to determine the degree to which the architecture supports the desired behaviors and to determine how the architecture can accommodate exception conditions that may occur. Tracing a scenario through a SysML behavioral diagram can be conducted as a "pencil and paper" exercise or as a simulation using a software tool that supports simulations of the diagram.

Kazman (1999) observes that three kinds of scenarios are effective:

- Use case scenarios of system/environment interactions;
- Growth scenarios for anticipated changes to a system; and
- Explanatory scenarios for extreme changes to a system.

Scenarios can also be developed to assess "what-if" situations such as robustness if one or more system elements fails, or the behavior that results from integrating a collection of systems into a system-of-systems architecture.

Scenarios to assess quality attributes can be constructed and traced through the elements of system architecture to identify the extent to which the architecture supports, or fails to support, desired quality attributes. It is important to assess quality attributes at the architectural phase of system development because the quality attributes of a system are largely determined by a system's architecture.

Different candidate architectures may provide the required functionality and behavior but it is typical that fewer candidates will support quality attributes such as usability, cyber security, ease of modification to make anticipated future changes, and maintainability. Scenarios for misuse cases can be developed to identify and explore potential security breaches in the architecture (Hope et al. 2004).

8.4 Architecture Evaluation Criteria

Candidate system architectures are evaluated for feasibility of design and implementation. Several aspects of feasibility must be examined, preferably using predefined evaluation criteria. Some of the evaluation activities may be performed in a preliminary manner during architecture evaluation and continued during analysis and verification and validation (V&V) of the system design definition.

The first step in developing the evaluation criteria for candidate system architectures is to characterize the design space, which is bounded by the prioritized system capabilities, system requirements, design constraints, design goals, and other constraints such as policies, infrastructure, schedule, and budget. Design constraints include quantified quality requirements and reflect the must-not and will-not capability statements (see Chapter 7). The design constraints may be numerous and include factors such as performance parameters, robustness, scalability, safety, security, sustainability, and survivability.

Additional evaluation criteria to be considered include the technologies that are available and affordable for implementing the candidate architectures, the infrastructure needed for system development, available resources (including kinds and number of engineers needed), and the schedule and budget allocated to system development. In addition, the evaluation criteria should include the ease of making anticipated changes to the architecture. And, Note 2 in Section 6.4.4.1 of 15288 indicates that a system's architecture should be as design-agnostic as possible to support flexibility in making design decisions during design trade studies; this evaluation criterion should also be included.

The chosen system architecture must also satisfy the constraints of legacy elements to be incorporated and certification processes that may be required when certifying a deliverable system.

Evaluation criteria should be prioritized because they are typically numerous and varied. For example, an architecture that optimizes safety and security may be somewhat different than an architecture that optimizes system performance. A system that must be delivered in 6 months may have a different architecture than one that is to be delivered in 18 months. Product-line architectures are different than "one off" architectures. Applying architecture evaluation criteria may require revisiting and revising stakeholders' requirements, the system capabilities to be delivered, and the resulting system requirements to be implemented.

In summary, evaluation criteria that could be considered in evaluating a system architecture definition are the following:

- Characterizing the design space for the candidate architectures;
- Evaluating technologies available and affordable for implementing the candidate architectures;
- Evaluating ability to incorporate required legacy elements into the candidate architectures;
- Evaluating ease of making anticipated changes to the candidate architectures;
- Determining the capability of candidate architectures to support a product line;
- Determining the capability of candidate architecture to support incorporation of the deliverable system into a system-of-systems;
- Determining the degree to which the candidate architectures are design-agnostic;
- Determining the capability of the candidate architectures to satisfy certify processes that may be required when certifying the deliverable system;
- Determining the ability of the candidate architectures to satisfy project constraints, including schedule, budget versus estimated cost, resources needed including numbers, and kinds of engineers and supporting personnel; and
- For each candidate architecture, determine the infrastructure requirements needed, to realize, verify, transition, and validate the deliverable system that would result.

Prioritized criteria can be used as the basis for trade studies and to evaluate the feasibility of alternative candidate architectures. However, it is important that the evaluation criteria be sufficiently comprehensive to avoid selecting an architecture that is too narrow in scope or is biased in some manner, while at the same time, the evaluation criteria should not be overly broad for the stated and anticipated needs.

The Goldilocks principle of architecture evaluation criteria is as follows:

> Not too narrow and not too wide, but just right.

8.5 Selecting the Architecture

Generating alternative candidate architectures is an important and sometimes neglected activity. The alternatives are (or should be) evaluated using prespecified evaluation criteria. It is important to include design rationales for the candidate architectures to document why the selected option was chosen and why the other options were rejected.

System engineers and software system engineers sometimes fail to document design rationale. Without design rationale to consult, they may modify the selected architecture at a later time, and then – at a still later time – regretfully realize why the original candidate architecture was chosen and why the modification was previously rejected. The design rationale must be updated when changes are made after reviewing the current design rationale and concluding that changes need to be made in the proposed manner.

One or more physical systems engineers and software systems engineers present to the decision makers the recommended architecture and the rejected options, including the design rationales for the recommended and rejected options and the trade-offs considerations for the recommended architecture.

Decision makers may include a system acquirer, one or more customers' business representatives, customer or customers' technical representatives, the supplying organization's managers, and the project manager. The decision makers choose the architecture that may be the recommended option, a different option, or an architecture that includes a combination of elements from the candidate architectures. The selection process is illustrated in Figure 8.3.

The selected architecture is typically reviewed and revised as necessary; this may involve revising the stakeholders' requirements, the system capabilities, and the system requirements. The selected architecture will be developed in more detail during the design definition phase.

Additional models and documentation are prepared for the selected architecture and additional V&V are performed, as needed. Verifying the

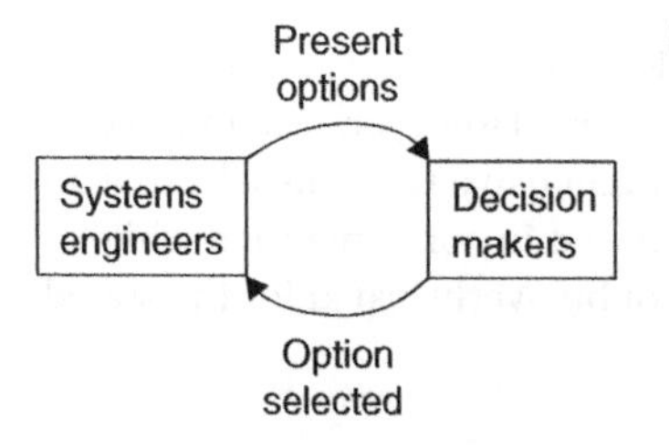

Figure 8.3 Selecting the system architecture.

documented architecture is concerned with determining the degree to which the architecture selected will, when designed and implemented, satisfy the system requirements, thus providing the system capabilities that are generated from the stakeholders' requirements. Validating the system architecture is concerned with determining the degree to which the documented architecture will serve the needs of the intended users of the architecture documentation. V&V of the selected architecture may result in revisions to the architecture and the work products on which it is based.

An integration plan is developed for the selected architecture. When using the I^3 development model, integration is merged with implementation because simulated elements are replaced with realized elements as those elements become available. The newly realized elements are verified and validated in combination with the previously realized system increment to form a new system increment, as illustrated in Figure 8.1.

A preliminary plan for packaging, delivery, transition, and system validation of the deliverable system (or any subsets of the final system) is also developed; this plan will be revised during system design and finalized during system implementation.

In summary, the following actions are performed following selection of the system architecture:

- The architecture is reviewed and revised as noted during the selection process;
- Additional V&V activities are performed, as needed; stakeholders' requirements, system capabilities, and system requirements are revised as necessary;
- Additional models and documentation are prepared for the selected architecture;
- An integration plan is developed for the selected architecture; and
- A preliminary plan for packaging, delivery, transition, and system validation of the deliverable system (or any subsets of the final system) is developed.

8.6 Principles of Design Definition

Sections 6.4.5 of ISO/IEC/IEEE Standards 15288 and 12207 provide guidance for the design definition process, which is distinct from but closely related to the architecture definition process (ISO 2015, 2017).

According to Subclause 6.4.5.1 of 15288 and 12207, the purpose of design definition is to

> provide sufficient detailed data and information about the system and its elements to enable the implementation consistent with architectural entities as defined in models and views of the system architecture.

Design definition provides the bridge from architecture definition to system implementation. Defining the system architecture and the system design are iterative and intermixed processes. System requirements, capabilities to be realized, and the stakeholders' requirements may also be revisited and revised during design definition.

During design definition, system elements defined during architecture definition are recursively decomposed as necessary and new elements are added. Requirements are allocated to the newly identified system elements and requirements allocated during architecture definition may be reallocated. Allocated requirements may be further decomposed and derived requirements may be added during design definition. Bidirectional traceability is maintained between the requirements allocated to each design element and to the corresponding architectural elements for which they are developed.

8.6.1 Design Evaluation Criteria

Major system (or subsystem) elements should be recursively decomposed into smaller elements until the following evaluation criteria can be satisfied with confidence.

- Areas of complexity and technical uncertainty have been identified.
- Previously identified implementation technologies have been confirmed or identified for revision.
- Conditions of use for each smallest system element have been specified (preconditions and postconditions including exceptions and side effects).
- Detailed design specifications have been prepared for the interfaces and connections that will be provided by each smallest system element.
- Interfaces and connections that will be needed by each smallest system element have been identified.
- System elements have been identified for procurement and for implementation.
- Previously identified infrastructure and resources needed for system implementation have been revised as necessary.
- The numbers and kinds of engineers in the various engineering disciplines needed to implement and oversee procurement of the lowest level elements in the system hierarchy have been determined.
- Cost and schedule estimates for system implementation, verification, validation, transition, and delivery have been updated with confidence.
- The risk management plan has been updated based on the satisfactory/unsatisfactory results of applying these evaluation criteria.

A large system may be decomposed into subsystems to avoid complexities created by excessive size; the architecture evaluation criteria in Section 8.4 should be applied to all subsystems at all levels of the functional system hierarchy, including the lowest level subsystems. The system and subsystem design teams should include systems engineers (both physical and software) who guide the design process and disciplinary subject matter experts who will provide expertise as needed. Application domain experts may also be needed.

8.6.2 Logical Design and Physical Design

Design definition includes logical design and physical design. Logical design is concerned with identifying system elements and defining the interfaces and other relationships among the system elements and interfaces to entities in the system environment, including flows of information, material, and energy both internal and external. Initial logical design is usually initiated during architecture definition and elaborated during design definition. Physical design is accomplished during design definition. Physical design artifacts may include items such as mechanical drawings, electrical schematics, and database schema.

Software design is primarily logical design but software elements may also be involved in physical design because physical elements such as analog-to-digital and digital-to-analog converters and the associated software may be needed to provide the interfaces to and interactions with physical system elements and external physical entities.

Nondevelopmental items (NDIs) are identified as candidates for incorporation into a system during system design. NDIs include commercial-off-the-shelf (COTS) items, reuse of relevant design elements, and acquirer-provided items. Legacy elements from a previous system may be identified for incorporation during system implementation. Sources of NDIs are determined, suitability for intended usage is evaluated, and procurement activities are initiated for selected NDIs. However, risks must be assessed and caution exerted when deciding to "buy or build" – Section 8.6.4 describes tradeoff considerations for buying or building. The following subsection summarizes the activities of design definition.

8.6.3 Activities of Design Definition

Table 8.2 summarizes the inputs, processes, and outputs of design definition for the I^3 system development model.

The inputs, processes, and outputs of design definition presented in Table 8.2 are elaborated as follows:

1. Stakeholders' requirements, system capabilities, system requirements, and system architecture are initially reviewed and revised as appropriate and then iteratively revised as needed during the integrated-iterative-incremental implementation subphases of the I^3 system development model.

Table 8.2 Design definition inputs, process, and outputs.

Inputs	Stakeholders' requirements, system capabilities, system requirements, system architecture, and constraints (technical and managerial)
	A preliminary plan for packaging, delivery, transition, and system validation of the deliverable system (or any subsets of the final system)
Process	The design definition activities described as follows are accomplished
Outputs	An initial system design definition, verified and validated
	An initial implementation plan for the I^3 system increments
	An initial plan for deliverable-system verification, packaging, delivery, transition, and validation

2. The system architecture and design definition are decomposed and revised to include additional system elements as necessary.
3. Behavioral scenarios are enacted to identify potential performance problems and other behavioral complexities (see Section 8.3.2.1).
4. Design artifacts and technologies initially identified during evaluation of candidate architectures are reviewed, revised, and expanded as necessary.
5. Buy-or-build trade studies are conducted.
6. Infrastructure needed for system realization is identified.
7. Internal and external interconnections and interfaces are designed and documented.
8. One or more design reviews are conducted; areas of complexity and uncertainty are identified.
9. Risk factors are identified and risk mitigation strategies are developed.
10. The number of I^3 incremental development subphases and the capabilities to be implemented during each subphase are initially specified (but may be revised later).
11. Detailed plans for design and realization of each system increment are made during the planning activity for each increment (see Figure 8.1).
12. Comprehensive design documentation is prepared initially and revised incrementally.
13. The initial system design is verified (both initially and incrementally) to determine the degree to which the design satisfies the preceding work products (stakeholders' requirements, system capabilities, system requirements, and system architecture). The system design is revised to ensure adequate coverage of those work products.
14. The initial system design is validated to determine the degree to which the design documentation will be useful for the intended users of the documentation. The documentation is revised as needed to ensure adequacy.
15. Detailed estimates of effort, cost, schedule, infrastructure, resources, and technologies needed for the technical activities of system implementation,

verification, transition, validation, delivery, and installation (and perhaps for operating and maintaining the system) that were determined during previous development phases are reviewed, revised, elaborated, and accepted.
16. Plans for deliverable system verification, transition, validation, delivery, and installation are finalized.

A system design definition and the resulting system realization can be compromised by failure to accomplish one or more of the indicated design definition activities.

8.6.4 Buying or Building

Procuring system elements versus fabricating physical elements or constructing software elements is a "buy or build" decision when system elements are available for procurement. The advantages of buying are sooner availability (perhaps) and the known quality attributes of proven units (perhaps). The disadvantages of buying are the following:

- Locating sources for the needed elements;
- Evaluating the functionality and quality attributes of available elements;
- Locating and evaluating may take more time than fabricating or constructing;
- Cost may be excessive as compared with fabrication (perhaps);
- Candidate elements may provide unwanted capabilities that will consume resources and must be masked to prevent unintended use;
- Including candidate elements may provide undesired emergent system capabilities that must be dealt with;
- Interfaces must be built to provide interconnections of the procured system elements to fabricated and constructed elements, and to the system environment;
- Building interfaces and V&V may be inhibited by lack of sufficient supporting documentation for the procured elements; and
- Failing to find acceptable candidate elements for procurement may result in descoping the requirements to match candidate elements that are available or fabrication of those elements.

These considerations may negate the potential benefits of buying rather than building one or more of the system elements.

The advantages of fabricating or constructing instead of procuring (building instead of buying) overcome some of the disadvantages of procuring:

- Locating and evaluating available elements is eliminated;
- Uncertainty about functionality, behavior, and quality attributes is reduced; and

- Requirements to be satisfied by the needed elements can be built into the system elements.

The primary disadvantages of building are the following:

- Resources and engineers need to build the system elements; and
- Time and expense to build compared with buying (perhaps).

Additional considerations for procuring (buying) software elements are the following:

- The soundness of the supplying organization from which software elements will be procured;
- The compatibility of updates the supplying organization may make to acquired software elements; and
- The ability of the acquirer who buys to make updates to the acquired software.

In some past cases, acquiring organizations have required the supplying organization to place in escrow a copy of the source code for the acquired software to be made available to the acquirer in case the supplying organization releases incompatible upgrades, discontinues supporting the software, is acquired by another organization, or goes out of business.

Supplying organizations are understandably reluctant to provide to the acquirer the source code for a software package. Most organizations that sell or license software to others are also hesitant to make changes to accommodate an acquirer. This is a major consideration that may tip the balance to construction rather than procurement of software to be incorporated into a system.

8.6.5 Verifying and Validating the System Design

Whether buying or building, or both, the system design definition must be verified to determine that the design definition adequately covers the selected architecture definition, system requirements definition, system capabilities, and stakeholders' requirements definition. The system design must also be validated to determine the extent to which the system design models and documentation will satisfy the needs of the intended users of the models and documentation; intended users of design documentation include the disciplinary and specialty engineers who will implement the system elements and interfaces, those who will verify and validate implemented system increments, those who will verify the deliverable system or deliverable subsets, those who will transition and validate a deliverable system in the operational environment, and those who will operate and maintain a delivered system.

Techniques for verifying a system design definition include the following:

- Traceability of coverage;
- Formal analysis of parametric constraints;
- Verification scenarios;
- Planning for deliverable system verification;
- Simulation;
- Prototyping;
- Inspections; and
- Reviews.

Techniques for validating a system design definition include the following:

- Reviews;
- Inspections;
- Analysis;
- "What-if" scenarios; and
- Planning for system implementation.

The accompanying box provides information on inspections and reviews (Figure 8.4).

Reviews and Inspections

Reviews and inspections are team efforts that can be applied to all work products of system development but are especially important for architecture and design. They can be elements of the 15288/12207 analysis process (see Clause 6.4.6, see also Fairley 2009, pp. 291, 292, and appendix 7B).

The following checklist includes issues to be investigated during an architecture or design review or inspection.

- *Architectural style*: Is there a better architectural style for this system?
- *Structure*: Can the architectural style be improved? If so, how?
- *Level of detail*: Is there too little or too much structural decomposition? If so, how can it be improved?
- *Internal interfaces and external connections*: Are any internal interfaces or external connections incorrect, incomplete, or missing? If so, what improvements are needed?
- *Diagrams, models, and documentation*: Are they appropriate and sufficient to define architectural structure and behavior? Will they provide for the needs of those who will use the architecture and design diagrams and models?
- *Inappropriate information*: Is inappropriate information such as schedule, budget, staffing, or implementation details included in the architecture and design definitions?

(Continued)

(Continued)

- *Unnecessary information*: Is information not needed to understand the architecture and design included?
- *Other*: Are there other issues to be addressed that are not included in this checklist?

Inspections and reviews differ in the following ways.

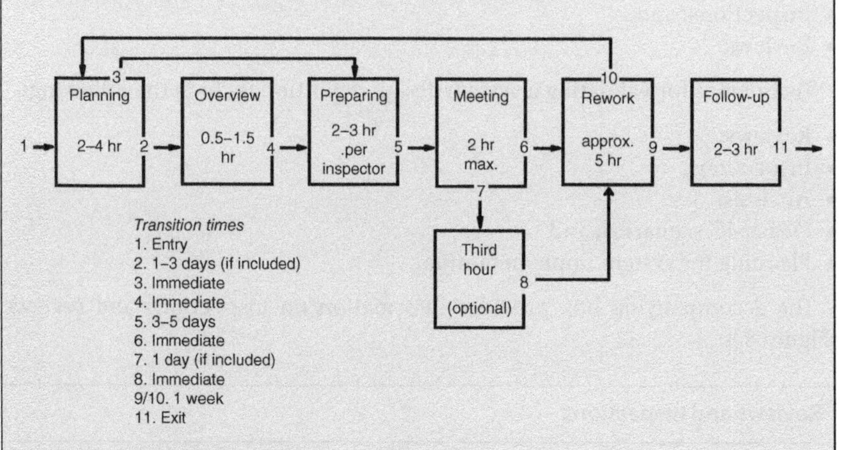

Figure 8.4 The inspection process.

A review is informal and can include any number of participants. An inspection is a formal process that includes a limited number of qualified and trained participants. Attendance is limited to not more than five or six participants.

Inspections include the following roles: a moderator, a presenter, a recorder, and one or more authors of the material being reviewed. The author cannot be the presenter (to provide objectivity) but the author can be the recorder (i.e. the note taker), so two is the minimum number of participants in an inspection – a moderator/presenter and an author/recorder. A checklist of items to review is used. Preparation is required. The maximum number of participants is limited to five or six to allow time for everyone to participate.

A review leader or leaders are informal moderators and presenters who are usually not trained in how to conduct reviews. Checklists are usually not used. Review notes may or may not be taken and the authors of the material may or may not be present. The number of participants in a review is usually not restricted.

Inspections include documented time spent in preparing for the inspection, the amount of time spent during the inspection, and the follow-up time spent. Checklists are used and forms are provided for recording issues that need further attention. Reviews may or may not include preparation time, checklists, and forms.

Action items are generated during inspections; they are assigned to individuals and tracked to closure. Reviews may or may not generate action items that may or may not be tracked to closure.

Documented results of inspections recently conducted are reviewed to identify recurring kinds of issues that may need systematic improvement and to determine ways in which the inspection process can be made more efficient and more effective.

8.7 RC-DSS Architecture Definition

The RC-DSS was introduced in Chapter 6. This section continues the RC-DSS case study presented in Chapters 6 and 7.

Various kinds of SysML diagrams that can be used to define the RC-DSS architecture (structure and behavior) are illustrated here.

A functional block diagram, depicted in Figure 8.5 as a SysML package that contains other SysML packages illustrates the five subsystems of the RC-DSS (Figure 6.1 repeated here as Figure 8.5).

Packages, **bdd**s, and **ibd**s are the primary SysML notations used to define system structure – SysML blocks are extended types of packages. A partially decomposed structure diagram for the architecture definition of the RC-DSS vehicle cabin is illustrated in Figure 8.6, depicted as a **bdd** and an associated **ibd** for the vehicle cabin.

As indicated in Appendix 8.A, bdds are black box representations of a system's structural elements. Ibds are white box representations; they provide the internal details of bdds. Ibds can also be used as separate from bdds, if desired.

The outer package diagram and the package type designator (i.e. [package]) in the bdd can be omitted and understood by default.

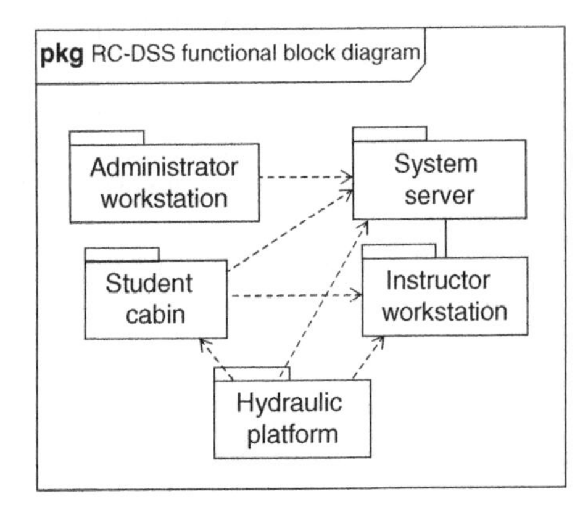

Figure 8.5 RC-DSS functional block diagram.

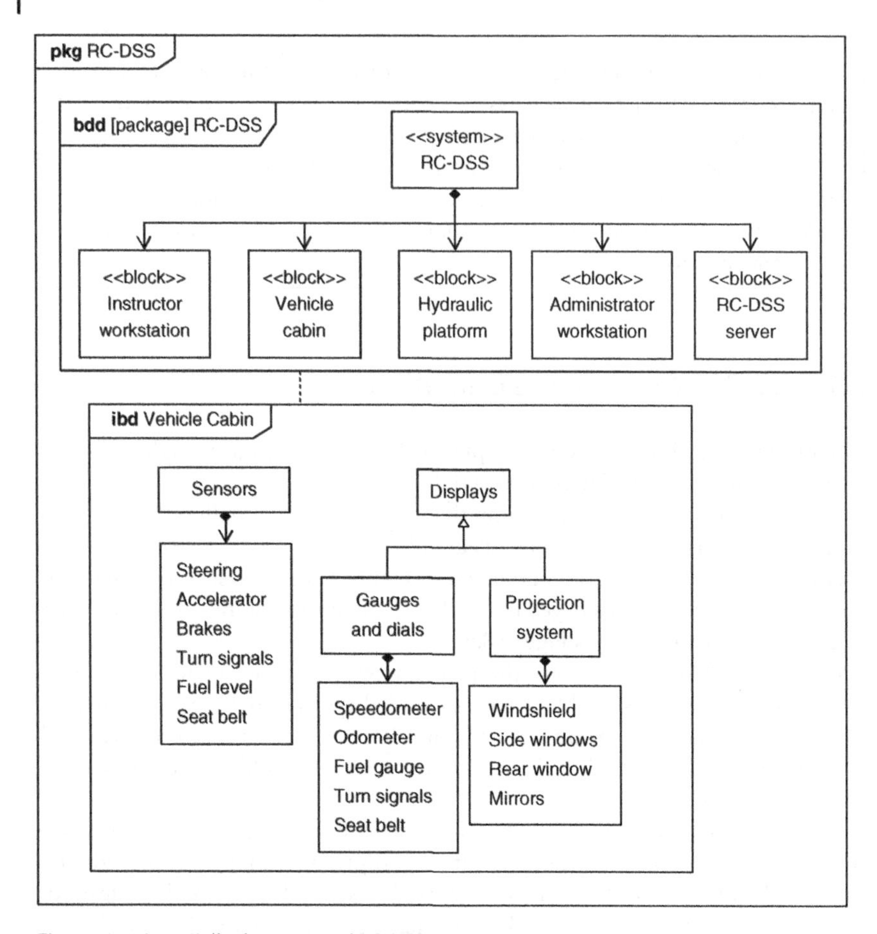

Figure 8.6 A partially decomposed RC-DSS system structure.

As stated above, the SysML behavioral diagrams are sequence, activity, state, and use case diagrams.

Figure 8.7 is a depiction of a SysML sequence diagram for authentication of an RC-DSS driving session.

Additional notations for sequence diagrams are illustrated and explained in Appendix 8.A.

Figure 8.8 is a SysML activity diagram. It provides additional scenarios for driving session authentication in addition to those in the sequence diagram. A sequence diagram can include only one scenario; three sequence diagrams would be needed to depict the scenarios in the activity diagram.

The activity diagram in Figure 8.8 does not include the fork and join concurrency constructs of activity diagrams; they are illustrated and explained in Appendix 8.A.

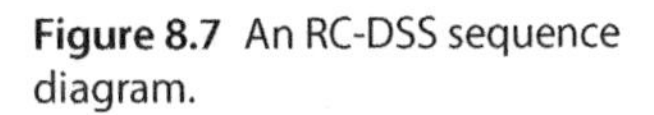
Figure 8.7 An RC-DSS sequence diagram.

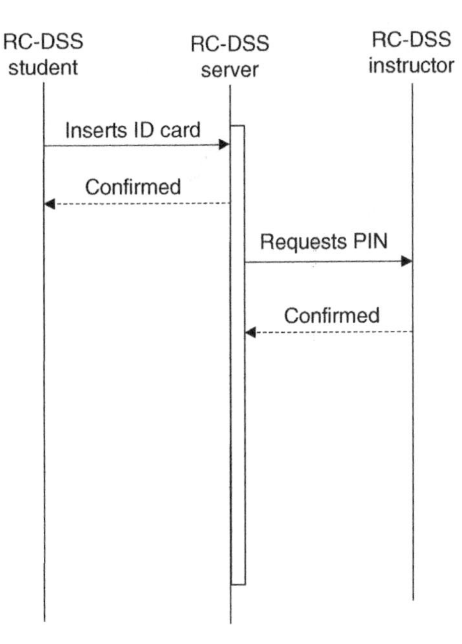

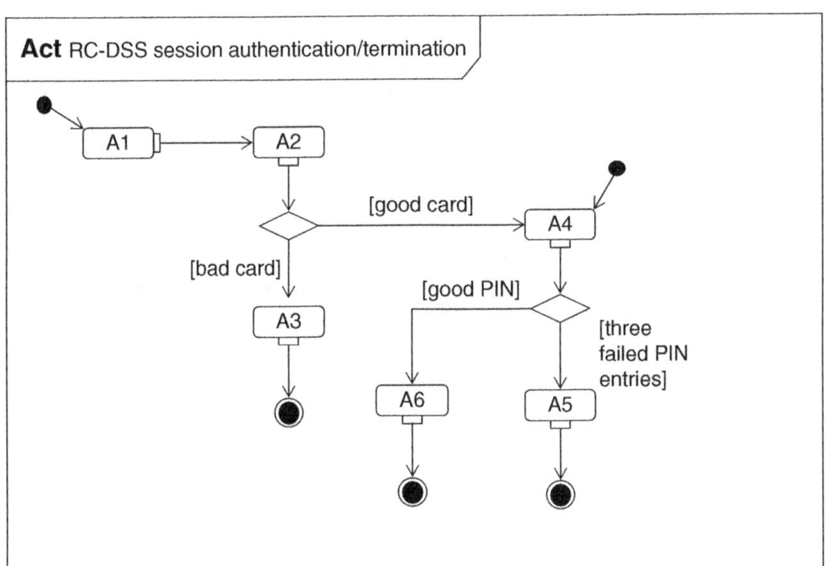

A1: Sturent inserts ID card
A2: System checks card
A3: Bad card ejected and session terminates
A4: Instructor enters PIN
A5: System shuts down after failed PIN entries
A6: System enters include nested state then terminates after the driving session

Figure 8.8 An RC-DSS activity diagram.

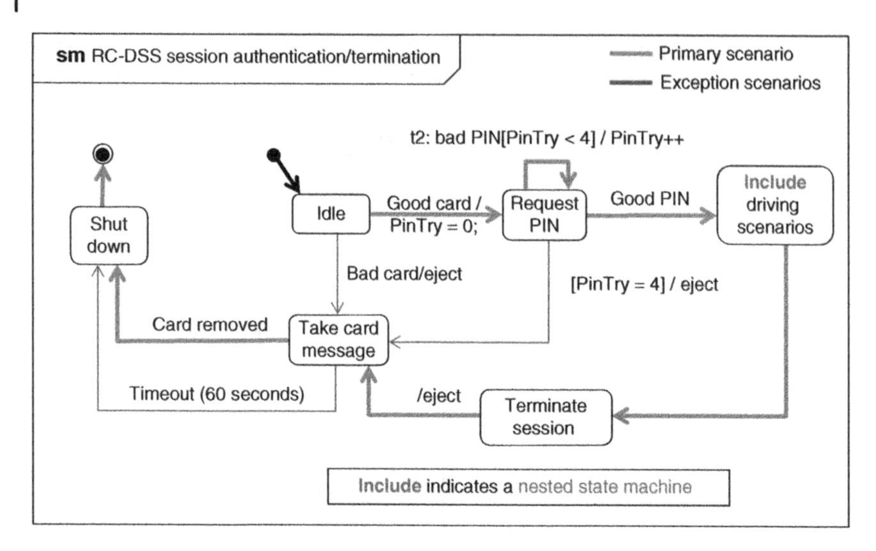

Figure 8.9 An RC-DSS state diagram.

An RC-DSS state diagram is illustrated in Figure 8.9.

The state diagram illustrates the primary scenario and the exception scenarios for RC-DSS authentication and termination, which might be indicated in an RC-DSS use case. The included state indicates embedded states for the driving scenarios. Observe that the state diagram illustrates multiple scenarios; a sequence diagram can depict only one scenario. Also note the state diagram does not indicate that the instructor enters the authentication PIN as do the sequence diagram and activity diagram.

Figure 8.10 illustrates an RC-DSS use case diagram that includes two actors and three use cases that depict a student enrolling for RC-DSS driving sessions.

The registrar (a secondary actor in a use case) assists the student as necessary. The ≪include≫ stereotype denotes a use case that is used to notify the student's sponsor the student has successfully enrolled or has encountered problems. The

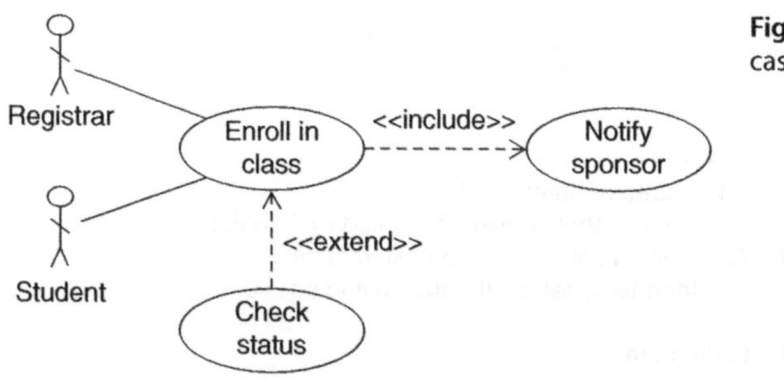

Figure 8.10 An RC-DSS use case diagram.

≪extend≫ stereotype is only used to check the student's enrollment status if problems are encountered.

More information concerning SysML diagrams is provided in Appendix 8.A.

8.7.1 RC-DSS Architecture Evaluation

Evaluation criteria for RC-DSS architecture options include the following:

1. A safe and secure environment for all hands-on RC-DSS users;
2. A realistic driving experience for student drivers;
3. A system architecture that will support a product line of simulators for various kinds of land-based vehicles having four or more wheels; and
4. A cost-effective implementation.

Architecture options for the DSS include the following:

A) A laptop-based joystick with visual display and audio output;
B) An unenclosed driver's seat, steering wheel, status display panel, and minimal controls and indicators (speedometer, accelerator, brakes);
C) A realistic enclosed representation of an automobile; and
D) A retrofitted flight-training simulator.

Table 8.3 indicates architecture options for the RC-DSS and the criteria satisfied by those options.

As noted in the table, evaluation criterion 1 could be satisfied by all four architectural options with appropriate design choices. Architectural options A and B are rejected because they do not satisfy criteria 2 and 3; criteria A and B could satisfy evaluation criteria 1 and 4 but they would not provide a realistic driving experience for student drivers and would not accommodate a product line architecture. Architectural option D is rejected because it does not satisfy criteria 2 and 4; sitting in the cockpit of a flight simulator would not provide a realistic experience of driving a land-based vehicle that has four or more wheels and the cost of acquiring and reprogramming a flight simulator would be prohibitive.

Architectural option C (a realistic enclosed representation of an automobile) satisfies evaluation criteria 1, 2, and 3; it satisfies criterion 4 in comparison to

Table 8.3 RC-DSS architecture options and criteria satisfied.

Criterion	1	2	3	4
Option				
A	X			X
B	X			X
C	X	X	X	X
D	X		X	

acquiring and retrofitting a flight simulator. However, using the flight simulator as an analogy, an architectural solution would be to mount realistic vehicle cabins on a hydraulic platform to provide the realistic feel of driving various vehicles on various road surfaces under various weather conditions. Modular and interchangeable real vehicle cabins will support product-line architecture.

8.7.2 RC-DSS Interface Definition

The interconnections and interfaces among system elements can be represented in a N^2 diagram (sometimes depicted as an $N2$ diagram), where N is the number of elements, as depicted in Table 8.4.

It is apparent from Table 8.4 that the system server provides output connections to the other four system elements (row 1) and that the other four elements provide input connections to the primary computer (column 1). In addition, the vehicle cabin has output connections from the hydraulic platform and the instructor workstation (row 2) and has input connections to the hydraulic platform and the instructor workstation (column 2). In addition to the system server connections, the hydraulic platform has an output connection to and an input connection from the primary vehicle cabin and the instructor workstation (row 3 and column 3). The administrator workstation has an output connection to and an input connection from the primary computer (row 4 and column 4). The instructor workstation has an output connection to and an input connection from the vehicle cabin in addition to the primary computer connections (row 5 and column 5).

Table 8.4 N^2 diagram for the RC-DSS system elements.

Interfaces to the RC-DSS server (E1)	Interfaces to the vehicle cabin (E2)	Interfaces to the hydraulic platform (E3)	Interfaces to the administrator workstation (E4)	Interfaces to the instructor workstation (E5)
E1: System server	E1 → E2	E1 → E3	E1 → E4	E1 → E5
E2 → E1	E2: Vehicle cabin	E2 → E3		E2 → E5
E3 → E1	E3 → E2	E3: Hydraulic platform		
E4 → E1			E4: Administrator workstation	
E5 → E1	E3 → E5			E5: Instructor workstation

Input → output.

Indirect connections are not shown on an N^2 diagram. For example, an instructor, using the instructor workstation, sends a request to the system server to send driving script signals to the hydraulic platform and the vehicle cabin. The instructor can also send driving intervention requests and receive responses. An N^2 diagram can be augmented with details of the interface connections during system design, if desired.

Note: N^2 diagrams are not included in SysML but are widely used in systems engineering.

Interchangeable vehicle cabins are chosen to support a product line of driving system simulators. The buy or build decision of buying and retrofitting real vehicles or building modular vehicle cabins will be made during the design definition phase of RC-DSS system development.

Design considerations for the vehicle cabin and other elements of the architecture depicted in Table 8.3 are presented in Section 8.8.

8.8 RC-DSS Design Definition

Architecture definition of the RC-DSS driving simulator is based on architecture definition Option C selected above and illustrated in Figure 8.10. The primary system elements are an enclosed vehicle cabin, a hydraulic platform, an instructor's workstation, an administrator's workstation, and a system server. Buy-or-build analysis determines that the server, hydraulic platform, and the instructor and administrator workstations can be purchased and that system architecture based on procuring these system elements will support a product line of driving system simulators by using interchangeable vehicle cabins. Review of the issues in Table 8.2 concerning potential downsides of buying these system elements does not reveal any significant risk factors.

However, interfaces among the systems elements will have to be developed and it has been determined that equations of motion for the hydraulic system control algorithms must be developed because a purchased hydraulic platform includes only basic control signals for the six degrees of platform movement. A control systems engineer (i.e. a specialty engineer) will be needed.

Options for enclosed vehicle cabins include the following:

1. Constructing vehicle cabin modules; and
2. Retrofitting real vehicle cabins.

Analysis has indicated that purchasing real vehicles and retrofitting the cabins with interfaces to the other system elements will have similar (perhaps less) cost compared with constructing cabin modules and will not require the facilities, tools, and construction workers needed to construct the cabin modules. Also, "buying plus retrofitting rather than building" will result in shorter development cycles while providing authentic vehicle interiors and instrumentation for student drivers. However, real vehicles will have to be instrumented to send and receive signals to and from the other subsystems, as indicated in Figures 8.5 and 8.6 and Table 8.4.

Additional design details for the vehicle cabin could be shown in the cabin ibd. Ibds could also be included for the other four system blocks.

As shown in Figure 8.6, instrumentation for the real vehicle cabin has been decomposed into its major elements: sensors and displays. Sensor data will be processed by the system server and relayed to the displays, the instructor workstation, and the hydraulic platform. Displays include the vehicle gauges and dials plus a 360° display of simulated driving environments projected through the instrumented windshield, side windows, rear window, and mirrors.

Design of the vehicle cabin instrumentation includes design of software interfaces that can be adapted to permit connections to the other system elements and internally to interconnect the sensors, displays, and projection system for the various kinds of vehicle cabins that may become RC-DSS systems.

A detailed design decision that has significant consequences is whether to reprogram a vehicle's ECUs to control the sensor and display indicators or whether to bypass the ECUs and instrument the sensor and display indicators with analog/digital (A/D) and digital/analog (D/A) converters and use them to sense and control the various indicators. It is determined that the latter design decision will be made because it may not be possible to obtain detailed ECU information from the vehicle manufacturers and the configurations of ECUs will likely differ for different kinds of vehicles. These considerations could present inhibitors to developing a product line of different RC-DSS configurations for different kinds of vehicles.

The design definition should be decomposed in sufficient detail to provide information needed to prepare the implementation plan, which will include the kinds and numbers of engineers and the infrastructure needed to implement an initial version of an RC-DSS. Details for preparing the implementation plan are provided in Chapter 10.

8.9 Controlling the Complexity of System Architecture and System Design

Complexity is the Achilles Heel of system development; simplicity is the antithesis of complexity. According to a quote attributed to Albert Einstein:

> Everything should be as simple as possible, but not simpler.

Simple is not the same as simplistic; simplistic system architecture and design definitions may be appealing but will not satisfy design parameters and other system evaluation criteria. The Einstein's quote expresses a design heuristic known as Occam's Razor. "Complexity" provides some background information on Achilles Heel and applying Occam's Razor to development of complex software-enabled physical systems.

Complexity

Achilles Heel

As told in ancient writings, Achilles was a Greek hero in the Trojan war. He accompanied Agamemnon to capture the city of Troy and gain the return of Helen to her husband Menelaus; a Trojan named Paris had abducted her. Helen had "the face that launched a thousand ships" according to Christopher Marlowe in his play Doctor Faustus, written in the late 1500s (Marlowe 2005).

There are various versions of the myth but all versions attribute Achilles' death to being shot by an arrow that severed his tendon that connects the calf muscle to the foot (the Achilles tendon). He was rendered helpless and was slain by the Trojan Paris. In one version of the myth, Achilles feet were the only part of him that was not protected by armor; in another version, Achilles was invulnerable in all of his body except for his heel because, when his mother dipped him in the river Styx as an infant, she held him by one of his heels that did not get immersed.

The term "Achilles Heel" is used colloquially to denote a weak point. It is used in system development to denote vulnerabilities in architecture and design caused by complexity that results in problems for system implementation, verification, validation, operation, and sustainment.

Occam's Razor

Occam was an English Franciscan friar, philosopher, and theologian who lived in the late thirteenth and early fourteenth centuries. One interpretation of Occam's Razor is that the simplest solution that solves the problem is to be preferred (but not a simplistic solution). In science, Occam's Razor is interpreted to mean the simplest hypothesis that explains the phenomenon under investigation should be accepted until new evidence requires reexamination of the theory and experimentation that confirms or rejects the hypothesis. Scientific elegance is achieved when the minimum amount of data to answer a question is used.

Occam's Razor is used in system development to metaphorically cut away complexity and reject unnecessarily complex architectures, designs, and implementations. Design elegance is achieved when the minimum number of system elements and interconnections will realize an objective.

Complexity makes systems difficult to design and implement, difficult to verify and validate, difficult to document, difficult to understand, difficult to evaluate, and difficult to modify.

Complex systems result from the following:

- Requirements that are incorrect and/or incomplete, and/or not consistent, and/or not concise, and/or unclear (i.e. do not satisfy the five Cs of requirements engineering: correct, complete, consistent, concise, and clear);
- Undocumented or inadequately documented system architecture;
- Undocumented or inadequately documented system design;
- Undocumented or inadequately documented system interfaces and connections (internal and external);
- Frequent, uncontrolled stakeholders' change requests;
- Undocumented legacy elements that must be incorporated;
- Policies, rules, and regulations that must be satisfied; and
- Lack of supporting documentation for stakeholders' requirements, system capabilities, system requirements, architecture, design, realization, operational concepts, traceability, and revision history.

Inadequate staffing, lack of needed skills, and/or insufficient time for proper system development are often the root causes of complexity.

Parsimony is the antidote for system complexity; parsimonious architectures and designs are simpler than the alternatives without being simplistic. As used in medicine, an "antidote" does not provide a cure but is used to counteract adverse physiological affects of a poison – Achilles didn't have an antidote to counter the effects of the poisoned arrow that killed him.

Parsimonious system design does not guarantee a cure for system complexity but it can counteract complexity.

Techniques of parsimonious design include the following:

- Abstraction;
- Composition;
- Containment;
- Encapsulation;
- Information hiding;
- Use of design patterns;
- Judicious naming of system elements;
- Hierarchical and associative relationships; and
- Multiple views of system architecture and design.

Abstraction is concerned with identifying and focusing on the major system elements and the interconnections among them while temporarily disregarding design details. Logical system design is conducted at an abstract level. The **bdd** of SysML provides a black box abstract view of a subsystem or system element (see Figure 8.A.2). Composition can be indicated using the bdd and ibd constructs of SysML in which system elements that are aggregated by the composing element "belong to" that element.

The package construct of SysML supports containment, encapsulation, and information hiding, but not composition (see Figure 8.2). A package can be used to group and contain any desired set of system elements, including other packages. Encapsulation prescribes boundaries around system elements. Information hiding conceals the internal details of system elements; visible interfaces provide access to the encapsulated and hidden design details and provide connections to other system elements. Package interfaces can be used to provide the connections among packaged system elements. Some packages may encapsulate physical elements with either physical or software interfaces and other packages may encapsulate software code and other cyber elements.

Design patterns are templates that specify solutions to commonly occurring problems that can be applied within specified contexts. Design patterns are based on experiences and good practices that provide simple and elegant approaches to controlling system complexity (Gamma et al. 1994).

Judicious naming provides names for system elements that are intuitive and informative. "Dynamo controller" is a more informative name than "D24701."

Hierarchical relationships among elements provide a top-down "divide and conquer" relationship among system elements in which lower levels of the hierarchy are parts of higher levels. Associative relationships are indicated for elements that are on the same level of abstraction (e.g. are at the same level in a hierarchy).

Multiple structural and behavioral views of system architecture and design are used because system architectures are usually too complex to be understood from a single architectural or behavioral view. As indicated above, SysML provides three structural diagrams and four behavioral diagrams that are used to define views of system architecture and design (see Figure 8.A.1).

8.10 Key Points

- System architecture defines system structure and system behavior.
- Multiple views of structure and behavior are often needed to understand a complex system.
- Two or more architectural alternatives should be defined and evaluated using predetermined evaluation criteria.
- Design definition provides the bridge from system architecture to system implementation.
- System design consists of logical design and physical design.
- Buy or build decisions for system elements should be evaluated in terms of risks and benefits of both options.
- Rationale for the selected architecture and design should be recorded for the selected architecture and design and for the rejected ones.

- SysML provides nine kinds of diagrams that can be used to define system architecture and system design.
- The RC-DSS case study provides examples of system architecture definition and system design definition.
- Complexity is the Achilles Heel of system development. Occam's Razor provides a metaphor for simplifying system architecture.
- Architecture and design should be as simple as possible, but not simplistic.
- Well-known techniques of system design can be applied to control system complexity.

Exercises

8.1. Subclause 6.4.4.1 of ISO/IEC/IEEE Standards 15288 and 12207 state that one of the purposes of architecture definition is to generate alternative system architectures. Briefly explain why it is important to generate more than one system architecture.

8.2. Briefly explain the differences between system structure and system behavior.

8.3. System architecture might be routine, innovative, or revolutionary.
(a) Briefly describe a system that probably has a routine architecture and explain why the architecture is probably routine (a system not mentioned in the text).
(b) Briefly describe a system that probably has an innovative architecture and explain why the architecture is probably innovative (a system not mentioned in the text).
(c) Briefly describe a system that probably has a revolutionary architecture and explain why the architecture is probably revolutionary (a system not mentioned in the text).

8.4. Briefly explain why simple system architecture is preferable to complex system architecture. Then briefly explain difference between a system architecture that is simple and one that is simplistic.

8.5. Briefly explain why architecture evaluation criteria should be prioritized.

8.6. Briefly explain the desired outcome of verifying system architecture.

8.7. Briefly explain the desired outcome of validating system architecture.

8.8. Briefly explain why using SysML diagrams (and others) are preferable to writing textual statements. Then briefly explain why diagrams alone are not sufficient to document system architecture and design.

8.9. SysML includes three structure and four behavior diagrams. Briefly explain why SysML includes more than one structure diagram and more than one behavior diagram.

8.10. ISO Standards 15288 and 12207 include separate processes for architecture definition and design definition (Clauses 6.4.4 and 6.4.5).

(a) Briefly explain why it is important to separately develop architecture definition and design definition.

(b) Briefly explain why it is important to develop architecture definition and design definition iteratively and concurrently.

8.A The System Modeling Language (SysML)

The SysML provides diagrams that can be used in various combinations to define system requirements, system architectures, and system designs. SysML is a subset and an extension of the Unified Modeling Language (UML 2017).

The leaf nodes of the taxonomy in Figure 8.A.1 depict the nine types of SysML diagrams. The open triangle notation indicates that the constructs at the tails of the lines are specializations of the constructs at the heads of the triangles.

As indicated in the figure, SysML includes three types of diagrams for defining system structure, four types of diagrams for defining system behavior, and two new types of diagrams (requirements and parametric diagrams) that are not included in UML (Friedenthal et al. 2009; SysML 2017, 2018). The open

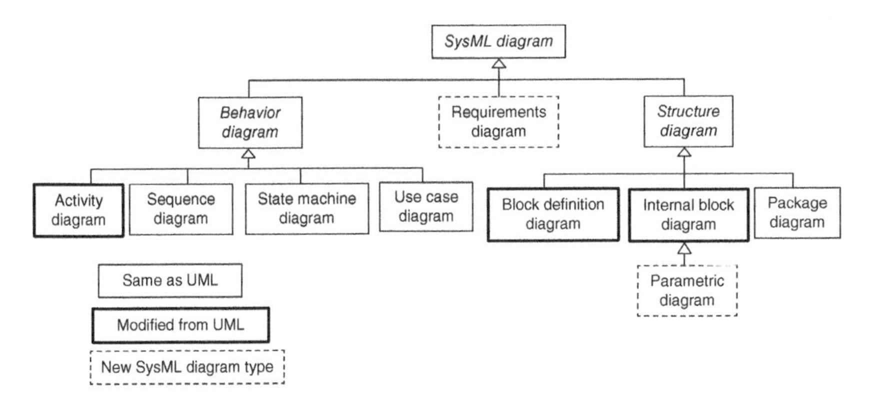

Figure 8.A.1 Nine types of SysML diagrams. Source: SysML (2018).

triangle notation in the figure indicates that the constructs at the tails of the lines are specializations of the constructs at the heads of the triangles.

Some diagrams familiar to systems engineers (functional flow block diagrams, N^2 charts, Ishikawa diagrams, and others) are not included in SysML. They are also used to define system requirements, architecture, and design.

8.A.1 SysML Structure Diagrams

As indicated in Figure 8.A.1, three types of SysML diagrams are used to define system structure:

- Package diagrams;
- Block definition diagrams; and
- Internal block diagrams.

The *package diagram* is unchanged from UML; it is the basic mechanism used to enclose SysML diagrams in various containment groupings. A SysML package can contain any arbitrary grouping of system model diagrams, including other packages. All system models defined using SysML are contained in an outer package, although the outer system package may not be explicitly included; it is understood by default. Figure 8.5 provides an example of a package diagram that contains other packages. It is the functional block diagram for the RC-DSS case study.

Dashed lines with open arrowheads, as in the Figure 8.5, define dependency relationships in SysML; the system elements at the tails of the lines are dependent on the elements at the heads of the lines. As depicted in Figure 8.5, four of the subsystems are dependent on the RC-DSS server. The hydraulic platform is dependent on the server and the instructor's workstation. The vehicle cabin is dependent on the server, the instructor's workstation, and the hydraulic platform.

Block diagrams are extended types of the UML/SysML package. They and packages are fundamental units of system structure in SysML. Block diagrams can be used to compose the functional-hierarchical structure of a system, or a collection of subsystems. A template for a bdd is illustrated in Figure 8.A.2.

The "sub" elements in Figure 8.A.2 may be subsystems or major system elements to be separately developed. The diamonds at the tails of the arrows indicate that the system is composed of the elements at the heads of the diamonds. The filled black diamonds indicate a "composition relationship" (aka "strong aggregation"). White filled diamonds would indicate a "reference relationship" (aka "simple aggregation"); it indicates that a subordinate element is part of its superior block but exists independent of the superior block. For example, a power supply could be a simple aggregate of a system (i.e. part of but existing

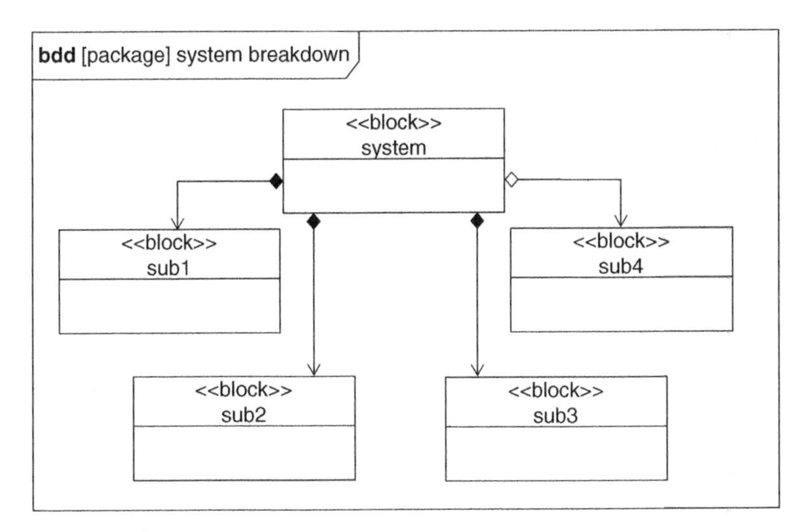

Figure 8.A.2 A template for a block definition diagram.

independent of the system). A SysML block structure diagram may include both black filled and white filled diamonds (SysML 2018).

ibds provide the internal details of bdds; they can be standalone or attached to the bdds for which they provide details. Bdds are black box representations; idbs are white boxes.

Figure 8.6 illustrates a bdd with an associated ibd for the RC-DSS case study. The containing package and the [package] type extension can be omitted from bdds, as shown in Figure 8.6, but if omitted are understood by default. A block must have a name and can encapsulate properties, operations, and constraints; all three are optional. Blocks can include provided and required interfaces, as desired, for both information and physical flows. Blocks can be recursively decomposed into other blocks.

Various forms of lines are used to connect the elements in and among pkgs, bdds, and ibds. They include but are not limited to associations of various kinds, containment, composition, dependency, and generalization/specialization (SysML 2017).

8.A.2 SysML Behavior Diagrams

The four behavior diagrams in SysML are the following:

- Sequence;
- Activity;
- State machine; and
- Use case.

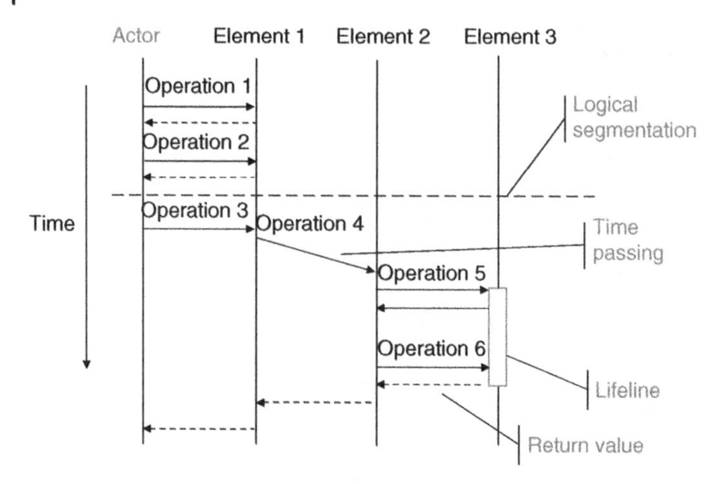

Figure 8.A.3 Template for a sequence diagram.

As indicated in Figure 8.A.1, the sequence, state machine, and use case diagrams are unchanged from UML. Activity diagrams are extended to allow flow of material and energy, in addition to data flows as in UML.

Sequence diagrams are often used to show external entities (called actors) that initiate sequences of interactions with system elements and receive responses. Sequence diagrams can be used to document scenarios in use cases. They can also be used to depict interactions among system elements. A template for a sequence diagram is illustrated in Figure 8.A.3.

As indicated in Figure 8.A.3, each participant has a "lifeline" of behavior through time that flows down the diagram. Arrows across the diagram depict the interactions among the participants. Messages and actions can be displayed on the arrows.

Figure 8.A.4 provides an example of a sequence diagram.

In the example, a hotel guest uses the telephone to request a wakeup call at 7:00 a.m. The wakeup call is received at 7:00 a.m. The wakeup application in the hotel server asks if the guest would like to hear the weather forecast. The guest replies "yes." The hotel server contacts the weather server. The weather forecast is returned to the hotel server and the hotel server returns the forecast to the hotel guest. Alternatively, the hotel wakeup application in the hotel server could be designed to connect the weather server directly to the hotel guest. In that case, the bottom two arrows in the example sequence diagram would be replaced with a single response arrow from the weather server to the hotel guest.

Activity diagrams are UML diagrams for which the concept of flow has been extended from data flow to include flows of material and energy, as desired. A template for an activity diagram is illustrated in Figure 8.A.5.

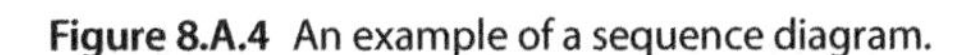

Figure 8.A.4 An example of a sequence diagram.

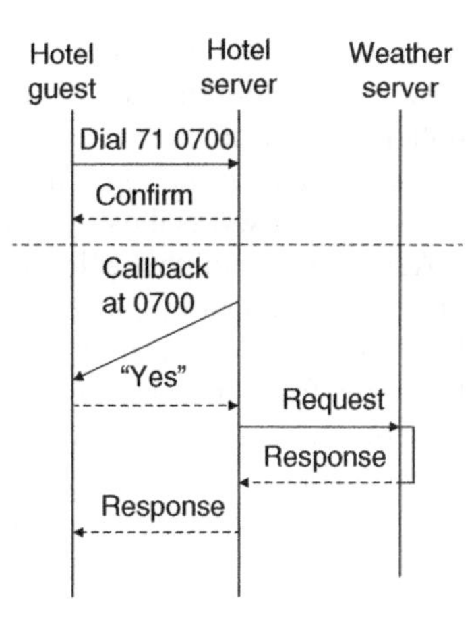

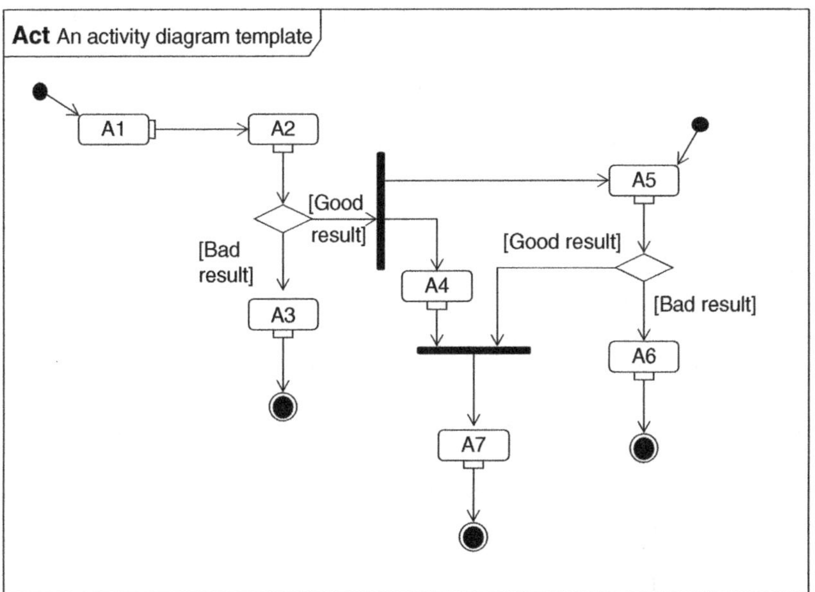

Figure 8.A.5 Template for an activity diagram.

As illustrated in Figure 8.A.5, activity diagrams are composed of nodes and connector lines. The lines, with arrowheads, indicate the direction of flow. The nodes in an activity diagram are actions that transform inputs into outputs, which in turn provide inputs to other nodes or deliver outputs. Activities can be documented using embedded activity diagrams, state diagrams, algorithms, equations, text, or any desired form of notation.

As illustrated, decision points can be used to direct flows. Concurrent flows can be initiated by the fork construct (as illustrated by the good result of activity

A2). Concurrent flows can be synchronized as illustrated by the join construct. Flow cannot continue to A7 until (or if) A4 is successful and a good result is obtained from A5.

A *state machine diagram* includes a set of states that are interconnected by transition arrows, as illustrated in Figure 8.9.

A system state might be "idle and waiting for input," "executing an algorithm," or "transforming an input into an output." Transformations while the system is in a specific state are called "activities." "Actions" can be specified during a transition from one state to another state; they are specified on the arrows. The state of a system at a point in time provides sufficient information to determine the next state at the next point in time.

The behavior of a system is determined by transitions through various connected states. Input stimuli can drive a system through a sequence of behavioral states, i.e. a scenario. State machine diagrams can include hierarchically nested state machines that can be recursively decomposed to any desired level. They can also be partitioned to depict concurrent behavior. Concurrent states can communicate; for example, a watchdog timer can trigger a reaction if a monitored state does not respond in a specified amount of time. State machine diagrams are based on the statechart notation developed by David Harel in the 1980s (Harel 1987).

Use case diagrams can be used to indicate the ways in which one or more actors interact with a set of use cases.

Figure 8.10 provides an illustration of a RC-DSS use case diagram that includes two actors and three use cases that depict a student enrolling for RC-DSS driving sessions. As indicated, the ≪include≫ and ≪extend≫ stereotypes are used to indicate relationships among the use cases.

The registrar (a secondary actor in a use case) assists the student as necessary. The ≪include≫ stereotype denotes a use case that is used to notify the student's sponsor the student has successfully enrolled or has encountered problems. Included user cases can be used whenever needed in conjunction with other use cases. An ≪extend≫ stereotype is used only in one use case; it can be thought of as an exception handler. In this example, ≪extend≫ is used to check the student's status if problems are encountered during enrollment.

These diagrams provide a high-level view of interactions between actors and use cases that are easily understood. A use case diagram can be thought of as a "visual table of contents" that can be linked to the use cases.

8.A.3 SysML Crosscutting Diagrams

Two SysML diagrams provide crosscutting views of system architecture:

- Parametric diagrams; and
- Requirements diagrams.

The views are crosscutting because they can be associated with structural or behavioral diagrams or can be used as standalone diagrams.

Parametric diagrams contain stereotyped blocks that are denoted by the ≪constraint≫ stereotype. Parametric blocks contain logical expressions and mathematical equations that constrain the parameters and other properties of the block with which they are associated.

A SysML requirements diagram (the ninth kind of SysML diagram) contains requirements text and icons denoted by the ≪requirement≫ stereotype. Icons can be included to illustrate a requirement or a collection of requirements. Each requirement diagram must include a name and some text. Other compartments in a requirements diagram can include modifiers such as derived, derivedFrom, refinedBy, and satisfiedBy. Comments such as "traced from" and "traced to" can be attached to requirements blocks using comment tags.

Requirement blocks can be organized in various ways to denote requirement hierarchy, derived requirements, and traceability links and to connect requirements to test cases and indicate other relationships among requirements. Various stereotypes are provided to denote the relationships. Requirements diagrams are crosscutting because they can include both structural and behavioral elements of a system and can be attached to all or part of a system model.

Stereotypes (≪xxx≫) are one of the three notations used to further define SysML constructs. The other two notations are tags (used to add comments) and constraints (used to state logical expressions and mathematical equations). SysML provides some additional predefined stereotypes; others can be locally defined.

References

Fairley, R. (2009). *Managing and Leading Software Projects*. Wiley.

Folmer, E., van Gurp, J., and Bosch, J. (2004). Software architecture analysis of usability. University of Groningen, the Netherlands. http://dl.ifip.org/db/conf/ehci/ehci2004/FolmerGB04.pdf (accessed 11 September 2018).

Friedenthal, S., Moore, A., and Steiner, R. (2009). OMG systems modeling language (OMG SysML™) tutorial. Object Management Group, September 2009. http://www.omgsysml.org/INCOSE-OMGSysML-Tutorial-Final-090901.pdf (accessed 3 July 2018).

Gamma, E., Vlissides, J., Johnson, R., and Helm, R. (1994). *Design Patterns: Elements of Reusable Object-Oriented Software*, 1e. Addison-Wesley Professional.

Harel, D. (1987). Statecharts: a visual formalism for complex systems. *Science of Computer Programming* 8 (3): 231–274. https://www.sciencedirect.com/science/article/pii/0167642387900359 (accessed 12 September 2018).

Hope, P., McGraw, G., and Anton, A. (2004). Misuse and abuse cases: getting past the positive. *IEEE Security and Privacy Magazine* 2 (3): 90–92.

ISO (2015). ISO/IEC/IEEE 15288:2015, Systems and software engineering – System life cycle processes. ISO/IEEE, 2015.

ISO (2017). ISO/IEC/IEEE 12207:2017, Systems and software engineering – Software life cycle processes. ISO/IEEE, 2017.

Kazman, R. (1999). Using scenarios in architecture evaluations. SEI Interactive, Software Engineering Institute. https://pdfs.semanticscholar.org/d05f/e163a11d4c8afd6a3e1c7e18d21b5bb400f5.pdf (accessed 12 September 2018).

Marlowe, C. (2005). *Doctor Faustus*. W.W. Norton.

Reisig, W. (2011). *Petri Nets: An Introduction*. Springer https://www.amazon.com/Petri-Nets-Introduction-Monographs-Theoretical/dp/3642699707 (accessed 3 July 2018).

SAE (2018). Taxonomy and definitions for terms related to driving automation systems for on-road motor vehicles J3016_201806. Society of Automotive Engineers, June 2018. https://www.sae.org/standards/content/j3016_201806/ (accessed 3 July 2018).

SysML (2017). OMG systems modeling language 1.5. Object Management Group. http://www.omgsysml.org/ (accessed 14 June 2018).

SysML (2018). What is SysML? Object Management Group. http://www.omgsysml.org/what-is-sysml.htm (accessed 31 July 2018).

UML (2017). The unified modeling language specification, Version 2.5.1. Object Management Group. https://www.omg.org/spec/UML/About-UML/ (accessed 13 June 2018).

9

System Implementation and Delivery

9.1 Introduction

System implementation occurs during Phase 5 of the integrated-iterative-incremental (I^3) system development model, which includes iterative-incremental subphases. During each subphase, software engineers can use iterative development processes, as desired, to construct and procure the software elements needed to realize the software elements of system capabilities and other disciplinary engineers can use incremental development processes, as desired, to concurrently fabricate and procure the physical system elements needed for the same system capabilities. Coordinated and concurrent implementation of software elements and physical elements needed to realize system capabilities is explained in the following sections.

System delivery occurs in Phase 6, which includes system transition from the development environment to the operational environment and system validation in the operational environment. The goal of system delivery is to gain acceptance of the delivered system by the appropriate system stakeholders.

Depiction of the I^3 system development model in Figure 5.2 is repeated here as Figure 9.1 for ease of reference. Phase 5 (implementation) and Phase 6 (system delivery) are highlighted in the figure.

This chapter provides readers with the opportunity to learn about iterative and incremental implementation of software-enabled physical systems. Key points are emphasized at the end of the chapter. References and Exercises are included to explore the issues in more detail.

9.2 I^3 Phases 5 and 6

The roles of systems engineers (both physical and software) during system implementation and delivery (I^3 Phases 5 and 6) are presented in Chapter 5. Their roles include coordinating system implementation among the implementation teams, maintaining the list of backlogged capabilities to be implemented,

Systems Engineering of Software-Enabled Systems, First Edition. Richard E. Fairley.
© 2019 John Wiley & Sons, Inc. Published 2019 by John Wiley & Sons, Inc.

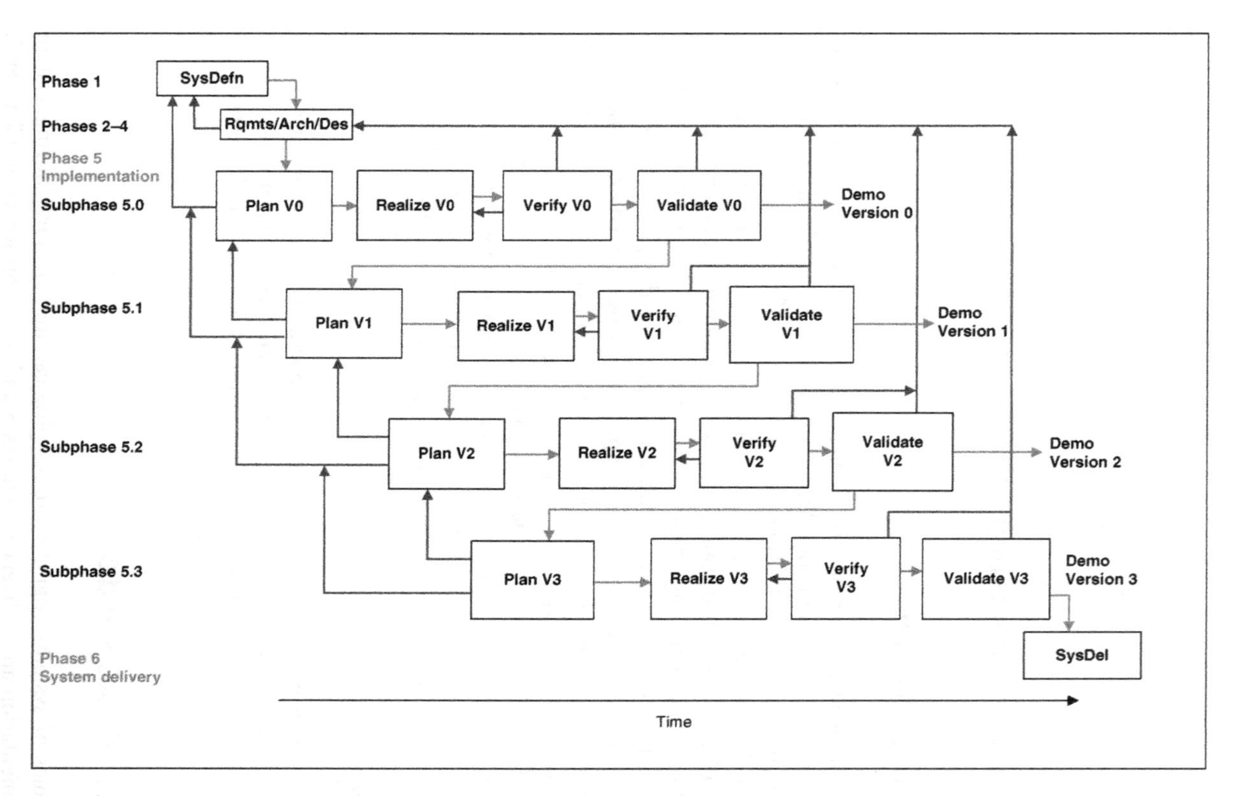

Figure 9.1 The I^3 system development model.

Table 9.1 I^3 phases and 15288/12207 processes.

I^3 development phases	15288/12207 Technical processes
Implementation	Implementation process (6.4.7) Integration process (6.4.8) Verification process (6.4.9) Validation process (6.4.11)
System delivery	Transition process (6.4.10) Validation process (6.4.11)

providing resources and tracking progress, evolving the system architecture and the system simulation model while ensuring the fidelity and integrity of the simulation, tracking progress, and providing ongoing communication with the project manager and the system stakeholders.

ISO/IEC/IEEE Standard 15288, augmented by ISO/IEC/IEEE Standard 12207, (herein referred to as 15288 and 12207) provides the basis for system implementation and delivery (ISO 2015, 2017). Table 9.1 illustrates the correspondence between the I^3 implementation and delivery phases and the 15288/12207 technical processes.

Executing the I^3 system development model involves concurrent realization of software elements and physical system elements that satisfy the requirements for the system capabilities. Realized system capabilities are integrated, verified, validated, and demonstrated in a sequence of subphases that produce successive versions of a system, culminating in a deliverable system.

Integration is merged with implementation when using the I^3 system development model because real elements replace simulated elements when they are realized. Verification and validation (V&V) occur informally when fabricated and procured physical elements replace simulated one, constructed and procured software elements replace simulated ones, and implemented real interfaces among system elements and real connections to the system environment replace the simulated ones. More formal V&V occurs at the end of each incremental subphase, as illustrated in Figure 9.1; V&V may be accomplished by the development engineers or by another functional unit, depending on the organizational structure, criticality of the system, and the acquirer's constraints, as stated in the statements of work (SOWs) or memos of understanding (MOUs).

The preliminary plan for packaging, delivery, transition, installation, and validation of the deliverable system (or any subsets of the final system) that was developed during system design is finalized during system implementation.

Phases 5 and 6 of system development are discussed in turn.

9.3 I^3 System Implementation

According to Clause 6.4.7.1 of 15288 and 12207, the purpose of the Implementation process is

> to realize a specific system element.

However, "implementation" is also used in 15288 and 12207 to denote realization of a process (e.g. the implementation process – see Clause 6 of 15288 and 12207). The term "implementation" is used in the I^3 system development model and throughout this text to denote the process of realizing a system increment (i.e. to "make real" a system increment). Each incremental subphase includes planning, realization, verification, validation, and demonstration (demonstration is a validation activity).

The inputs, processes, and outputs of I^3 system implementation for systems engineering are indicated in Table 9.2.

Previous I^3 development phases that have supplied the supporting work products for system implementation and the corresponding 15288/12207 processes are listed in Table 9.3.

Stakeholders' needs, system capabilities, system requirements, system architecture, design definition, and the results of system analysis (including risk factors and mitigation strategies) may be revised during planning of incremental subphases.

During Subphase 0 of I^3 implementation a model-based systems engineering (MBSE) approach is used to develop a simulation model of the system architecture and the connections to the system environment; simulation models are developed for each of the subsystems of a large system and a simulation of

Table 9.2 Implementation of inputs, processes, and outputs.

Inputs	Current versions of work products generated and modified during previous development phases including a prioritized list of system capabilities to be implemented
Activities	A simulation model is developed and the integrity and fidelity of the simulation is certified during Subphase 0
	The number of implementation subphases is initially determined and capabilities are allocated to each subphase, subject to constraints of schedule and resources initially determined
	The number of subphases and allocated capabilities are revised as necessary during planning of each incremental subphase
	During each subphase planning, realizing, verifying, validating, and demonstrating are coordinated
Outputs	A deliverable system with accompanying documentation and supporting infrastructure as specified in the SOW or MOU

Table 9.3 Previous I^3 phases and 15288/12207 technical processes.

I^3 development phases	15288/12207 Technical processes
SysDefn	Business or mission analysis process (6.4.1)
	Stakeholder needs and requirements definition process (6.4.2)
	Project planning process (6.3.1)
Rqmts/Arch/Des	Project planning process (6.3.1)
	System requirements definition process (6.4.3)
	Architecture definition process (6.4.4)
	Design definition process (6.4.5)
	System analysis process (6.4.6)

the connections among the subsystems is developed. One or more specialist engineer may needed to develop and certify the integrity and fidelity of the simulation or simulations.

Subsequent phases use an iterative approach to realization of software elements, as desired and an incremental approach to realization of physical elements, as desired. Some software elements may be constructed "from scratch," some may be developed by a subcontractor, some may be instantiated from library elements, and some may be procured from one or more vendors or from public archives (i.e. open sources). Physical elements may be fabricated "from scratch" by project engineers, some may be fabricated on contract by a vendor or subcontractor, some may be modifications of existing elements, some may be provided by the acquirer, and some may be procured as commodity items.

The following aspects of the I^3 implementation phase differ from conventional models for system implementation, especially for software-enabled physical systems in which software provides some (perhaps most) of the system functionality, behavior, quality attributes, interfaces among and coordination of the physical elements, and connections to the system environment:

1. Each iterative-incremental subphase produces a working version of the combined real and simulated elements of the system. The initial version (Version 0 in Figure 9.1) comprises simulated architectural elements, simulated interfaces among the architectural elements, and simulated connections to the system environment. Subsequent versions incrementally replace simulated elements, interfaces, and connections. Initial versions will have limited simulations of capabilities that will increase in subsequent subphases as the architecture evolves and real system elements replace simulated elements.

   Ideally, the deliverable version of a system will result when all simulated elements have been replaced by real elements. In some cases, the deliverable system may include one or more simulated elements, or it may include some elements that do not provide all of the specified capabilities. Alternately, a delivered system may include the highest priority capabilities but not some of the lowest priority capabilities. In these cases, the delivered version may be replaced by a subsequent version when the desired capabilities become available or are improved.

2. Integration occurs continuously. Real elements replace simulated elements as they become available. Concurrent development of the software elements and physical elements for a system capability requires continuous and detailed communication between the software engineers and the disciplinary engineers who develop the physical element(s). A systems engineer (physical or software) or a senior team member will be assigned the task of coordinating development of software and physical elements, the interfaces among system elements, and connections to the simulated system environment.

   "Sandbox" copies of the current system version are used as test harnesses for experimenting with, developing, and integrating new system elements to replace simulated elements. In some cases, development of software elements may lead development of physical elements for a system capability. Software elements can be developed using the interfaces and responses of the simulated hardware elements. Real software elements that have been completed can be used to provide interfaces and behaviors for the physical elements being developed; iterative modifications to software elements that have been completed and physical elements being developed are to be expected.

   In other cases, software development may lag development of physical elements, perhaps because the software to be developed for a system capability is complex and the physical elements are readily procured. A similar approach to that presented in the previous paragraph can be used.

   If necessary, interface mismatches among physical elements and between software elements and physical elements can be accommodated using software bridges to transform the outputs of physical elements into inputs needed by software elements and other physical elements, and conversely to transform outputs generated by software into inputs for physical elements. Analog-to-digital (A/D) converters may be needed to convert physical analog signals into digital form for processing in the software elements.

   Digital-to-analog (D/A) converters may be needed to convert processed digital data into analog form to provide analog inputs to physical system elements. A/D and D/A converters and software bridges may also be needed to connect software elements and physical elements to the system environment.

   "Adapter and Bridge Design Patterns" describes two approaches to building interfaces that provide linkages between two software elements, between two physical elements, or between a software element and a physical element.

Software engineers and physical-element engineers spend the time waiting for their counterparts to complete the system elements need to realize a system capability by making modifications to their completed elements to better accommodate the counterpart elements being developed, improving implemented system capabilities, clearing technical debt for their completed system elements, making improvements to their methods and tools, and making plans for the upcoming subphase.

They may work ahead on system elements needed for the next capability to be implemented but caution must be observed to avoid getting too far ahead, which can result in avoidable rework if the assumptions made become invalid when the counterpart engineers develop real elements to replace the simulated elements used to implement the next set of system elements.

Adapter and Bridge Design Patterns

In general, a design pattern is a template that provides a solution to a design problem within a context. The pattern concept was introduced by architect Christopher Alexander and has been adapted by other disciplines, notably by computer science and software engineering (Alexander 1977; Gamma et al. 1994). A software design pattern is composed of classes that must be instantiated to provide solutions to specific design problems.

The software adapter pattern is designed to accept digital input data, perhaps from a physical element A/D converter, transform it as needed, and provide it as an output, perhaps to a physical element D/A converter, or perhaps the input is from a software element and the output is to another software element.

A template for software adapter patterns is illustrated in Figure 9.2.

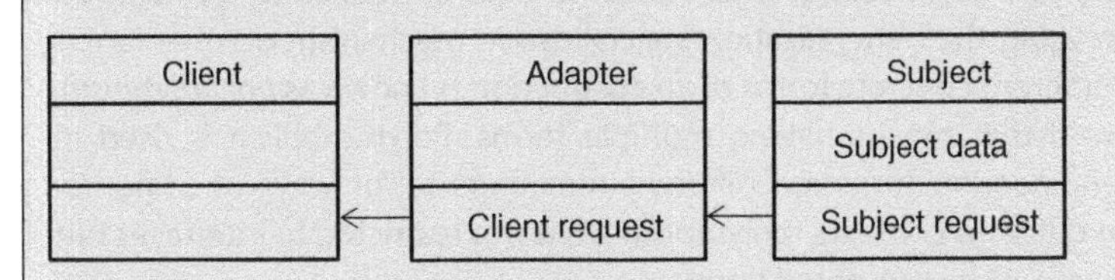

Figure 9.2 A template for adapter patterns.

In this example, a client (a software element or a physical element, perhaps connected by an A/D converter) requests data from a subject that has the needed data but the data is in an incompatible format. For example, the subject

(Continued)

(Continued)

has a list of data items sorted in ascending order; the client needs the list sorted in descending order.

The dependency arrows indicate that the client is dependent on the adapter and the adapter is dependent on the subject.

The pattern works as follows:

1. The client invokes the adapter using the client request interface of the adapter (the client request is an adapter *access method*).
2. The adapter invokes the subject using the subject request access method.
3. The subject sends the requested data to the adapter.
4. The adapter reformats the data (e.g. resorts the list) and sends it to the client.

The client could bypass the adapter, directly request the subject data list, and resort the list before using it. Alternatively, the subject could resort the list before sending it directly to the client. But these approaches would require modifying either the client software or the subject software. The adapter pattern provides the needed interface without modifying either the client or the subject.

Another design pattern, the Façade pattern, is similar to the adapter but it does not transform the data. Software facades, like building facades, hide complexities behind less complex interfaces. For example, a subject might have four access methods to be used for different purposes by different clients (for example, create a new list, add an item to the list, delete an item from the list, find an item in the list and retrieve it) but a client might need only one of the interface methods (e.g. find an item in the list and retrieve it). The client does not need to use, and perhaps should not be allowed to use, the other four interface methods – perhaps based on security issues. The Façade would provide the one interface method for the client to use.

The bridge pattern, like the adapter pattern and the Façade pattern, is widely used in software engineering. It can provide different concrete forms of an abstraction using the generalization/specialization mechanism of inheritance. Providing different concrete forms of an abstraction is known as *polymorphism*, a Greek word that means having multiple forms. Polymorphism is used in software engineering to assign different meanings to, or ways of using, an interface in different contexts (polymorphism is not restricted to interfaces but other uses will not be presented here).

An example of a bridge pattern is presented in Figure 9.3.

The diamond-shaped icon indicates the superclass/subclass semantics of inheritance. The superclass (machine controller) generalizes the subclasses by providing properties and methods common to the subclasses. The subclasses (machine queries) inherit and can use the superclass properties and methods; they are specialized by adding the specific properties and access methods for each subclass.

The machine controller is a software program that controls the machines in an automated factory (see "Smart Factory" in Section 1.2). The controller periodically queried the mechanical status of each machine to determine rate of throughput, whether the machine is performing satisfactorily, whether it is malfunctioning, or if the machine is due for periodic maintenance. A/D converters are used if the machines are not designed to provide digital data.

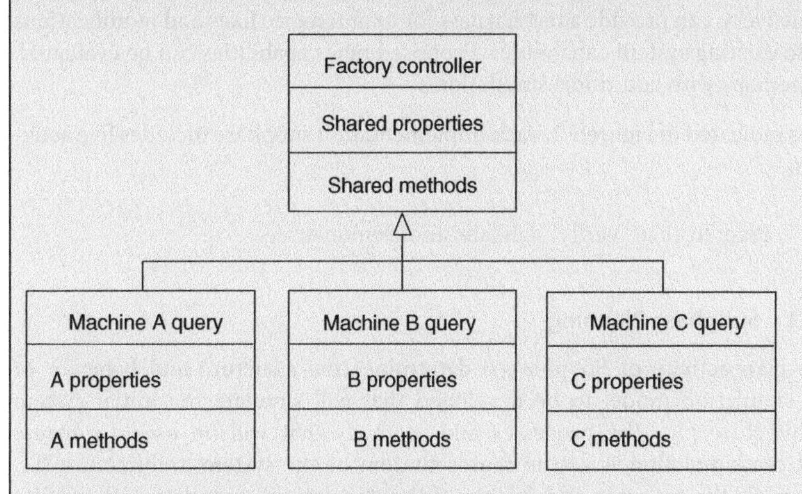

Figure 9.3 An example of a bridge pattern.

Some of the machines may be of different kinds; all machines have common properties and methods and each different kind of machine has specific properties and methods. The machine controller queries are all of the same form but the interface details for each of the different kinds of machines are different (i.e. the queries are polymorphic). Queries are of the following form:

> Query.MachineA.method1; Query.MachineB.method1; Query.MachineC.method1; and so forth.

The shared properties and shared methods of the machine controller are inherited and available for use in each of the machine query classes. When a new or different machine is added, the unique properties and methods are encapsulated in a new machine query class and the class is attached to the machine controller superclass.

3. V&V occur continuously in the sandbox and as distinct activities at the end of each implementation subphase when newly available system elements

replace simulated elements in the previous version to produce a next version of the evolving system that includes one or more newly realized system capabilities.
4. Emergent and nondeterministic behaviors may appear incrementally as newly realized elements replace simulated elements. Desired behaviors can be incorporated and undesired behaviors can be ameliorated.
5. The infrastructure used for system implementation, if retained after system delivery, can provide a test harness for maintenance fixes and modifications to existing system capabilities. Proposed new capabilities can be evaluated, perhaps with additional simulations.

As indicated in Figure 9.1, each implementation subphase includes five activities:

Plan; Realize; Verify; Validate; and Demonstrate.

9.3.1 Subphase Planning

The plan activity of Subphase 0 determines the structure and behavior of the simulation model to be developed that will simulate the initial system architecture plus the processes and methods that will be used to ensure that the simulation is a true representation of the system architecture (i.e. to ensure the integrity and fidelity of the simulation). Simulation specialists may be involved in Subphase 0 and may periodically be involved in evolving the simulation model and certifying the fidelity and integrity of the modified models.

An important activity during planning Subphase 0 is determining the priority order by which system capabilities will be implemented during the subphases of Phase 5 (criteria for prioritizing capabilities are presented in Table 5.2). The prioritized list of capabilities is sometimes called the implementation backlog (see Figure 4.10). The backlog may be revised and reprioritized as desired by the system implementers, with the concurrence of the stakeholders' representative (i.e. the customer in Figure 4.10). However, the capabilities to be implemented during a subphase should not be changed after the subphase commences, to avoid implementation thrashing. A major change of mission or system requirements may of course necessitate abandonment of the current incremental implementation subphase.

In general, plan is a one- to two-day activity during which the implementation team determines the work to be accomplished, the work assignments, and perhaps, changes to the implementation process to improve communication, coordination, efficiency, and effectiveness. The plan activity for Phase 0 may require more time than planning during subsequent subphases of system implementation.

The planning activity also includes a retrospective analysis of lessons learned during the recently completed subphase (sometimes called a postmortem review); it may be determined that different methods, techniques, and/or tools will be incorporated in the current subphase or it may be determined that a "time out" subphase is needed to integrate a new development tool or to clear accumulated technical debt (Fairley and Willshire 2017).

The duration of subphases is fixed, usually at one month but not longer than three months. Shorter subphases provide more frequent evidence of desired progress and early warning of problems to be addressed before they become serious problems.

The number and kinds of engineers that will be involved is determined and fixed for the duration of a subphase. Fixing the duration and resources determines the amount of work that can be accomplished during a subphase and facilitates tracking of progress by using two controlled variables to determine the third one – the amount of work that can be accomplished. Techniques such as technical debt and burn down can be used to track progress – these and other techniques for tracking progress are presented in Chapter 11.

There may be multiple implementation teams, perhaps one for each subsystem or multiple teams for a large subsystem, that have, perhaps, been decomposed into smaller parts. "Team Size" illustrates that multiple small teams, with single points of contact between each pair of teams, is more efficient than one large team.

Team Size

Analysis has shown that individual productivity of software developers decreases with increasing team size because of the need for increased communication and coordination among the increasing number of team members; the same decrease probably occurs for any team engaged in closely coordinated intellect-intensive work, including teams of physical systems engineers (Boehm 1981, 2000).

One of the constructive cost model (COCOMO) estimation models developed by Dr. Boehm and colleagues at Southern California University uses the following equation to estimate effort as a function of system size (Boehm 1981):

$$\text{Effort} = 3.6 \times \text{size}^{1.2}.$$

The exponent represents the nonlinear increase in effort required for increased system size, which is attributed to loss of productivity of individual engineers when more engineers are involved. The equation was derived from empirical data collected on software projects.

(Continued)

(Continued)

The following effort estimates can be calculated, assuming effort is in engineer-months and size is in units such as the number of use cases:

$$\text{Effort1} = 3.6 \times 20^{1.2} = 131 \text{ engineer-months}$$

and

$$\text{Effort2} = 3.6 \times 40^{1.2} = 301 \text{ engineer-months.}$$

Doubling the system size requires 301 engineer-months of effort, as compared to 131 engineer-months for a system of half the size, which is 39 more engineer-months than would be required by linear scaling (131 to 262).

Assuming the project is to be completed in 15 months, the number of engineers will be approximately

$$131/15 = 9 \text{ engineers, or}$$

$$301/15 = 20 \text{ engineers.}$$

The nonlinear scaling of engineers (20 instead of 18) compensates for the loss of productivity by individual engineers caused by the increased number of communication paths.

The number of communication paths among N engineers is $N(N-1)/2$, which is the number of links in a fully connected graph. For the example:

$$9 \times 8/2 = 36 \text{ communication paths for system size} = 20, \text{ and}$$

$$20 \times 19/2 = 190 \text{ communication paths for system size} = 40.$$

The number of communication paths for the second team (190) can be reduced by dividing it into two teams of 10 engineers each with a single two-way communication link between the two teams, as indicated by the first term in the equation:

$$2 \times 1/2 + 2(10 \times 9/2) = 91.$$

Finally, note that the number of communication paths for one team of 20 is decreased by ~50% (91/190 = 0.48) when the team is divided into two teams of 10 each with one communication link between them. This is the primary reason multiple teams of 10 or fewer members each is preferable to larger teams.

9.3.2 Realization, Integration, Verification, and Validation of System Capabilities

The realization/integration activity requires most of the time and effort during a subphase. V&V at the end of a subphase will typically require one or two days for V&V and one day or less for stakeholder demonstrations.

Brief meetings (sometimes called standup meetings) are conducted each day during a subphase to review progress; make plans for the day; and identify problem areas, roadblocks, and other progress inhibitors. Daily standup meetings are conducted by each team and include all members of the team: hardware, software, and other disciplinary engineers plus specialty engineers.

9.4 I³ System Delivery

System delivery involves transitioning a deliverable system into the operational environment and validating that the system provides the specified capabilities.

Transition may involve packaging and shipping one or more copies of a system by a secure method of transportation to one or more installation sites. Installation of a large software-enabled physical system may be preceded by site preparation, which may be the responsibility of the supplier or the acquirer.

The purpose of the validation process is to determine the degree to which the installed system satisfies the stakeholders' requirements. The goal of system validation in the operational environment is to gain system acceptance by the stakeholders and other appropriate personnel. Demonstrations of capabilities allows the system stakeholders to determine the degree to which the system satisfies their requirements.

It is not uncommon that a contracted SOW will include a 30-day burn-in period during which the system supplier will remedy shortcomings of the delivered system.

9.5 Key Points

- When using the I³ system development model, implementation is concerned with developing system increments in the subphases of Phase 5, as illustrated in Figure 9.1.
- Each subphase of Phase 5 includes planning, realization, verification, validation, and demonstration.
- Software engineers and other disciplinary engineers concurrently implement and integrate the software and physical system elements needed to realize system capabilities.
- Real system elements replace simulated elements after they are realized, verified, and validated. Incremental versions of a system include simulated and real system elements.
- The deliverable system is the result of the final incremental subphase of Phase 5.
- System delivery (Phase 6) includes the transition and validation processes of ISO/IEC/IEEE Standards 15288 and 12207.

- The transition process moves the deliverable system from the simulated operational environment to the real operational environment.
- The goal of Phase 6 validation is to gain acceptance of the delivered system by the system stakeholders.

Exercises

9.1. System implementation happens in Phase 5 of the I^3 system development model. Phase 5 supports iterative development of software elements and incremental development of physical elements.
(a) Briefly explain how iterative implementation differs from incremental implementation.
(b) Briefly explain how iterative implementation and incremental implementation are similar.

9.2. I^3 Phase 5 supports concurrent development of software elements and physical elements needed to realize a capability.
(a) Briefly explain how concurrent development of software elements and physical elements occurs during Phase 5.
(b) Briefly explain some of the issues that have to be handled during concurrent development of software elements and physical elements.

9.3. Phase 6 of I^3 system development involves transition and validation.
(a) Briefly explain the transition process.
(b) Briefly explain some of the issues that may have to be resolved during system transition.

9.4. Validation occurs during I^3 Phase 5 system implementation and during I^3 Phase 6 validation.
(a) Briefly explain how validation during Phase 5 is different from validation during Phase 6.
(b) Briefly explain how validation during Phase 5 is similar to validation during Phase 6.

9.5. Briefly explain the roles of systems engineers as follows.
(a) Briefly explain the roles of systems engineers during I^3 Phase 5 system implementation.
(b) Briefly explain the roles of systems engineering during I^3 Phase 6 system delivery.

References

Alexander, C. (1977). *A Pattern Language: Towns, Buildings, Construction*. Oxford University Press.

Boehm, B. (1981). *Software Engineering Economics*. Prentice Hall.

Boehm, B. (2000). *Software Cost estimation with COCOMO II*. Prentice Hall.

Fairley, R. and Willshire, M. (2017). Better now than later: managing technical debt in systems development. *IEEE Computer* 81–87.

Gamma, E. et al. (1994). *Design Patterns: Elements of Reusable Object-Oriented Software*. Addison-Wesley.

ISO (2015). ISO/IEC/IEEE 15288:2015, Systems and software engineering – System life cycle processes, ISO/IEEE, 2015.

ISO (2017). ISO/IEC/IEEE 12207:2017, Systems and software engineering – Software life cycle processes, ISO/IEEE, 2017.

Part III

Technical Management of Systems Engineering

Part III of this text includes Chapters 10–12. They cover management of the technical work activities for a systems engineering project. The topics in Part III are as follows: Chapter 10 covers planning and estimating, Chapter 11 covers assessing and controlling, and Chapter 12 covers organizing, leading, and coordinating.

Systems Engineering of Software-Enabled Systems, First Edition. Richard E. Fairley.
© 2019 John Wiley & Sons, Inc. Published 2019 by John Wiley & Sons, Inc.

10

Planning and Estimating the Technical Work

10.1 Introduction

The term "technical management," as used in this text, is concerned with managing the technical work activities that facilitate development of complex software-enabled physical systems. The scope of technical management is clarified in the introduction to Clause 6.3 of ISO/IEC/IEEE Standards 15288 and 12207:

> This set of technical management processes is performed so that system-specific technical processes can be conducted effectively. They do not comprise a management system or a comprehensive set of processes for project management, as that is not the scope of this standard.
>
> (ISO 2015, 2017)

The key elements of technical management, as pursued by one or more physical systems engineers and one or more software systems engineers, working in a cooperative and coordinated manner, are developing a technical management plan; executing the plan to include assessment, analysis, and control of the work processes and work products; revising the plan periodically and as warrant by the situation; and leading and coordinating the work activities of the system development team.

This chapter is concerned with planning the technical work. Chapter 11 covers assessment, analysis, and control, and Chapter 12 covers leading and coordinating the work activities of a system development team.

Planning the technical work needed to develop a complex software-enabled physical system involves applying the planning processes of project management to plan the technical work activities (Fairley 2009; Fairley et al. 2013; PMBOK 2017).

This chapter provides readers with the opportunity to develop an understanding of the methods, techniques, and tools used to prepare estimates and

Systems Engineering of Software-Enabled Systems, First Edition. Richard E. Fairley.
© 2019 John Wiley & Sons, Inc. Published 2019 by John Wiley & Sons, Inc.

the other essential elements of a technical management plan. References and Exercises are provided for accessing more information and gaining additional understanding.

This chapter is keyed to the I³ system development model but the methods and techniques can be applied to other kinds of system development.

10.2 Documenting the Technical Work Plan (SEMP)

A systems engineering management plan (SEMP) is prepared to address the overall approach to managing the engineering activities of a systems project. An SEMP is subordinate to, and closely integrated with, the project management plan (PMP) developed by the project manager with the assistance of the systems engineers. It may be a part of the PMP but it is usually a standalone and closely coordinated document (the SEMP). The SEMP may be called a system management plan (SMP). It may be called a software engineering plan or a software development plan (SDP) for software-intensive systems that incorporate available hardware elements.

In some cases, the term "system management plan" is used interchangeably with SEMP, and in other cases, as in contracting, the acquirer may prepare an SMP and the supplier will prepare an SEMP in response to the SMP. The focus here is on the SEMP, which provides guidance for conducting the technical work activities of a systems project; the SEMP also provides information to be used when preparing the PMP. A technical management plan for a systems program is based on the collective SEMPs of the projects that constitute the program.

Various formats for SEMPs are available and can be found by Internet search. ISO/IEC/IEEE Standard 24748-4 provides a template for SEMPs (ISO 2016). The following material provides a checklist of items typically contained in an SEMP. Some of the items may be "business as usual," for example, configuration management and quality assurance (QA) provided by separate organizational units. Other items may be contained in other documents, for example, the PMP or the system requirements definition. Reference to and methods of access to these other items should be included in the SEMP.

The checklist of items to be included in SEMPs, either directly or by reference, include the following:

1. Introductory material: A cover or title page, a history of revisions, a table of contents, and introduction (purpose, intended audience, key terms).
2. An executive summary that includes a brief statement of the business or mission that needs to be satisfied and the project scope.
3. References for and linkages to the statement of work (SOW), the PMP, the system requirements, and other relevant documents.

4. The development process model, or models, to be used that incorporate(s) appropriate engineering methods and tools, as tailored for the project.
5. The current estimate for technical work to be accomplished based on a standard estimation template (see Section 10.3.1), plus a link to previous baselined estimates, if any.
6. A work breakdown structure (WBS) and schedule network.
7. A staffing plan for the number of system developers and supporting staff by types of skills and skill levels, when they will be needed, how many will be needed, for how long they will be needed, and how they will be acquired.
8. A plan for the engineering methods and tools to be used by system developers for the technical processes derived from Clause 6.4 of ISO/IEC/IEEE Standards 15288 and 12207 and how they will be acquired, if necessary. The list of the technical processes is provided in the accompanying sidebar (ISO 2015, 2017).

Technical Processes of 15288 and 12207

The following technical processes are listed and described in Clause 6.4 of the 15288 and 12207 standards. Not all systems projects will execute all of these processes. Some projects may start with system requirements definition (based on previously executed business or mission analysis and stakeholder needs and requirement definition) and conclude with the transition and validation processes. Other projects may start by conducting business or mission analysis and conclude with architecture definition; still others may start with transition and validation and provide systems engineering support for operation and maintenance throughout the operational life cycle of a system, concluding with system retirement and disposal. Annex A of 15288 and 12207 provides guidance for tailoring the technical processes for specific projects.

Some or all of 14 technical processes to be accounted for in an SEMP are the following:

1. Business or mission analysis process;
2. Stakeholder needs and requirements definition process;
3. System requirements definition process;
4. Architecture definition process;
5. Design definition process;
6. System analysis process;
7. Implementation process;
8. Integration process;
9. Verification process;
10. Transition process;
11. Validation process;

(Continued)

(Continued)
12. Operation process; 13. Maintenance process; and 14. Disposal process.

9. Organization of the systems engineering team and the system development teams along with a list of job descriptions for key project members, to include competencies, roles, responsibilities, and when needed.
10. A plan for acquiring and utilizing consultants and subcontractors to address staffing shortfalls and special needs.
11. A plan for acquiring needed resources in addition to personnel.
12. A training plan that includes training for engineers and others as needed for the processes, methods, tools, and relevant aspects of the application domain; who will receive the training; who will provide it; when and where it will be delivered; and a process for certifying trained individuals.
13. A plan for the technical infrastructure and development environment (or infrastructures and environments), including the manufacturing infrastructure and environment, if applicable.
14. A plan for preparing and releasing intermediate system builds, if appropriate.

10.2.1 Documenting Other Technical Processes

An SEMP should also include plans for other technical management processes, as indicated in Clause 6.3 of the 15288 and 12207 standards; they include (ISO 2015, 2017) the following:

1. Project assessment and control;
2. Decision management;
3. Risk management;
4. Configuration management;
5. Information management;
6. Measurement; and
7. Quality assurance.

1. Project assessment and control is addressed in Chapter 11.
2. *Decision management*: Decision-making occurs constantly on an informal basis and on a formal basis, as warranted. According to Clause 6.3.1.1 of ISO/IEC/IEEE Standards 15288 and 12207, the purpose of the decision management (DM) process is to

> provide a structured, analytical framework for objectively identifying, characterizing and evaluating a set of alternatives for a

decision at any point in the life cycle and select the most beneficial course of action.

The DM process is used to resolve technical issues (and technical management issues) that need to be decided using structured deliberations. Trade studies and engineering analysis are often used to assess the impact of proposed decisions on factors such as cost, schedule, performance, quality, and risk. A decision is reached by ranking proposed decisions using a selection model. Assumptions made and the decision rationale are noted and maintained to document the decision made and to provide guidance for future decisions that may be made. If necessary, the system analysis process may be used to guide a formal analysis of the situation (see Clause 6.4.6 of ISO 15288).

3. The risk management process is addressed in Chapter 11.
4. *Configuration management*: The purpose of engineering configuration management is to establish and maintain control of configuration items (CIs) for a system being developed and then during deployment throughout the system life cycle. A CI is any work product or item of technology that is treated as an individual unit. CIs can be items such as the SEMP, requirements definition, architecture definition, design definition, realized system elements, system increments, test plans, test cases, and test results. Related items may be grouped into a CI, such as the system elements that provide a system capability or the system elements that constitute a system release.
   CIs are placed under version control and cannot be changed without adhering to a change management process that typically involves approval by a change control board (CCB). Generating approved changes creates an evolutionary log of how a CI is changed over time. A depiction of the change management process is illustrated in Figure 10.1.
   The activities of configuration management include identification of CIs, change management, status accounting, and CI audits. Various kinds of tools support the processes of configuration management; they range from spreadsheets to databases to version control tools. A version control tool is often used in an informal manner during system development to track and control changes to evolving system elements and system increments.
5. *Information management*: The information management process is concerned with planning, executing, and controlling the dissemination of appropriate information to various system stakeholders, including those within the system development organization such as the systems engineers, system developers, the project manager, and other organizational elements plus external stakeholders as appropriate.
6. The measurement process is presented in Chapter 11.

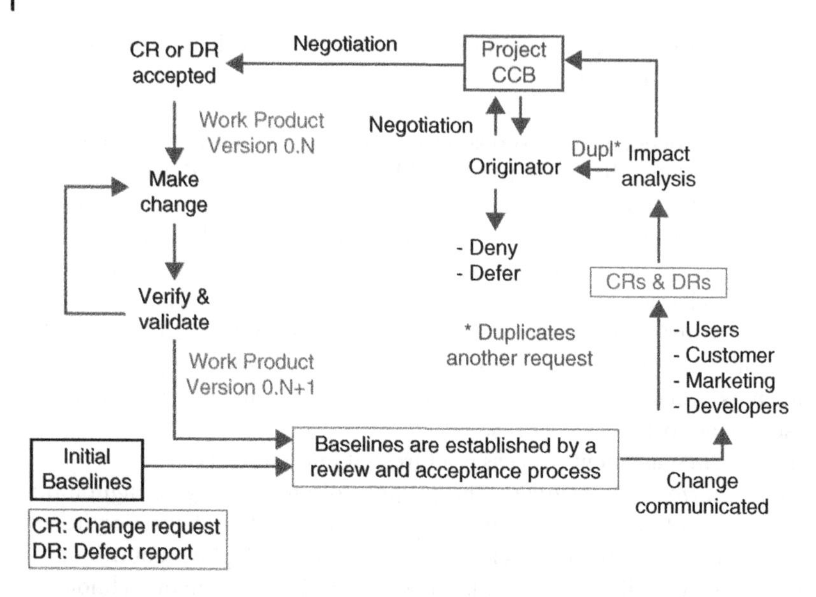

Figure 10.1 The change management process. Source: Fairley 2009, figure 7.4 (p. 283).

7. *Quality assurance*: QA and quality control are often confused, particularly in software engineering where "testing" is often used interchangeably with "quality assurance" rather than correctly being regarded as a quality control mechanism. QA is an engineering management process used to assure that regulations, policies, procedures, guidelines, and plans that impact the quality of system elements, subsystems, system increments, and the final system are being followed. A diagram for QA of software and systems engineering projects is illustrated in Figure 10.2, which is repeated from Chapter 2.
   As illustrated in the figure, QA is, or should be, an oversight process. To correct an observed discrepancy, the first action taken by QA personnel is to meet with the project manager and lead systems engineer(s) to discuss possible remedies. As indicated in the figure, discrepancies that cannot be resolved at the project level are reported to the decision makers with recommended actions.

In one organization, issues and recommended actions are reported to the decision managers as green (proceed), yellow (caution), and red (stop work). In a project familiar to the author, the project manager and lead engineer

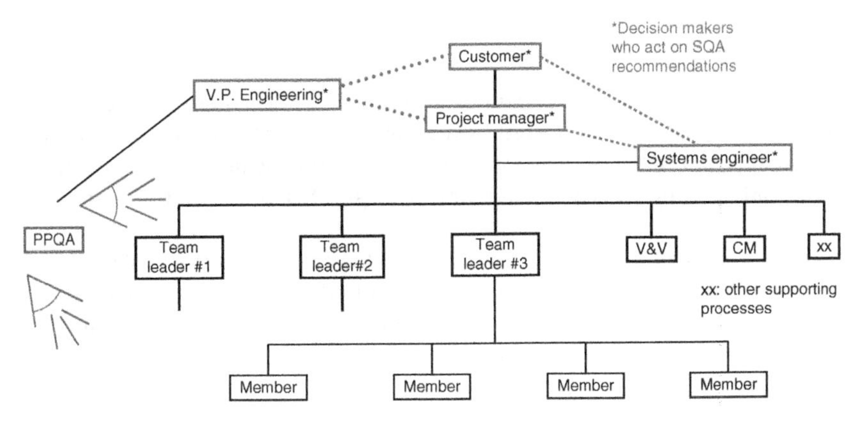

Figure 10.2 Quality assurance for systems engineering projects.

requested that QA report their project as "yellow" because they could not otherwise obtain the attention of the decision makers to resolve the issue. In another instance, the project manager/systems engineer requested that a QA person be transferred to the project because that person had a broader overview of the project than any of the engineers assigned to the project. After being transferred, she became a system developer; she no longer played the role of QA observer.

Although sometimes regarded as a foe of systems engineering, the role of QA is, or should be, to play an advisory role in recommending solutions for issues that can inhibit production of a quality system in an efficient and effective manner. Referring issues to decision makers that cannot be resolved at the project level intentionally limits the authority of QA, which improves interactions of QA personnel with project managers and systems/software engineering leaders by removing the image, and perhaps the reality, of QA as the "projects police."

As indicated above, this extensive list of systems engineering processes will be tailored for each project. Some of the planning activities may be "business as usual"; others may be conducted by other elements of the organization (e.g. configuration management and QA). In those cases, the plan for communicating and coordinating, along with the respective roles and responsibilities of systems engineers and others, should be specified. In other cases, some or all of the planning activities may be the responsibility of systems engineering personnel. These cases include small organizations, small noncritical projects in larger organizations, projects conducted at remote locations, and "Skunk Works Projects."

Skunk Works Projects

Skunk Works projects (also known as Skunkworks projects) are research and development projects conducted by a small group of people. These projects are usually conducted at a remote site, free of the constraints of the parent organization. They are conducted with a sense of urgency and a need for secrecy.

The first known use of the term "Skunkworks" occurred in the 1940s when Lockheed Corporation established a small team, led by aeronautical and systems engineer Clarence "Kelly" Johnson, to develop the P-80 Shooting Star jet fighter, which was the first jet fighter to be operationally deployed by the U.S. Army Air Forces. The P-80 is illustrated in Figure 10.3.

Figure 10.3 The P-80 Shooting Star jet fighter. Source: USAF – U.S. Air Force photo.

Design of the P-80 started in 1943 and the first P-80 jets were deployed 143 days later (approximately 4.5 months). The U.S. Air Force and Navy used the T-33 aircraft, a variant of the P-80, as a trainer for many years. The Lockheed Martin Skunk Works later developed many other famous aircraft.

A corner of the modern Lockheed Martin Skunks Works aircraft hangar that depicts the LM Skunkworks logo is illustrated in Figure 10.4. The hangar is located in Palmdale, California.

Figure 10.4 The Lockheed Martin Skunk Works logo. Source: Picture taken by Bernardo Malfitano of AirShowFan.com.

The Kelly Johnson team adopted the name Skunkworks for their facility because it was located in a circus tent next to a paint factory that smelled like skunks.

10.2.2 Developing the Initial Plan

An initial version of an SEMP will contain whatever information is available. A plan for evolving the SEMP should be developed and followed (i.e. a plan for planning). After an SEMP is stable, it will typically be updated at intervals of one to three months, more frequently in the early phases of a project and less frequently as a project stabilizes.

Development of an SEMP should occur as a coordinated activity with the project manager in conjunction with development of the PMP. Small projects may not need a separate SEMP; the plan for technical activities can be included in the PMP. However, two or more systems engineers (a minimum of one physical and one software systems engineer) should be involved in developing the SEMP or technical management elements of a PMP for the technical management activities of a software-enabled physical system project.

The essential activities of developing the initial version of a technical plan for a systems project (i.e. an SEMP) are the following:

- Reviewing and clarifying available technical information.
- Available documentation may include the outcomes or one or more of business or mission analysis, stakeholders' needs and requirements, system capabilities, system requirements, and (perhaps) a skeletal proof-of-concept

architecture. These documents may be in early stages of development and may lack sufficient information to support development of a definitive plan.
- Making initial estimates of the required schedule, personnel and other resources, technology, and infrastructure for the technical work activities.
- An initial estimate should be made without regard to constraints; it will likely be modified when consulting with the project manager who will provide the constrained allocations for technical work activities. This may result in reviewing and revising the stakeholders' requirements with the appropriate stakeholders. In some cases, a project may be redefined, deferred, or cancelled.
- Identify known technical risk factors.
- A major risk factor (i.e. a potential problem) is that the initial plan will be used as the basis for making project commitments, especially on contracted projects where responses to requests for proposals (RFPs) and subsequent negotiation of an SOW is often accomplished based on insufficient knowledge.

Estimating schedule, budget, resources, technology, and infrastructure for the technical work of a software-enabled systems project is one of the first tasks to be undertaken when developing an SEMP. Estimates are revised as more is understood about the project.

10.3 The Estimation Process

An estimate is a prediction of future events based on historical data and current data adjusted for differences between the historical plus current data and what is known or assumed about the future.

Historical data provides the basis of estimation (BOE) for a project.

The goal of estimating technical work to be accomplished is determining a set of parameters that provide a high level of confidence an acceptable system can be developed and delivered within the bounds of the project constraints. The parameters and constraints to be considered are the following:

- System features;
- Quality attributes
- System capabilities;
- Requirements and architecture;
- Development infrastructure;
- System size;
- Technical effort;
- Schedule;
- Skills and skill levels;
- Other resources;

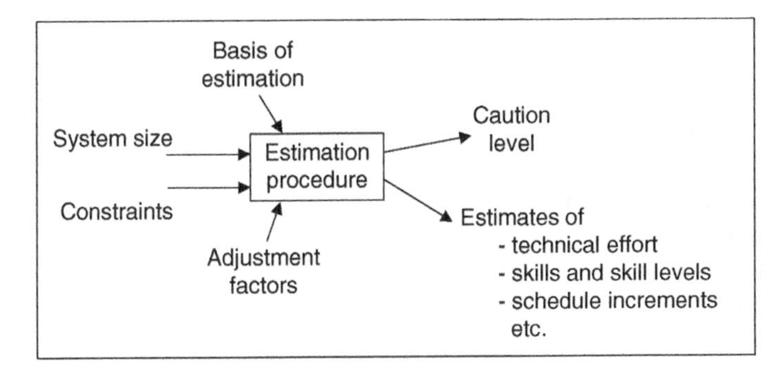

Figure 10.5 An estimation process. Source: Fairley 2009, figure 6.1 (p. 210).

- Budget; and
- Technology.

Some of these parameters will be inputs to an estimation procedure (the currently known information) and others are the estimates produced as outputs of the procedure. For example, the input parameters might be estimated system size, the constraints that must be satisfied, and the BOE. The outputs may be estimated technical effort, number of personnel by skill and skill level, schedule increments, and the caution level for the estimate as compared with the input constraints. An example is provided in Figure 10.5.

The estimation procedure for the process in Figure 10.5 can be applied manually or sometimes algorithmically, depending on the estimation procedure used. System size, for example, might be stated as 150 use cases, where each use case is in the range of 10(±2) steps in the behavioral scenario and the number of exceptions to be handled is limited to three or fewer (thus providing a consistent size measure).

The caution level in the figure results from comparing the estimates to the constraints. It may be ranked (low, medium, high), (green, yellow, red), or some similar indicator. For example, a project estimated as requiring 180 staff-months of technical effort and 18 months (thus 10 engineers for 18 months) might result in a low or green caution level if the project constraints indicate that the technical work should not require more than 300 staff-months and 20 months (more than the estimate), or the caution level might be high or red if the estimate is 180 staff-months and 18 months but the project constraints are not more than 120 staff-months and 15 months (less than the estimate).

Adjustment factors applied to an estimate typically account for attributes such as (Boehm et al. 2000):

- Personnel factors
  - Skill levels

  - Experience
  - Team cohesion
- Comprehension factors
  - System domain familiarity
  - System requirements familiarity
  - System architecture familiarity
  - Development process familiarity
- Complexity factors
  - Domain complexity
  - Design complexity
  - Interface complexity

Adjustment factors are typically in the numeric range of 0.5–1.5. A factor of 1.2 for personnel skill would indicate very low skill levels that would increase the estimated effort by 20%; a factor 0.8 would indicate very high skill levels and reduce estimated effort by 20%. On the other hand, an adjustment factor of 0.7 for complexity would reduce estimated effort by 30% and a complexity factor of 1.3 would increase estimated effort by 30%.

All estimates are based on assumptions that various project factors are accurate. An accurate estimate depends on the accuracy of the inputs used to make the estimate. It may be assumed that system size, the BOE, and the adjustment factors are accurate representations for the system being estimated, or it may be acknowledged that an early estimate is likely to be inaccurate because the size and adjustment factors are not accurately known.

In general, an estimate cannot be more accurate than the parameters on which it is based.

> A well-known saying: garbage in = garbage out.

Three additional points:

1. The scope of effort being estimated must be understood. The BOE may include only the implementation effort for past projects or it may include all of the technical work, including the work performed for all of the development phases from eliciting stakeholders' requirements to deliverable system transition and validation, including (or not) effort for supporting processes such as configuration management and QA.
2. The estimated caution level is advisory. It may be overridden in the documented estimate by additional factors; for example, a preliminary estimate of project feasibility may be rated low in estimated caution but adjusted to medium or high when documenting the caution because the input parameters for an initial feasibility estimate are not sufficiently understood to instill confidence in an estimate based on those parameters, or an estimate may not account for other factors such as a new or difficult customer.

Alternatively, an estimate may be rated high in caution level but low on documented caution because the customer and the system to be developed are familiar, the BOE is valid, and the estimate is based on stable system requirements and architecture.

3. Systematic estimation procedures support "what-if" analysis of estimates that result from varying the input parameters. What-if analysis can expose risk factors to be encountered during the project and can indicate remedial actions that must be taken to reduce risk. For example, reducing staffing risk and increasing confidence in the estimate may result from increasing the skill level rating of the team members from medium to high. Additional training and/or recruitment of key personnel may be needed to increase the skill level rating to increase confidence in the estimate.

   What-if analysis may indicate an algorithmic estimation procedure being used is especially sensitive to variation in some of the input parameters, for example, estimated system complexity or variations in the schedule constraint. Estimation sensitivities that result from the estimation algorithm being used may be discounted if the sensitivity seems inappropriate, or the sensitivity may be justified, which indicates that special attention is needed for ensuring the accuracy of the sensitive parameter, or parameters.

These considerations can be summarized in the following principles of estimation:

Estimation principle #1:

A project estimate is a projection from past and present to the future, adjusted to account for differences between the past-present and the future.

Estimation principle #2:

All estimates are based on a set of assumptions that may or may not be true and a set of constraints that may or may not be satisfied.

Estimation principle #3:

Projects must be reestimated periodically as understanding grows and aperiodically as project parameters change.

10.3.1 An Inverse Estimation Process

The estimation process in Figure 10.5 can be inverted: given a set of project constraints, an estimate of system content can be developed. For example, how many system elements can be developed for a given schedule and a set of resources? The inverted estimation process is illustrated in Figure 10.6.

For example, inputs of 200 staff-months of effort, adjustment factors for medium skill levels, high system complexity, medium confidence level, and five schedule increments might produce an estimate of 180 use cases for estimated system size. The estimated number of use cases that could be implemented for

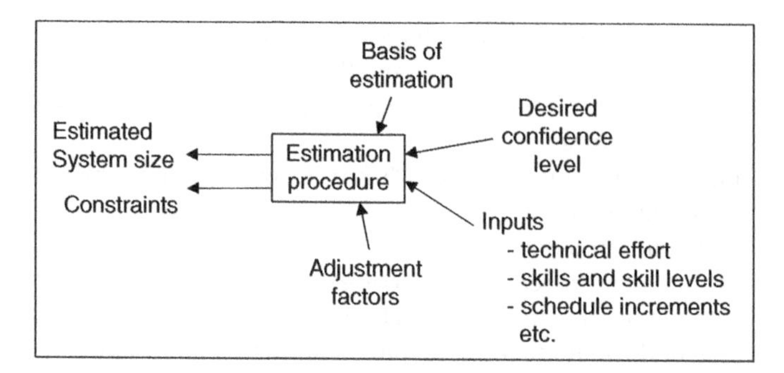

Figure 10.6 An inverted estimation process. Source: Fairley 2009, figure 6.3 (p. 215).

the input parameters would each be sized to 10 or fewer interaction steps in the behavior scenarios and 3 or fewer exceptions to be handled, which would be consistent with the BOE used to make the estimate.

The estimated project constraints might include an effort constraint of 240 staff-moths and a schedule constraint of 20 months, thus a constraint of not fewer than 12 engineers (240/20) for the input confidence level.

10.3.2 Making an Initial Estimate

An initial estimate should be prepared to assess the feasibility of a proposed project. Current information should be reviewed and clarified to the extent possible. Documentation developed during Phases 1 and 2 of the I^3 system development model should be reviewed and clarified to the extent possible: business or mission analysis, stakeholders needs and requirements, system capabilities, system requirements, and a proof-of-concept architectural model, if available. Relevant historical data should also be consulted. Risk factors, uncertainties, and the known-unknowns should be identified and documented as a basis for stating a confidence level along the initial estimate.

Risk Factors, Uncertainty, and Known-Unknowns

A risk factor is a known potential problem that, should it become a real problem, will inhibit progress and may prevent successful completion of a project without violating the project constraints. Composite risk is the set of known potential problems (i.e. risk factors). Each risk factor is characterized by an estimate of the probability and impact should the potential problem become a real problem plus the mitigation strategy (i.e. the remedy) to be applied if the potential problem becomes a real problem.

Risk factors are tracked and mitigating actions are initiated when the potential problem becomes a real problem, i.e. when (if) an objective indicator crosses a predetermined threshold.

Risk management processes and methods are addressed in Chapter 11.

A potential problem may be categorized as an uncertainty because there is insufficient information to estimate, with confidence, the probability and/or impact should the potential problem become a real problem.

The known-unknowns are aspects of technical work processes or technical work products that have been identified as issues that caused problems in past projects but it is unknown if the current project may or may not have the same issues because of factors such as a different customer; a different project manager; a different development team; or new/modified processes, methods, or tools.

Uncertainties and known-unknowns are placed on a watch list and periodically reviewed. Some of them may become risk factors and be placed on the list of risk factors to be mitigated as the project evolves and more information becomes available concerning uncertainties and known-unknowns.

As previously mentioned, care must be taken that initial estimates based on preliminary requirements and a proof-of-concept architecture do not become the de facto estimates for making project commitments.

Initial estimates and subsequent estimates can be made using the estimation techniques described in the following sections.

10.4 Estimation Techniques

Estimation techniques include rule of thumb (ROT), analogy, expert judgment, Delphi, wideband Delphi, regression-based estimation models, and Monte Carlo estimation.

10.4.1 Rule of Thumb Estimation

Practical experience is the basis for rule of thumb (ROT) estimates. Some ROTs are generally accepted, industry-wide. For example,

> It costs 100 times more to fix a problem after system deployment than to fix it prior to system deployment.

Some ROTs are based on local experience; for example,

> In our organization it costs 75 times more to fix an interface problem post-release as to fix it pre-release.

A development ROT, for example, might be

> On average, it takes one engineer-week to implement a system capability based on a use case with 10 or fewer steps in the primary behavioral scenario and not more than 3 exception conditions to be handled.

An estimate for a system having 200 capabilities-based use cases would require 200 engineer weeks to implement, perhaps 5 engineers for 40 weeks or 15 engineers for 20 weeks (the latter to account for loss of individual productivity based on increased communication and coordination). The estimate might be adjusted upward or downward based on particular factors for a project, such as a new engineering team or knowing the engineers who will implement the system.

This ROT might be applicable for a particular kind of system in a particular domain, such as an enterprise information system for the insurance industry. Analysis might reveal that past projects have exhibited a similar productivity metric so that the average of one engineer-week per capability is a reliable metric or the productivity metric for past projects may vary widely, thus making the average value an unreliable ROT. Or, the ROT might be based on two past projects, making the metric statistically invalid. And, applying the ROT for other kinds of systems in other domains would not be valid (e.g. an embedded real-time system project).

Estimates based on reliable ROTs are useful to provide rough order of magnitude estimates when doing feasibility studies and "what-if" scenarios.

However, it is a risky endeavor to base a project estimate on ROTs alone. In general, estimates that will result in commitments should be made using two or more independent estimates using two or more different estimation techniques.

10.4.2 Estimation by Analogy

Analogy is a widely used technique for estimating project attributes in the engineering disciplines, including systems engineering. The goal of analogy-based estimation is to find one or more analogous projects for which the attributes of interest are known. The closer the analogy, the more confident will be the estimate. An ROT, for example, can be used with greater confidence if it is based on projects that are analogous to the one you are estimating.

Analogy-based estimates can be quite simple (e.g. a similar project required 10 people for 18 months to accomplish the system implementation phase) or it can be quite sophisticated. In the latter case, an organization might have compiled a relational database of past projects. Each row in the data schema would contain data for a completed project. Each column would record an attribute of past projects, such as

- The customer;
- The kind of system;
- The scope of activities included;
- System size and the size measure used;
- Adjustment factors applied;
- The system development model used;
- Development tools used;
- Deliverable system produced;
- Estimated and actual project duration;
- Estimated and actual effort;
- Estimated and actual cost;
- Prerelease and postrelease problems encountered; and
- Lessons learned.

To make an estimate, the known and estimated characteristics of the project to be estimated are entered in a query that retrieves a list of projects that match the current project within a specified range, for example, all projects that developed software-enabled physical systems of high complexity with medium staffing skills built by engineers using the I^3 system development model in the domain of medical instrumentation and that are within ±10% of the estimated system size.

The primary strength of analogy-based estimation is

Strong analogies provided a good BOE for a project.

The primary weakness is

Weak analogies produce inaccurate estimates.

An estimate is thus no better or worse than the analogies on which it is based.

10.4.3 Expert Judgment

Expert judgment involves asking one or more experts for their estimates of project attributes such as effort, time, skills and skill levels needed, and risk factors. Referring to Figure 10.5, the BOE is in the heads of one or more experts. The adjustment factors they apply may include subjective factors such as knowing the people who will manage and do the work, the politics of customer relations, and frictions that may exist among internal elements of the development organization. System attributes include whatever information is available for the experts to examine (e.g. desired features and quality attributes, system capabilities to be realized, system requirements).

Experts might decide that the requirements are too vague and incomplete for them to render an opinion (which is useful to know). At the other extreme, different kinds of experts may be able to provide estimates for different elements of the system architecture for the envisioned system (e.g. the vehicle cabin, hydraulic platform, server and data repository, and the instructors' and administrators' workstations for the driving system simulator described in the sidebar in Chapter 6).

The primary strengths of expert judgment are

Different kinds of experts can provide estimates for different kinds of system elements or subsystems;

and

Experts can include subjective and political factors that are not typically recorded in databases of past projects.

The primary weaknesses of expert judgment are

Experts may be overly optimistic in estimating the time and resources needed for them to do the work rather than the time and resources needed by less-expert developers;

and

Their recall of past experiences may be incorrect or incomplete.

10.4.4 The Delphi Method

The Delphi method is a variation on expert judgment in which a group of experts make their individual estimates and document their rationales in isolation from one another. The method is based on the assumptions that a group meeting can bias individual estimates but that group estimates are more accurate and nuanced (if unbiased) than individual estimates.

A facilitator provides the documented attributes of the project to be estimated and an estimation form and then collects, collates, and summarizes the resulting estimates and rationales. Names are removed and the summaries are collated and provided to the experts for a second round of estimation. The experts may change their estimates and rationales based on the summaries of the other estimators or they may maintain their previous estimates and strengthen their rationales.

Each expert's estimates tend to stabilize and the group estimates may converge somewhat after two or three rounds of anonymous estimation. If wide disparities in the estimates remain after three rounds of estimation, the experts meet and express their concerns, which may or may not be resolved. Disparities in estimates are based on each expert's experiences (i.e. his or her BOE) and provide information for risk management.

The Delphi estimation method is named after the ancient Greek Delphi oracles, as described in "Greek Delphi."

Greek Delphi

The name "Delphi" is based on the oracles of Delphi in ancient Greece. The oracles were typically virgin women who were thought to speak for the various Greek gods. Temples were erected as sites for worshiping the gods and as a place for the oracles to live and be consulted; the women often lived in groups within the temples.

The oracles were famous for delivering ambiguous prophecies. In one instance, a Greek king sought the advice of an oracle about the outcome of an approaching battle with the Persians. The oracle replied: "If you cross the river, a great empire will be destroyed." The king thought the response was favorable but the battle was lost and the Persians destroyed his empire.

In another proclamation, an oracle advised that the best approach for an upcoming battle was: "Only the wooden palisades may save you." Some thought this meant that a wooden fence should be erected around the Acropolis and a stand taken there. Others thought it meant that the battle should be taken to the sea, and others thought it meant that the Athenian Greeks should flee by ship to Italy.

It was believed that Delphic prophecies were infallible. It was thought that prophetic failures were failures to correctly interpret the responses, not an error of the oracle.

10.4.5 Wideband Delphi Estimation

The wideband Delphi estimation method is a variant of the Delphi method developed by Barry Boehm and John Farquhar in the 1970s (Boehm 1981). Wideband Delphi proceeds as follows:

1. A facilitator distributes the project documentation and an estimation form to each expert.
2. After a brief period of study, the experts meet as a group with the facilitator to discuss estimation issues.
3. The experts fill out the estimation forms anonymously while sitting together.
4. The facilitator collects, collates, and distributes summaries of the estimates and rationales.
5. The experts discuss the summary results.

Steps 3–5 are repeated until each expert's estimate and rationale stabilize or until it is determined that consensus cannot be reached. The repetitions may occur on different days.

An electronic blackboard on which the experts' estimates and rationales are anonymously displayed when entered on their laptop computers or meeting room terminals facilitates wideband Delphi estimation.

Delphi teams and wideband Delphi teams usually include three to seven experts and a knowledgeable facilitator.

10.4.6 Regression-Based Estimation Models

Regression-based estimation models are based on equations derived from historical data collected from past projects. The equations are derived by regression analysis, which derives "best fit" equations from historical data (Chaatterjee and Simonoff 2012). The equations provide the BOE.

In most regression-based estimation models for systems and software projects, the primary estimation equation is based on regression analysis of the relationship between a single independent variable (e.g. size) and a dependent variable (e.g. effort), as in

$$E = a \times S^b,$$

where E is the effort, S is the size, and a and b are the constants derived by regression analysis.

For example, if size S is the number of system capabilities and E is the effort in staff-months, the equation might be

$E = 2.5 \times S^{1.2}$.

Thus, $E \sim 40$ when $S = 10$ and $E \sim 91$ when $S = 20$.

Note the nonlinear increase in effort when size is doubled.

Some measure of size is often used as the independent variable in regression-based estimation equations because size can be determined more objectively than other product attributes and size measures such as number of use cases or system capabilities can be estimated more accurately than other system attributes, especially in the early phases of a project.

Referring to Figure 10.5 as it relates to a regression-based estimation model,

- The BOE is the size-effort regression equation derived from historical data;
- Size is the primary input parameter; and
- An effort adjustment factor (EAF), which is the product of the collective adjustment factors in Figure 10.5, is applied as a multiplier in the effort–size equation:

$$E_{adj} = a \times S^b \times \text{EAF}.$$

Dr Barry Boehm and his colleagues developed the best-known regression-based estimation models based on large sets of collected data (USC 2018). He termed his first published models, developed for estimating software projects, "COCOMO," which is an acronym derived from the phrase "constructive cost model." The COCOMO models were termed "constructive" because Dr Boehm illustrated, by examples in his textbook, how to construct and validate a regression-based estimation model.

The EAF in the first-published COCOMO models is the product of 15 cost drivers, each of which, when set to value 1.0 (i.e. nominal) has no effect on the estimated effort. Experts using the wideband Delphi technique developed the set of cost drivers. The ranges of values for each cost driver were selected using statistical analysis and expert judgment applied to past projects.

Dr Boehm published the original set of COCOMO models in his textbook *Software Engineering Economics* in 1981 (Boehm 1981). The USC Center for Software and Systems Engineering, under Dr Boehm's leadership, has developed several other estimation models, including COCOMO II and the constructive systems engineering model (COCOMO II 2018; COSYSMO 2018). The SoftStar Systems web site provides access to a suite of regression-based estimation tools, including original COCOMO, COCOMO II, COSYSMO, and a tool for calibrating regression models using local historical data (SoftStar 2018).

A word of caution: The equations for regression-based estimation, as distributed in the estimation models, are calibrated using historical data available to the developers of those models. The models should not be used without validating them by comparing the estimates produced by the models to historical data for projects conducted in your organization. If an estimation model, as provided, cannot be locally validated, the model must be recalibrated by deriving new regression equations using local data. The referenced estimation models support recalibration by allowing changes to the parameters in the models (e.g. original COCOMO constants and exponents) and the values of the cost drivers.

10.4.7 Monte Carlo Estimation

The Monte Carlo estimation technique can be used to produce probability distributions for estimates. When using a regression-based estimation model, probability distributions for system size and cost drivers can be randomly sampled, repeatedly. Each sampled combination of inputs is used to produce an estimate. The process of developing a probability distribution of estimates is illustrated in Figure 10.7.

If the process is repeated a few hundred or a few thousand times, a histogram of probable effort can be generated, as illustrated in Figure 10.8.

The simulation that produced the histogram in Figure 10.8 was run 300 times. If 12 of 300 estimates are computed to have the same value of effort, E, the probable effort will be 0.04 (12/300). The probability that a project can be completed with an amount of effort less than or equal to a stated value, E, is determined by summing up all of the probabilities for values of effort less than or equal

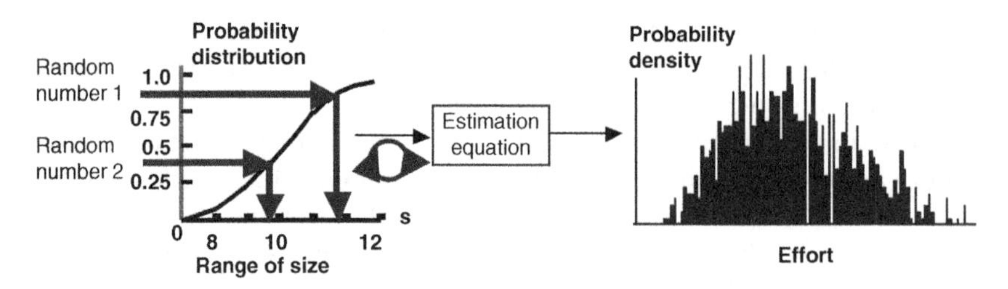

Figure 10.7 Illustrating Monte Carlo estimation. Source: Fairley 2009, figure 6.13a.

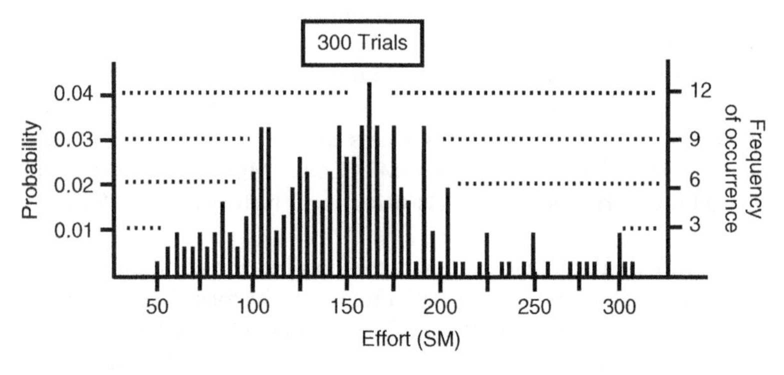

Figure 10.8 An effort histogram. Source: Fairley 2009, figure 6.13b.

to E. In the histogram in Figure 10.8, for example, it is 80% probable that the project can be completed with 200 staff-months of effort or less because 80% of the calculated values of effort are at, or to the right of, $E = 200$ SM. Adjustment factors can be applied, as desired.

Monte Carlo

Monte Carlo simulation is named for the principality of Monte Carlo, located in France. It is famous for gambling casinos that are, of course, based on probable outcomes. For many years, Monte Carlo simulation, as applied to business, science, and engineering, was the realm of specialists but the advent of powerful personal computers and applications software based on spreadsheets has made simulations easy to perform.

The negative side of easy use is that, to be correctly applied, Monte Carlo simulation requires an understanding of the underlying theory, how to specify input probability distributions, and how to interpret the resulting probability densities and distributions. As the saying goes

> A fool with a tool is still a fool.

A simulation tool, such as the Oracle Crystal Ball, can be used to specify probability functions for input parameters such as the estimated size and effort multiplier probability distributions in the input cells of a spreadsheet; i.e. probability functions are entered rather than single values (ORACLE 2018). The spreadsheet tool uses the equations provided by the user to repeatedly sample the probability distribution functions and compute solutions that are saved in output spreadsheet cells. A variety of graphs can be produced from the values in output cells, such as the histogram in Figure 10.8.

10.4.8 Bottom-Up Estimation

Bottom-up estimation is based on making estimates for technical work by aggregating the estimates for a set of system capabilities or system requirements at the lowest level of decomposition of an early architecture diagram, an architecture definition, or a WBS. Lower level estimates are rolled up to provide estimates for various elements in the hierarchy, including the top level.

Any of the methods previously addressed can be used to make bottom-up estimates for individual capabilities or individual system elements (ROT, analogy, expert judgment, Delphi, wideband Delphi, regression-based estimation, or Monte Carlo estimation). Different estimation methods can be applied to different subsystems. The lowest level estimates can be rolled up to any desired level in a hierarchy.

A WBS depicts hierarchical groupings of related work activities and the work products that will be developed by each activity. A partial WBS for an automated teller machine (ATM) project is illustrated in Figure 10.9.

As illustrated, a WBS is a tree structure that can also be represented as an indented list. Element 1 in Figure 10.9 "Manage Project" could be presented as follows:

1. Manage project
   - 1.1 Initiate project
     - 1.1.1 Identify stakeholders
     - 1.1.2 Develop/clarify requirements
     - 1.1.3 Prepare initial estimates
     - 1.1.4 Prepare initial project plan
     - 1.1.5 Obtain commitment to the plan
   - 1.2 Conduct project
     - 1.2.1 Measure and control the project
     - 1.2.2 Lead and direct the personnel
     - 1.2.3 Communicate and coordinate
     - 1.2.4 Manage risk
   - 1.3 Closeout project
     - 1.3.1 Obtain product acceptance

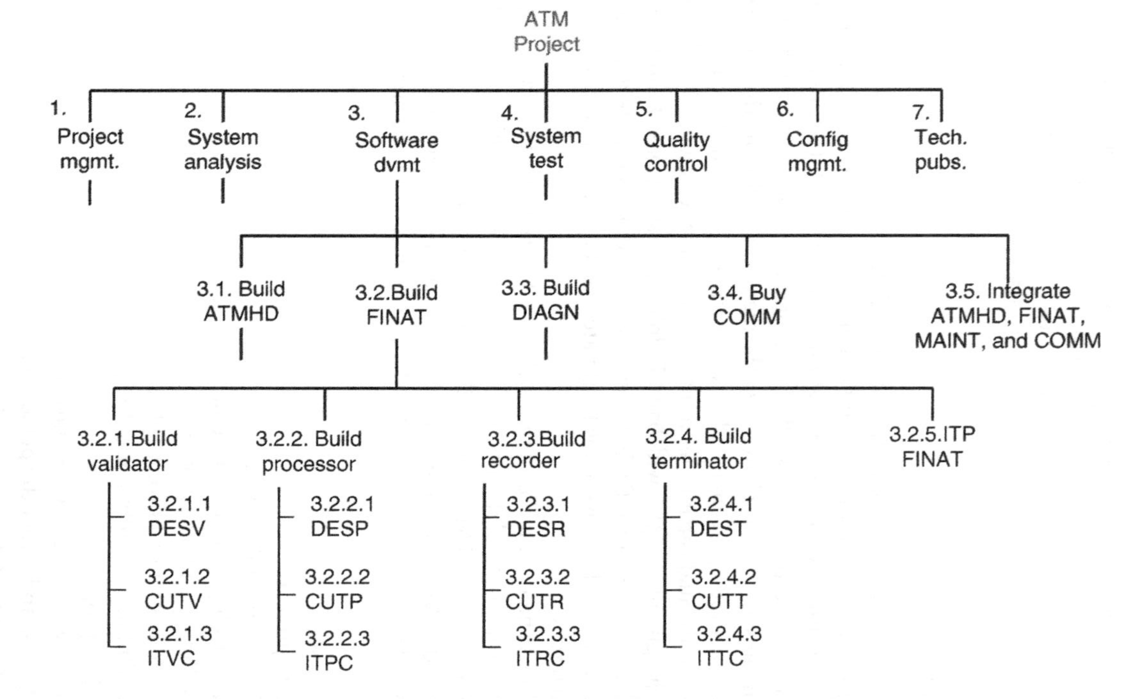

Figure 10.9 A partial WBS. ATMHD, hardware drivers; FINAT, financial transactions; DIAGN, diagnostics; COMM, communications; DES, design; CUT, code and unit test; ITP, integrate and test. Source: Fairley 2009, figure 5.5a.

1.3.2 Conduct a postmortem review
1.3.3 Prepare and distribute a lessons-learned report
1.3.4 Assist in reassigning project personnel

The PMI Practice Standard for Work Breakdown Structures provides information on developing and using a WBS (PMI 2006). Section 5.4 of the managing and leading software projects (MLSP) textbook provides 15 guidelines for designing WBSs (Fairley 2009).

Bottom-up estimation is perhaps the most accurate approach to estimation because of the increased level of detail at which various estimation techniques can be applied and because positive and negative variations in inaccuracies at lower levels may "average out" when aggregated at higher levels. Critical path and program evaluation and review technique (PERT) estimates can be prepared for scheduling of work activities using the estimated schedules for each work element. The critical path method is presented in Section 5.5.1 of the MLSP text and the PERT method is presented in Section 5.5.2 of MLSP.

The primary strength of bottom-up estimation is

- The increased accuracy of estimates that result from an increased level of detail and the accompanying level of understanding.

The primary weakness is

- Lack of sufficient knowledge or time to prepare and decompose system capabilities, system requirements, a system architecture, or a WBS early in a project. However, bottom-up estimation should be used as soon as capabilities, requirements, or architecture begin to emerge; periodical reestimation should occur as the system evolves.

10.5 Documenting an Estimate

Organizations should have a standard template for documenting estimates, so that all projects include the same factors, which will facilitate development of lessons learned and a database of consistent estimates can be developed to provide or improve a BOE. The items that should be included in an estimation template are the following:

- Project identifier;
- Version number and date of the estimate;
- Total estimated effort;
- Total estimated schedule;
- Name(s) of the estimator(s);
- Amount of time and effort spent in making the estimate;
- Rationale for the estimate (i.e. why is this estimate being made? feasibility, initial estimate, periodic update, aperiodic update, etc.);

- Elements changed (for updates to an estimate);
- Estimation methods and tools used;
- The BOE for each method or tool used (industry averages, expert judgment, local historical data, etc.);
- A list of assumptions made for each method or tool used;
- A list of constraints observed in making the estimates;
- A list of inputs used for each method or tool used (e.g. size, constraints, adjustment factors);
- Estimates provided by each estimation method or tool (e.g. total effort, schedule, project milestones, effort for various project activities by project phase, estimated prerelease and postrelease defects, estimated reliability at product delivery, total life-cycle costs);
- A range of estimates for effort, schedule, resources, cost, and quality attributes with associated probabilities for each method or tool used;
- Identified risk factors for the project;
- The estimator's level of confidence in the accuracy of the estimate (0–10, low, medium, high); and
- Information, resources, and time needed to make an improved estimate.

Estimation, like all engineering processes, should be conducted in accordance with a well-defined procedure (i.e. a set of steps to be followed). A multistep estimation procedure is listed and discussed here.

1. Determine the purpose of, and required confidence in, the estimate.
2. Determine the information needed and sources for it, to include estimated size, complexity, and required quality attributes.
3. Plan the schedule, resources, and responsibilities for developing the estimate.
4. Develop the system capabilities, system requirements, and system architecture in as much detail as possible and as warranted.
5. If warranted and possible, develop the size, complexity, and required quality attributes for each system requirement and each element in the system architecture.
6. Select two or more estimation techniques to be used.
7. Supply any additional factors required by the estimation techniques to be used. Prepare estimates using the selected estimation techniques.
8. Conduct sensitivity analyses on the estimates.
9. Reconcile differences in the estimates.
10. Document risk factors exposed by the estimation process.
11. Prepare a plan for updating the estimate at periodic intervals and as aperiodic events occur.
12. Prepare and implement a plan for baseline retention of estimation data, the documented estimates, and on-going updates to the estimates.

13. Document the estimate using a standard template for estimates, to include the information similar to that in the template provided in Section 10.5.1.

As with all processes, the procedural steps listed above should be tailored to fit the needs of the situation. If Step 1 (determine the purpose of and required accuracy of the estimate) reveals that the estimate is a "ball park" estimate to determine the feasibility of a contemplated project, a quick ROT calculation may be sufficient. If Step 1 reveals that the estimate is for the organization's next major system, on which the survival of the company may depend, two or more independent estimates using different estimation techniques should be conducted with great care and may involve feasibility studies, prototyping, and analysis of the competition.

Several steps in the estimation procedure use the phrase "in as much detail as possible and as warranted." Depending on the purpose and criticality of the estimate, development of system architecture before making a commitment estimate may or may not be warranted.

Step 6 indicates that more than one estimation technique should be used and Step 10 calls for reconciling the difference in the estimates produced by multiple techniques. Again, depending on the purpose of the estimate and the information available for making the estimate, use of multiple estimation techniques may or may not be warranted; however, you should use multiple techniques if you are preparing to make commitments for a project based on the estimates.

To reconcile differences produced by different estimates, Step 9 in the estimation procedure calls for conducting a sensitivity analysis on each resulting estimate. Sensitivity analysis is concerned with determining the sensitivity of variations in the estimated outputs based on variations in the estimation inputs. Large variations in estimated outputs that result from small variations in the inputs indicate that the estimation technique is sensitive to those input parameters. Knowing that the estimated values are sensitive to certain input values may result in closer examination of those inputs and may help to explain why two estimates produced by two different techniques do not agree.

For example, product size is the primary input parameter to many estimation methods and is thus a sensitive input parameter. In COCOMO II, the combined effect of the personnel effort multipliers is the second most sensitive parameter. The combined effect of the range of values in personnel effort multipliers, as specified in the COCOMO II textbook, can cause variations of approximately 10 : 1 in effort estimates for software projects, which is consistent with the observations of others (DeMarco and Lister 1999).

Step 11 (document risk factors exposed by the estimation process) provides information to be included in the documented estimation; also, it can be used to prepare the risk management plan (see Chapter 11). The estimation procedure may have revealed, for example, that the requirements are too vague to support an accurate estimate, that the schedule constraint results in unacceptable risk to successfully completing the project within the constrained duration, or that the system developers do not have sufficient knowledge of the new development environment to successfully use it on the project.

Step 12 (prepare a plan for updating the estimate at periodic intervals and as aperiodic events occur) is concerned with preparing a plan to keep the estimate current as understanding of the project increases, and as conditions change. Many organizations update project estimates on a monthly basis. Changes in the requirements, reduction of the budget, or loss of a key person are examples of aperiodic events that would warrant revision of an estimate.

Step 13 (prepare and implement a plan for baseline retention of estimation data, the documented estimate, and on-going updates to the estimate) is concerned with version control of documented estimates, the data on which each estimate is based, and updated versions of the estimate that are created periodically and aperiodically. And as with all revisions to baselined work products, the following information should be recorded for each version of an estimate:

- Date of the revision;
- Who made the revision;
- Reasons for the revision;
- Data used to make the revision;
- Elements changed; and
- Who was notified of the change.

Baseline control of documented estimates removes ambiguity as to which estimate is the current one and creates an audit trail of how and why the estimates changed over time.

The final step in the estimation procedure, Step 14 (document the estimate using a standard template for estimates), should be based on a standard template for recording estimates that is used throughout the organization. It is important to use a standard template to document estimates so that all estimates for all projects record similar information, from which consistent historical data across all projects can be accumulated. The template should provide for recording the information listed in the following section.

10.5.1 An Estimation Template

Organizations should have standard templates for recording estimates for the overall project and for the technical work activities. The template should support documenting and reporting of the following information:

- Project identifier;
- Version number and date of the estimate;
- Project domain and organizational unit;
- Name(s) and organizational unit(s) of the estimator(s);

- Amount of time spent in making the estimate;
- Scope of work included in the estimate;
- Total estimated effort;
- Total estimated schedule;
- Rationale for the estimate (i.e. why is this estimate being made? feasibility, initial estimate, periodic update, or aperiodic update);
- System elements and/or work elements changed for the update (after the initial estimates);
- Estimation methods and tools used;
- The BOE for each method or tool used (industry averages, expert judgment, local historical data, regression equations, etc.);
- A list of inputs for each method or tool used (e.g. size, adjustment factors, constraints);
- A list of assumptions made for each method or tool used;
- A list of constraints observed in making the estimates;
- Estimates data provided by each estimation method or tool (e.g. total effort, schedule, project milestones, effort for various project activities by project phase, estimated reliability at product delivery, total life-cycle costs);
- A range of estimates for effort, schedule, resources, cost, and quality attributes with associated probabilities for each method or tool used;
- Risk factors for the project;
- The estimator's level of confidence in the accuracy of the estimate (1–5; low, medium, high); and
- Information, resources, and time needed to make an improved estimate.

10.6 Key Points

- An SEMP provides plans for the technical work activities to be accomplished.
- The SEMP is subordinate to and closely integrated with the PMP.
- For small projects, the items in an SEMP may be embedded in the PMP.
- The SEMP should be updated periodically and aperiodically as events dictate.
- A standard template should be used when preparing an SEMP. The template may be tailored as appropriate.
- A project estimate is a projection from past and present to future, suitably adjusted to account for difference between past and future.
- Accurate estimates are based on a set of assumptions that must be realized and a set of constraints that must be satisfied.
- Estimation methods include ROT, analogy, expert judgment, Delphi, wide-band Delphi, and regression-based estimation tools.
- Regression-based estimation models must be calibrated using local data.
- Estimate should be based on a stable BOE.

- Size is the primary variable in most estimation models. Adjustment factors are applied to adjust for the difference between the BOE and the project being estimated.
- Projects must be reestimated periodically as understanding grows and aperiodically as project parameters change.
- Estimates should be prepared using at least two different methods.
- Estimates should be documented using a standard template.

Exercises

10.1. Technical management is based on the methods and techniques of project management.
(a) Briefly describe the similarities of technical management and project management.
(b) Briefly describe the difference between technical management and project management.

10.2. Project management plans (PMPs) are developed by project managers and systems engineering management plans (SEMPs) are developed by systems engineers.
(a) Briefly describe the similarities of PMPs and SEMPs.
(b) Briefly describe the differences between PMPs and SEMPs.

10.3. The 14 topics to be included in an SEMP are listed in Section 10.2. Search the Internet to find one or more templates for SEMPs.
(a) Briefly explain the similarities between the SEMP topics in Section 1.2 and the topics you found in other SEMPs.
(b) Briefly explain the differences between the SEMP topics in Section 1.2 and the topics you found in other SEMPs.

10.4. An estimate is a prediction of future events based on historical data and current data adjusted for differences between the historical plus current data and what is known or assumed about the future.
(a) Identify and briefly describe two areas of modern society other than engineering where estimates are made using this definition of estimation.
(b) Briefly describe the historical and current data that is used to make estimates or each of your identified areas.

10.5. Some adjustment factors for systems engineering projects are listed in Section 10.3. Identify and briefly describe three additional adjustment factors that could be used when making estimates for systems engineering projects.

10.6. Estimation principle #2 in Section 1.3 is: All estimates are based on a set of assumptions that may or may not be true and a set of constraints that may or may not be satisfied.
  (a) Identify three assumptions that may not be true for technical management of systems engineering projects.
  (b) Identify three constraints that may not be satisfied for technical management of systems engineering projects.

10.7. Estimation methods for engineering projects include rule of thumb and analogy.
  (a) Briefly describe a rule of thumb estimate you use in your day-to-day life.
  (b) Briefly describe an analogy you use in your day-to-day life.

10.8. Estimation methods for engineering projects include Delphi and wideband Delphi. Identify and briefly describe two areas of modern society other than engineering where the Delphi or wideband Delphi estimation method could be used.

10.9. Briefly describe two reasons a regression-based estimation equation could be misapplied when making an estimate.

10.10. Briefly describe two reasons a standard estimation template should be used for all project estimates.

References

Boehm, B. (1981). *Software Engineering Economics*. Prentice Hall.

Boehm, B., Abts, C., Brown, A. et al. (2000). *Software Cost Estimation with COCOMO II*. Prentice Hall.

Chaatterjee, S. and Simonoff, J. (2012). *Regression Analysis*, 1e. Wiley.

COCOMO II (2018). Constructive cost model II. http://csse.usc.edu/csse/research/COCOMOII/cocomo_main.html#downloads (accessed 30 July 2018).

COSYSMO (2018). SystemStar: The COSYSMO estimation tool. http://www.softstarsystems.com/oversys.htm (accessed 24 July 2018).

DeMarco, T. and Lister, T. (1999). *Peopleware*, 2e. Dorset House.

Fairley, R. (2009). *Managing and Leading Software Projects*. Wiley.

Fairley, R. et al. (2013). *Software Extension to the PMBOK® Guide (SWX)*, 5e. Project Management Institute https://www.pmi.org/pmbok-guide-standards/foundational/pmbok/software-extension-5th-edition (accessed 9 June 2018).

ISO (2015). ISO/IEC/IEEE 15288:2015, Systems and software engineering – System life cycle processes, ISO/IEEE, 2015.

ISO (2016). ISO/IEC/IEEE 24748-4:2016, Systems and software engineering – Life cycle management – Part 4: Systems engineering planning, ISO/IEEE, 2016.

ISO (2017). ISO/IEC/IEEE 12207:2017, Systems and software engineering – Software life cycle processes, ISO/IEEE, 2017.

ORACLE (2018). Oracle crystal ball. https://www.oracle.com/applications/crystalball/index.html (accessed 30 July 2018).

PMBOK (2017). *A Guide to the Project Management Body of Knowledge*, 6e. Project Management Institute https://www.pmi.org/pmbok-guide-standards/foundational/pmbok (accessed 15 June 2018).

PMI (2006). *Practice Standard for Work Breakdown Structures*. Project Management Institute.

SoftStar (2018). SystemStar and Costar – Tools for software estimation and systems engineering estimation. http://www.softstarsystems.com/index.html#How%20to%20Contact%20Us (accessed 30 July 2018).

USC (2018). The Center for Systems and Software Engineering. http://csse.usc.edu/csse/about/ (accessed 30 July 2018).

11

Assessing, Analyzing, and Controlling Technical Work

11.1 Introduction

Managing the technical activities of a systems project, as practiced by physical systems engineers and software systems engineers, involves planning and estimating (Chapter 10); assessing, analyzing, and controlling (this chapter); and leading and coordinating the work activities of the system developers (Chapter 12).

Plans and estimates provide objective targets against which progress of the work activities and the resulting work products can be determined. Determining project status involves assessing expended effort, schedule, and budget; the progress of work product development; and comparing plans to actuals. Relationships among project attributes are also examined. For example, it is possible to make good schedule progress while expending more effort than planned, or to make good cost and schedule progress while developing poor quality system elements. The goal of technical management is to keep effort, schedule, and progress in balance.

Progress reports (generated periodically and aperiodically as events dictate) indicate the project attributes that are conforming to plans and those that need to be investigated for possible corrective action. If, in a given reporting period, effort is as planned but personnel cost is higher than planned, for example, this indicates that more expensive (i.e. more highly skilled) system developers than planned are being utilized, perhaps because the work is more difficult than anticipated, thus justifying more highly skilled system developers, or perhaps because the work is being done by highly skilled, and expensive, system developers but the work is complex and is taking longer than planned.

Other costs should also be measured and compared with plan. Travel cost may be higher (or lower) than planned; equipment costs may similarly deviate from plan. In any case, deviations from plan indicate that the deviations need to be investigated and corrective action should be applied, if needed.

Project control is exerted by applying corrective action when the relationships among project attributes becomes unbalanced (as above), or when one

Systems Engineering of Software-Enabled Systems, First Edition. Richard E. Fairley.
© 2019 John Wiley & Sons, Inc. Published 2019 by John Wiley & Sons, Inc.

or more dimensions of progress deviate from plan by more than an acceptable amount; for example, a delay of two days in achieving a major milestone may not require corrective action but being two weeks late may constitute an unacceptable delay for which corrective action must be taken. Similarly, a 2% overrun of allocated memory for an incremental build of an embedded software-enabled subsystem may be acceptable but a 20% overrun is probably not acceptable.

Reference material for this chapter include the following:

- *A Guide to the Project Management Body of Knowledge* (PMBOK 2017)
- ISO/IEC/IEEE 15288:2015, Standard, Systems and software engineering – System life cycle processes (ISO 2015)
- ISO/IEC/IEEE Standard 12207:2017, Systems and software engineering – Software life cycle processes (ISO 2017)
- *Managing and Leading Software Projects* (Fairley 2009)

This chapter provides the opportunity for readers to develop an understanding of principles, methods, and techniques used to assess, analyze, and control the technical work activities and the resulting work products of a system development project. This chapter is keyed to the I^3 system development model but the methods and techniques can be applied to other kinds of system development.

References and Exercises are provided to gain access to more information and to gain additional understanding.

11.2 Assessing and Analyzing Process Parameters

The first consideration for assessing and analyzing technical work activities is: what should be assessed and analyzed?

At minimum, the following attributes of technical work being performed should be assessed and analyzed:

Process parameters to be assessed along with the relationships among them include the following:

- *Effort*: Amount of work expended for various work activities;
- *Schedule*: Achievement of objectively measured milestones;
- *Cost*: Expenditures for various kinds of resources, including effort;
- *Progress*: Work products completed, accepted, and baselined; and
- *Technical risk*.

Product parameters include the following:

- *System capabilities*: Capabilities implemented and demonstrated;
- *Quality attributes*: Those visible to system users and other external stakeholders, and those visible to system developers (see Table 7.2);
- *System performance parameters*;

- *Rework backlog*: Work needed to clear work product discrepancies; and
- *Stability of system requirements.*

Process attributes are measured against the product attributes. Among the process attributes, schedule may (or may not) be more important than effort or cost, and security may be a more important system attribute than performance. Depending on the relative importance of the various process and product attributes, more effort may be expended on measuring and assessing some attributes than on measuring and controlling others.

Product and process measures are, or should be, byproducts of the procedures, methods, tools, and techniques used to develop a system. Excessive time, effort, and cost spent in obtaining, analyzing, and acting on product and process measures is a symptom of ineffective development and management processes that need to be improved.

Various methods can be used to collect, analyze, and display process assessment results, including

- Effort assessment;
- Binary progress assessment;
- Technical debt assessment;
- Earned value (EV) assessment; and
- Risk assessment.

Each is addressed in turn.

11.2.1 Assessing and Analyzing Effort

As illustrated in Figure 11.1, there are three categories of effort (Fairley and Willshire 2005):

- Original work;
- Evolutionary rework; and
- Avoidable rework (retrospective and corrective).

Original work is the effort expended in establishing initial baselines of work products (e.g. capabilities, requirements, design, realized system elements, test

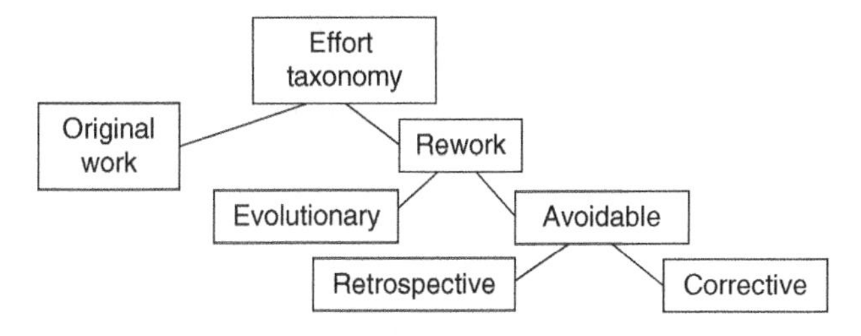

Figure 11.1 A taxonomy of project effort. Source: Fairley 2009, figure 8.1 (p. 337).

plans). Evolutionary rework is spent to change baselined work products in ways that add value to an evolving system. For example, the system design and some system elements may have to be reworked because of a change in requirements that results from a change of mission for a customer's mission-critical system.

Avoidable rework is rework required to make changes to baselined work products developed during original work or evolutionary rework. In principle, avoidable rework (as the name implies) is work that should not be done. But because humans are not infallible, some small amount of avoidable rework is in fact unavoidable. As indicated in Figure 11.1, there are two kinds of avoidable rework: retrospective and corrective.

Retrospective rework occurs, for example, when one or more system elements must be reworked during iterative or incremental development to accommodate the elements to be developed during the next iteration or incremental phase, for example, reworking an interface or the system structure. Excessive amounts of retrospective rework may indicate the need for more attention to interface design, for example.

Corrective rework occurs when an error in a work product has to be fixed. Because corrective rework is the result of mistakes made by humans, and because humans are not infallible, a certain amount of avoidable rework is to be expected. A large percentage of total effort devoted to avoidable rework, like a high fever in a sick person, is a problem in itself, but more significantly, it is an indicator of other serious problems. In the case of systems projects, unacceptably high levels of avoidable rework are a symptom of inefficient and ineffective work processes, skills, and/or tools.

Refactoring (i.e. restructuring) of software during iterative development provides an example of the distinctions among evolutionary rework, retrospective rework, and corrective rework. Evolutionary rework occurs when refactoring of the system structure is done to accommodate a new requirement that could not be foreseen. Retrospective rework occurs when a feature that is known to be needed as a basis for the next system increment is not included in the previous iteration and thus has to be added now. Corrective rework occurs when mistakes are discovered in a work product that have been previously reviewed, tested, demonstrated, and accepted.

Avoidable rework (retrospective and corrective) is the bane of most system development organizations. Avoidable rework of 5% or less of total effort is attained in the best software organizations; 20% or less of total work attributable to avoidable rework is a modest goal that can be achieved by most system development organizations (Fairley and Willshire 2005).

Unfortunately, many organizations do not separately assess original work, evolutionary rework, retrospective rework, and corrective rework and are thus unaware of how project resources are being spent. In many organizations, avoidable rework is a large percentage of total project effort;

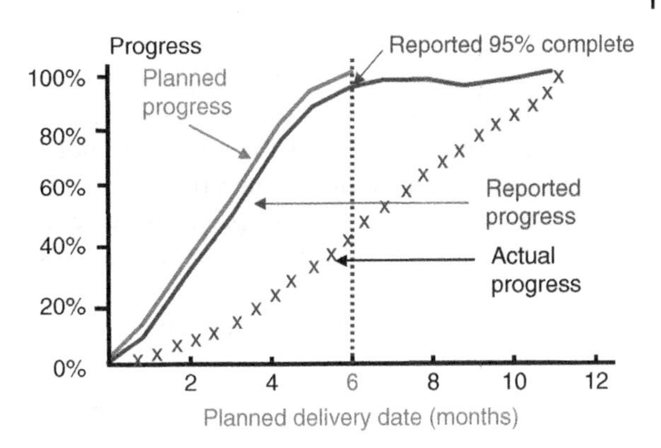

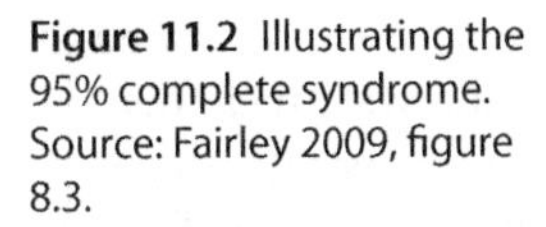
Figure 11.2 Illustrating the 95% complete syndrome. Source: Fairley 2009, figure 8.3.

these organizations would probably initiate corrective actions if they were aware of wasted effort.

11.2.2 Assessing and Analyzing Binary Progress

Progress of technical work can be accurately assessed using binary assessment. The concept was termed "binary deliverables" by DeMarco and separately termed "binary tracking" by Fairley, both in the early 1980s; it is named "binary assessment" in this text (Fairley 1985; DeMarco 1986). When using the I^3 system development model, verification, validation, and demonstration of system capabilities are based on objective completion criteria for realization of system capabilities. Using binary assessment, implementing a system capability is counted as 0% complete until the associated completion criteria are satisfied. The capability is then counted as 100% complete; assessment is thus binary.

Binary assessment of progress helps to avoid the 95% complete syndrome of systems projects:

> Assessing progress using "guesstimates" often results in systems being reported as 95% complete as the planned completion date approaches and remain reported as 95% complete for a long period of time.

Binary assessment is depicted in Figure 11.2, where each "X" represents progress based on binary assessment of realized system capabilities and provides accurate assessment of actual progress. Also note in the figure that actual progress deviated from planned progress early in the project. Corrective action should have been taken no later than the second month of the project.

Binary completion criteria can also be applied at a lower level of system development, for example, when implementing system elements needed to realize a system capability. Successful execution of test cases that were developed before implementing a system element (i.e. test-driven development, as described in Section 4.5.2) provides objective completion criteria for the system element.

Note that Figure 11.2 is intentionally exaggerated to illustrate the 95% completion syndrome. This situation would not (should not) occur if the I^3 capabilities-based incremental iterative-increment system development process, with binary assessment of progress, were used. Unfortunately, systems projects that do not utilize binary assessment sometimes display the 95% syndrome illustrated in Figure 11.2.

Figure 11.2 illustrates the maxim of binary assessment:

> It is better to be 100% complete with 90% of the work and know it is true (binary assessment at month 10), than to think that 100% work is 90% complete and hope it is true (the guesstimate at month 5).

Because binary assessment does not give credit for work in progress, assessment accuracy can be improved by decomposing large capabilities into smaller units. For example, suppose Capability A is decomposed into three capabilities: capabilities A1, A2, and A3. For simplicity of the example, it is assumed that Capabilities A1, A2, and A3 each will require equal amounts of effort; weighting factors could be applied otherwise. Binary completion of Capability A1 will indicate 33% completion of Capability A, binary completion of Capabilities A1 and A2 indicates 66% completion of Capability A, and so forth.

In general, smaller capabilities (i.e. smaller units of required effort) increase the accuracy of assessment but at the risks of micro-management and excessive project resources devoted to assessing and controlling. To avoid these problems, effort should not be planned and assessed in units smaller than one staff-week. Individual engineers may decompose their work into small units; for example, a software engineer may complete several internal construction iterations during a weekly iteration cycle while constructing a software element for a software-hardware capability, but for purposes of assessing and measuring the effort and binary assessment of work completed should not be sized smaller than one staff-week.

After an implemented capability satisfies its acceptance criteria, the capability work package is closed and never reopened. Subsequent modifications of capabilities are documented in change requests and defect reports, which facilitates assessment of rework, as described in Section 11.2.1.

The lower limit on assessing effort and the resulting work products using binary assessment should not be less than one staff-week. The upper limit on effort sizing for a capability is determined by considerations of risk. The fundamental purpose of assessment and control is to provide frequent demonstrations of progress and early warning of problems when progress is not as planned or expected. If a capability is scoped to require a large amount of effort, and completion of the task is not as planned, early warning of problems will not be obtained. A general guideline (i.e. a rule of thumb):

> Each capability to be implemented should be sized to require between 1 staff-week and 1 staff-month of effort. Ideally, capabilities will be sized for completion within one to two weeks.

Because effort is the product of people and time, two people might implement a two staff-week capability in one week, or one person in two weeks, thus providing frequent assessment of progress. The effort sizing includes the combined efforts of all engineers, of various disciplines, who will be involved in concurrently realizing a capability. Large capabilities will need to be decomposed to the indicated size.

Suppose, for example, that the technical effort required to develop the capabilities for a system is estimated to require 10 people for 15 months (150 staff-months of effort). If implementing each capability were estimated to require, on average, two staff-weeks, the technical effort would consist of 300 units of effort, with 20 capabilities completed each month, on average. Planning and assessment at this level of detail is best accomplished in an iterative-incremental manner, as in the I^3 system development model.

In some cases, completing implementation of a capability may be delayed because it will be dependent on implementation of another capability or perhaps dependent on availability of hardware or software that is being procured.

Larger projects (e.g. 200 people for 30 months) should be organized into teams of 10 or fewer staff-members per team. Each team leader, working with her or his team members, will coordinate planning of work units, assure that binary acceptance criteria for work products are satisfied, and use binary assessment to measure and to report progress (or lack thereof) to the physical systems engineer(s) and software system engineer(s), their staffs, and other appropriate stakeholders.

11.2.3 Estimating Future Status

Table 11.1 provides an example of using binary assessment to determine the status of technical work in progress; the table indicates that 900 of 1000 system capabilities are sufficiently covered by the system architecture, detailed design specifications for 750 capabilities have been developed, 500 of the 1000 capabilities have been realized, 200 capabilities have been successfully verified, and 28 of 200 capabilities have been validated.

Clearly, the product is being developed in an iterative and incremental manner because 14% of the realized capabilities have been validated while the design of 250 capabilities remain to be completed. The "percent complete" column is accurate because binary assessment of work products is being reported. Work

Table 11.1 An example of project status using binary assessment.

Status	Percent complete (%)
900 of 1000 capabilities traced to the architecture	90
750 of 1000 capabilities designed	75
375 of 750 capabilities realized	50
75 of 375 capabilities verified	20
28 of 200 capabilities validated	14

Source: Fairley 2009, table 8.7.

Table 11.2 Percent of effort for various work activities.

Work activity	Percent of development effort (%)
Architecture definition	17
Design definition	26
Capability implementation	35
Verification	10
Validation	12

Source: Fairley 2009, table 8.8.

required for the remaining 30 requirements may be 80% or 90% completed but they are not counted as progress measurements until they are 100% complete.

Table 11.2 is used to determine overall percent complete for the technical work. The table contains typical percentages of effort for various types of work activities in the organization where the project is being conducted.

Using Tables 11.1 and 11.2, it can be determined that 17% of the work is 90% complete (architecture definition), 26% of the work is 75% complete (design definition), and so forth. Combining the two tables provides the overall percent complete for the project:

$$(90 \times 0.17) + (75 \times 0.26) + (50 \times 0.35) + (20 \times 0.10) + (14 \times 0.12) = 56\%\text{complete.}$$

The project is 56% complete and thus 44% of the work remains to be completed.

Suppose the technical work is at the end of its seventh month and 75 staff-months of effort has been expended thus far. If 75 staff-months are 56% of total project effort, the remaining 44% will require 60 staff-months of effort ([44/56] × 75). Average staffing level has been approximately 11 system developers (75/7); 60 staff-months of remaining effort using 11 system developers will require ~5.5 additional months (60/11). It is therefore estimated that the project will require a total schedule of 12.5 months (7 + 5.5).

The project could be completed in 11 months if 4 engineers could be added in a manner that did not slow progress while assimilating the new developers and if the new developers were instantly as productive as the present 11 system developers. It is highly unlikely that the project could be completed in 9 months total time (another 2 months) by adding 19 system developers (60 developer-months/2 months = 30 developers; 30 = 11 present + 19 new).

A caution: The above calculations assume that the remaining work to be done for each capability is at the same level of granularity and difficulty as work completed and that a constant rate of progress will prevail for the remaining work activities. It may be that the most difficult parts are completed; i.e. the 250 remaining capabilities are all simple, and the project is more than 56% complete. Conversely, the 250 remaining capabilities might be the most complex ones and the project is less than 56% complete.

The accuracy of the estimate-to-complete can be improved as follows: first, decompose each capability and the associated requirements until

- complexities and risk factors are exposed;
- opportunities for reuse of existing system elements are identified; and
- effort to implement each capability can be estimated.

Then, weighting factors (say, on a scale of 1–5) can be applied to each remaining capability to be developed based on relative estimated effort. It may be, for example, that some capabilities are estimated to require three times or five times the amount of effort compared with some of the other capabilities. If a capability is estimated to be a "20" in relative effort compared with other capabilities on a scale of 1–5, that capability clearly needs to be decomposed into a set of smaller, derived capabilities. The remaining effort and schedule to complete the project can be calculated more accurately using the weighting factors.

11.2.4 Assessing and Analyzing Technical Debt

Technical debt is incurred when some of the work needed to develop a system is deferred until later, either knowingly or unknowingly (Fairley and Willshire 2017). Technical debt that is not "paid" can result in a delivered system that has inadequate functionality or behavior, has unacceptable quality, lacks supporting documentation, is hard to use or operate, and perhaps has other undesirable attributes.

The term "debt" is appropriate because not confronting a technical issue until later typically incurs a high rate of interest in the form of increased effort, time, resources, technology, and/or cost to correct a problem that could have been avoided if a different decision had been made earlier. And, like a rate of financial interest, the incurred debt increases with increasing time because the interest is compounded. Debt cascades because deferred work can result in problems in subsequent work. Accumulating technical debt is akin to placing a mortgage on future endeavors that must be repaid later.

Technical debt is often incurred because of constraints that cannot be violated and assumptions that are later shown to be false. Constraints on requirements, design, schedule, budget, staffing, resources, or technical parameters can inhibit development of a satisfactory technical solution because of lack of time, resources, and other constraints that prevent doing what needs to be done when it needs to be done. Assumptions that are believed to be true (or believed will be true) and are later shown to be false can result in accumulation of technical debt. For example, technical debt may be incurred because initial system requirements were assumed to be stable and feasible but were later shown to be volatile or infeasible.

And sometimes, technical debt is unknowingly incurred when skills, processes, methods, or tools are inadequate to avoid technical debt. This may occur when project managers and system developers have inadequate skills or processes, or when an organization enters a new line of business without sufficient preparation or contingency planning, or when innovative extensions to an existing system are pushed beyond limits of technology, knowledge, or ability.

In some cases, technical debt is passed on to others (i.e. users, customers, operators, maintainers, other stakeholders). Those others may unknowingly accept a less-than-acceptable system or they may knowingly accept the system and pay the debt to improve it. However, delivery of a system that carries technical debt often results in negative consequences for the system development organization (e.g. loss of reputation or future business). Other consequences of unpaid technical debt may be manifest for system stakeholders as loss of monetary units, degradation of equipment or data, or injury to and loss of human life.

Assessing technical debt requires a plan for progress and periodic comparison of actual progress to planned progress. Figure 11.3 (similar to Figure 11.2) illustrates planned and reported progress versus actual progress for the technical work of a systems project. The vertical bars depict technical debt as a deviation between reported progress and actual progress. Well-managed technical work would have confronted accumulating technical debt early in system development if actual status had been accurately reported using binary assessment. Also depicted is the well-known 95% complete syndrome cited earlier.

Reducing or eliminating technical debt may require rework of existing artifacts (see Section 11.2.1). Rework to clear technical debt may create the risk of

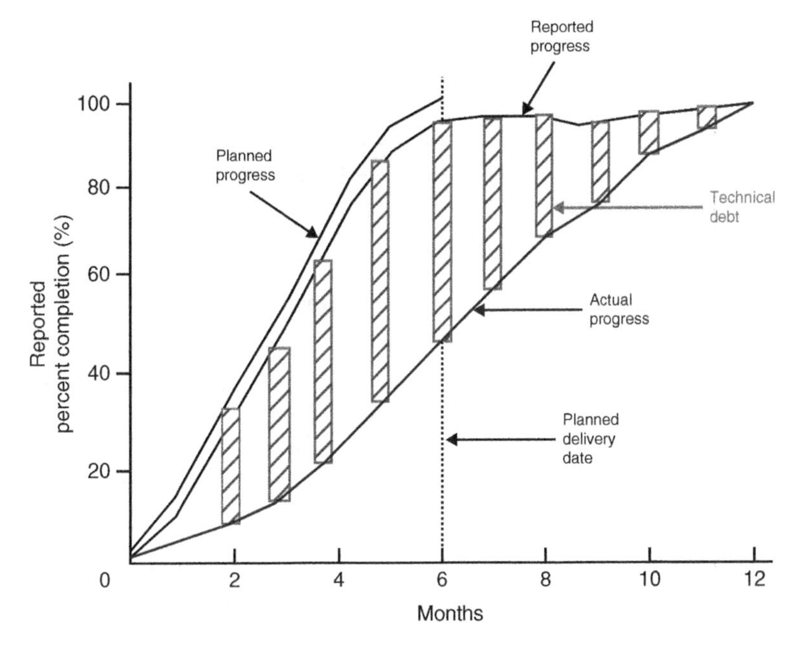

Figure 11.3 Illustrating accumulation of technical debt. Source: Fairley and Willshire 2017, IEEE Computer, May 2017 (p. 81).

accumulating future technical debt if the effort, schedule, and resources used to clear present technical debt results in violation of project constraints when accomplishing future work.

Well-managed technical work will accumulate small amounts of technical debt and discharge it during the subsequent iterative-incremental subphases of the system implementation phase (i.e. Phase 5 of the I^3 system development model). Figure 11.4 (a burndown chart) illustrates an iteration cycle of 15 work-days duration (i.e. three work-weeks) during which 27 of 30 planned tasks were completed. The remaining three tasks will be incorporated for completion in the plan for the next iteration cycle.

An occasional "time out" iteration cycle or incremental phase may be needed to clear accumulated technical depth that cannot be cleared routinely.

11.2.5 Assessing and Analyzing Earned Value

Earned value is widely used to assess and analyze the progress of technical work for a systems project. Binary assessment of task completions can provide accurate assessments. The EV process for assessing and analyzing technical work is as follows:

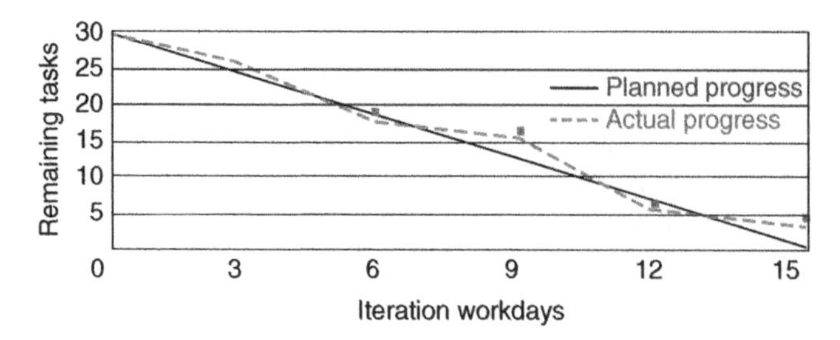

Figure 11.4 Burndown chart for an iteration cycle. Source: Fairley and Willshire 2017, IEEE Computer, May 2017 (p. 85).

1. Specify the planned effort and duration for each task to be completed during the coming iterative-incremental subphase of the implementation phase (Phase 5) of the I^3 system development model (see Chapter 5).
2. When a task is assessed as completed using binary assessment, the planned effort for the task is credited as its "earned value" (i.e. *earned value* is the planned effort that is "earned back" when a task is completed).
3. For all tasks completed to date, compare the sum of the earned values (i.e. the planned effort for all of the completed tasks) to the sum of the actual effort for the tasks.
4. For all tasks completed to date, compare the number of tasks that should have been completed to those that have been completed.

If the cumulative actual cost for all completed tasks to date is greater than the planned cost for those tasks (Step 3), the technical work is over budget; conversely, if actual cost is less than planned cost, the project is under budget. Similarly, if fewer tasks have been completed to date than planned for completion, the technical work is behind planned schedule; if more tasks have been completed than planned, the technical work is ahead of planned schedule (Step 4). Binary assessment assures that the technical work as reported to be complete is actually complete because binary tracking indicates that the work products have passed their acceptance criteria.

The terminology of earned value is summarized in Table 11.3.

From the formulas in Table 11.3, it can be seen that a CPI (cost performance index) > 1 means the actual cost is greater than the budgeted cost. A SPI (schedule performance index) > 1 means the technical work is behind planned schedule; conversely, a CPI < 1 indicates the technical work is ahead of planned schedule. A CPI < 1 and an SPI < 1 would mean that the technical work is under budget and ahead schedule. All four combinations of CPI and SPI are possible.

A caution: The formulas for CPI, SPI, estimated actual cost (EAC), and estimated completion date (ECD) in Table 11.5 are sometimes inverted, as follows:

$$\text{CPI} = \text{BCWP/ACWP},$$

Table 11.3 Earned value terminology.

Term	Definition	Explanation
BCWP	Budgeted cost of work performed	Cumulative amount of the budget for all tasks completed to date (i.e. the earned value)
ACWP	Actual cost of work performed	Actual cost of all tasks completed to date
BCWS	Budgeted cost of work scheduled	Planned cost of all tasks scheduled for completion to date
BAC	Budgeted actual cost	Planned cost of the total project
SCD	Scheduled completion date	Planned completion date of the project
EAC	Estimated actual cost	EAC of the project based on progress to date
ECD	Estimated completion date	ECD based on progress to date
CV	Cost variance	CV = ACWP − BCWP
SV	Schedule variance	SV = BCWS − BCWP
CPI	Cost performance index	CPI = ACWP/BCWP
SPI	Schedule performance index	SPI = BCWS/BCWP
CVC	Cost variance at Completion	CVC = EAC − BAC
SVC	Schedule variance at completion	SVC = ECD − SCD

EAC = BAC × CPI and ECD = SCD × SPI.

$$\text{SPI} = \text{BCWP}/\text{BCWS},$$

which makes

$$\text{EAC} = \text{BAC}/\text{CPI}$$

and

$$\text{ECD} = \text{SCD}/\text{SPI}.$$

In this case, CPI < 1 and SPI < 1 would mean the technical work is over budget and behind schedule. Both sets of formulas produce the same values of EAC and ECD if they are consistently applied.

An example of an earned value report is provided in Figure 11.5. In this example, the technical work is over budget because, using the formulas in Table 11.3, the CPI is greater than 1 (i.e. A > C) and behind schedule because the SPI is greater than 1 (i.e. B > C).

All eight combinations of A, B, and C are possible, as illustrated in Table 11.4. The relationships among A, B, and C in Figure 11.5 are those in column 1 of Table 11.4.

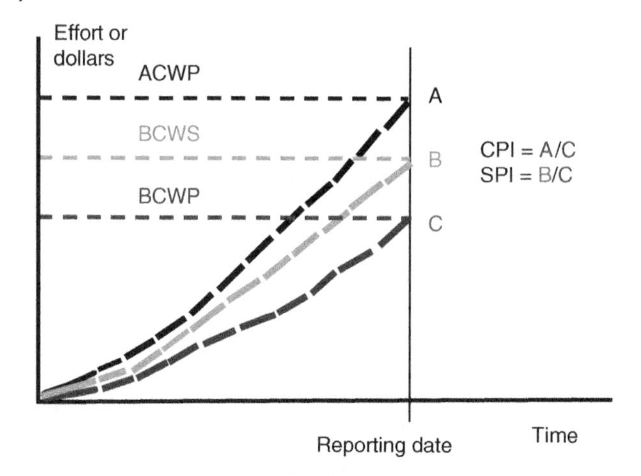

Figure 11.5 An example of earned value assessment. Source: Fairley 2009, figure 8.6 (p. 349).

Table 11.4 Earned value relationships.

Orientation Condition	A B C	A C B	B A C	C A B	B C A	C B A
Cost overrun	X		X		X	
Cost savings		X		X		X
Schedule delay	X	X	X			
Schedule advance				X	X	X

A, ACWP; B, BCWS; C, BCWP.

The CPI and SPI can be used to calculate the EAC and ECD, using the formulas below Table 11.3. An example is provided in Figure 11.6.

Because the CPI and SPI will vary from one reporting period to the next, the EAC and ECD will also vary from period to period. For example, a project may be over budget but ahead of schedule in one reporting period, back on planned budget and schedule in the next, and on budget but behind schedule in the next reporting period.

The examples in Figures 11.5 and 11.6 are exaggerated for the purposes of illustration. On a well-run project, the ACWP and BCWP lines will closely track the BCWS line; i.e. the project will remain roughly on schedule and on budget throughout the project. Said another way, the schedule variance (SV) and cost variance (CV) will be near 0 and the CPI and SPI will be near 1 in each reporting interval; the cost variance at completion (CVC) and schedule variance at completion (SVC) will be near zero and the engineers will deliver an acceptable system on schedule and within budget. Assessment results similar

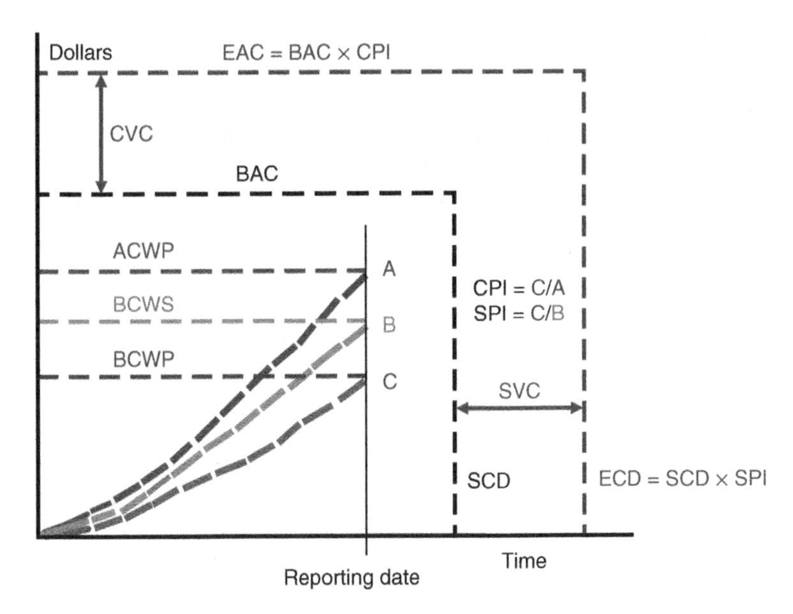

Figure 11.6 Earned value projections of estimated actual cost and estimated completion date. Source: Fairley 2009, figure 8.7 (p. 350).

to those in Figures 11.5 and 11.6 can be generated using spreadsheets to do the calculations and prepare the graphs.

The previous examples were presented using technical effort as the "cost" variable. Actual cost in monetary units can be used in place of effort or to account for effort cost and other costs.

In summary, EV assessment is a concise way of presenting

(a) the current status of cost (effort plus other costs), schedule, and progress on work products;
(b) on-going (e.g. monthly) estimates of the final cost of a project (EAC) and the delivery date of the system (ECD) based on the on-going current status of the technical work; and
(c) trends over time.

Necessary conditions for accurate assessment of earned value include the following:

1. Specification of work units to be completed during the upcoming incremental subphase of the implementation phase (Phase 5) of the I^3 system development model;
2. Binary assessment based on objective acceptance criteria for the work products;

3. Iterative-incremental development and assessment of work products at frequent intervals to provide demonstrations of progress and early warning of problems;
4. Accurate reporting of effort and time required for the work tasks;
5. Version control of baseline work products;
6. Standard methods, tools, and formats across the organization for capturing and reporting earned value status; and
7. Use of earned value assessment to forecast EAC and ECD.

In this section, the principles of earned value assessment are related to the I^3 system development model and binary assessment of progress. However, the principles can be applied when using other system development models; see Chapter 8 of the cited reference (Fairley 2009).

For earned value reporting to be effective, effort and time (item 4) must be accurately reported. Time cards prepared on a weekly basis (e.g. each Friday afternoon) are usually not accurate because the amount of time spent on various work activities during the week will not be accurately recalled. Pleasant tasks will be recalled as being much shorter than was actual time spent and unpleasant or difficult tasks will be recalled as taking much longer than actually spent, or perhaps they will not be reported.

Alternatives to weekly time cards include the following:

(1) Electronic templates, completed at the end of each work day;
(2) Templates for recording work activities that are attached to planned units of work; and
(3) Templates attached to the version control system that require entry of data on checking in accepted work products.

It is the responsibility of each team member to report time and effort data for each task in a timely manner; it is the responsibility of each team leader to ensure that the team members are entering accurate time and effort data. It is the responsibility of quality assurance personnel to periodically audit the reporting system to ensure that time and effort data are accurate and timely.

A final caution on collecting and reporting project data (both product and process):

> Reporting of performance data should be at the level of teams (3 to 7 members) and aggregations of teams.

Performance should *never* be related to individuals, except in private one-on-one meetings where the emphasis should be "what can I do, as your team leader to help you do a better job?" and not "how could you be so stupid as to have made so many mistakes?" Nothing will kill a measurement program faster or more effectively than public disclosure of data related to individual productivity and individual quality of work.

11.2.6 Assessing and Controlling Risk

The goal of risk management is to identify and mitigate potential problems with sufficient leadtime to prevent adverse impacts (i.e. to assess and control risk). In mission-critical situations, risk may be expressed as the potential for loss of human life or the potential for significant loss of information or property and the negative impact if human life, information, or property is lost. During system development, risk management focuses on assessing and controlling project factors such as budget, schedule, resources, and cost and on deliverable system functionality, behavior and quality attributes such as safety, security, and reliability that could have adverse impact on those who will use or be affected by the delivered system.

Risk management has been, and is, extensively studied, written about, and practiced. This section provides a brief overview of risk management. Chapter 9 of *Managing and Leading Software Projects* provides an extensive discussion of risk management for software projects that is equally applicable to systems projects (Fairley 2009).

Informally, risk can be characterized as the chance a bad thing might happen and the negative consequences if it does happen. More formally, risk can be characterized by probability, p, where $0 < p < 1$, and potential loss, L. Probability $p = 0$ indicates that the negative thing will never happen and probability $p = 1$ indicates that the negative thing will happen with certainty. For systems projects, potential loss L is usually expressed in monetary units, on an ordinal scale of (low, medium, high), or in dimensionless units of utility, which is usually measured on a scale of 0–100 to express relative value within a context. A glass of water, for example, has greater utility for a person lost in a desert (perhaps 100) than for a person in the comfort of his or her home (perhaps 10).

Specific risks are characterized as “risk factors.” Some commonly occurring factors for systems projects are listed in Table 11.5.

Each specific risk factor can be characterized by probability, p, and potential loss, L. In some cases, it may be possible to quantify risk factors. In those cases, risk exposure is the product of probability and potential loss:

$$\mathrm{RE} = p \times L.$$

A risk factor with probability $p = 0.25$ and a potential loss of 100 000 USD has a risk exposure of

$$0.25 \times 100\,000 \text{ or } 25\,000 \text{ USD}.$$

Alternatively, a risk factor with probability $p = 0.25$ and utility of 50 has a risk exposure of 12.5 dimensionless units.

Risk exposure can be used to compare the exposure of different risk factors.

Table 11.5 Some commonly occurring risk factors for systems projects.

Risk factors	Examples
Schedule	Inadequate calendar time
Budget	Insufficient funds
Requirements	Infeasible, unstable, incorrect, incomplete, inconsistent
Personnel	Recruitment, ability, retention
Process	Inefficient and/or ineffective procedures
Resources	System development environment and tools; supporting organizations
Technology	Platform for implementation and domain of application
Geography	Multiple development sites
External factors	Vendors, subcontractors, and procurement
Operational risks	Missing features, inadequate performance
Quality	User and customer dissatisfaction
Maintenance	Error correction; needed improvements to features and quality

Source: Fairley 2009, table 9.2 (p. 366).

Risk leverage factors (RLFs) can be used to evaluate the leverage gained by investing in effort to reduce probability or potential loss:

$$RLF = (RE_{before} - RE_{after})/RR_{cost},$$

where RE_{before} is the risk exposure before risk reduction (RR) and RE_{after} is the risk exposure after risk reduction; RR_{cost} is the expense of implementing the risk reduction activity. RLF is a dimensionless quantity that can be used to evaluate various risk reduction investments.

One way to determine investment strategies is to calculate and compare RLF for various investment strategies for highest priority risk factors. RLF is calculated by calculating the risk exposure before mitigation, the risk exposure after mitigation, and dividing the difference by the cost of mitigation:

$$RLF = (RE_{before} - RE_{after})/(\text{cost of mitigation}).$$

Suppose, for example, that an investment of \$25 000 (i.e. 25 000 USD) is being considered to reduce probability from 0.4 to 0.1 for a risk factor with potential impact of \$500 000. Risk exposure before mitigation is \$200 000 (0.4 × 500 000), risk exposure after mitigation would be \$50 000 (0.1 × 500 000), and the cost of mitigation is estimated to be \$25 000. The RLF is thus

$$RLF = (200\,000 - 50\,000)/25\,000 = 6.0.$$

RLF is a dimensionless number that can be used to evaluate different risk reduction investments. Larger RLFs indicate better investment strategies.

There is no guarantee, of course, that the investment will reduce the probability from 0.4 to 0.1, nor is it certain that the risk factor will become a problem if it is not mitigated; the probability is "only 40%." However $500 000 would be a severe financial loss to an organization and a severe loss of reputation for you, as the principal physical systems engineer or principal software systems engineer for a project. If you do not spend $25 000 to mitigate the risk factor and the problem occurs with a loss of $500 000 USD, you may be removed as lead systems engineer (i.e. technical manager). If you do spend $25 000 and the problem does not occur, you (and others) will never know if it would have occurred without spending $25 000.

In this regard, investing in risk mitigation is akin to investing in automobile insurance. You may be one of the fortunate ones who has never had a serious automobile accident and it could be said that you have wasted a lot of money paying for auto insurance (especially if you are my age); however, the potential financial impact created by not buying insurance is so great that most rational people will continue to buy insurance, even though the probability of an accident based on historical evidence is low, and even if it were not required by law.

In some cases, it may not be possible to quantify risk probability and potential loss. Qualitative determination of risk exposure, based on subjective factors, can be assessed as illustrated in Table 11.6.

There will never be enough time, money, or resources to perform adequate mitigating actions for all identified risk factors because it is very easy to generate a long list of risk factors using the risk identification techniques listed below. Risk factors that have the highest priorities, as determined by risk exposures, risk leverage, qualitative analysis, and subjective factors, should receive the majority of limited risk management resources.

The procedures of risk management include the following:

- Risk identification;
- Risk analysis and prioritization;
- Risk mitigation; and

Table 11.6 Qualitative determination of risk exposure.

Potential loss	**Low**	**Medium**	**High**	**Very high**
Probability				
Low	Low	Medium	High	Medium
Medium	Low	High	High	High
High	Medium	High	Very high	Very high
Very high	Medium	High	Very high	Extremely high

Source: Fairley 2009, table 9.1B (p. 564).

- Assessing and controlling risk.

In general, many different methods and techniques can be used to identify risk factors for a systems project. They include the following:

- Checklists;
- Brainstorming;
- Expert judgment;
- Strengths, weaknesses, opportunities, and threats (SWOT) analysis;
- Lessons-learned files;
- Cost modeling;
- Schedule analysis;
- Requirements triage;
- Assets inventory; and
- Tradeoff analysis.

Identified risk factors can be analyzed and prioritized using the following:

- Expert judgment (individual, Delphi, Delphi wideband – see Chapter 10);
- Monte Carlo calculations (see Chapter 10);
- Analysis of assumptions and constraints;
- Risk exposure and risk leverage; and
- What-if analysis.

Risk mitigation strategies for identified risk factors are listed in Table 11.7.

1. *Risk avoidance*
 Risk avoidance is concerned with changing a situation to reduce the probability of a potential problem. If a timing constraint in a real-time system is of concern, perhaps the timing constraint can be relaxed or perhaps a faster hardware processor can be used; if there is insufficient time to complete the project, perhaps the schedule can be extended, thus avoiding the risk of late delivery.

Table 11.7 Risk mitigation strategies.

Strategy	Approach
Avoidance	Change the system or the project
Transfer	Reallocate the source of the potential problem
Acceptance	Use a watch list
Immediate action	Reduce probability and/or loss
Deferred action	Activate a delayed action, if warranted
Crisis management	When risk management fails

Source: Fairley 2009, table 9.8 (p. 383).

Risk factors are often created by project constraints and can sometimes be avoided by modifying the constraints. Process constraints (schedule, budget, resources), system constraints (features and quality attributes), and technology constraints (processor speed, available memory) should be examined. Some constraints may be essential to a successful outcome but it may be possible to relax or remove some other constraints on closer examination.

2. *Risk transfer*

Risk transfer involves reassigning the work needed to implement a system capability for which a risk factor has been identified to another organizational unit that can better handle the risk factor. Data compression, for example, may have to be implemented in a special purpose hardware chip if there is significant risk that a data compression algorithm cannot be executed rapidly enough in software. Or, a subcontractor or consultant group might be used if a system capability to be realized will be more complex than thought and the subcontractor or consultant has special expertise. For example, a specialist in optics or data compression might be used to develop some elements of a space-borne telescope.

Care must be taken that transferring a risk factor does not create other unintended risks. The time and expense required to design and develop a special purpose chip may create unacceptable risks to cost and schedule. Managing a new subcontractor may represent a greater risk to success than the learning curve required for your team. There may be a risk factor that a project will fail if a subcontractor fails to deliver acceptable system elements within an acceptable time frame.

3. *Risk acceptance*

Risk acceptance is the third strategy for mitigating a risk factor. Acceptance involves acknowledging the risk factor but taking no action at the present time other than placing the risk factor on a watch list. Although risk acceptance does not result in a specific mitigation activity, each risk factor on a watch list is frequently reexamined on a periodic basis to determine if the level of probability, impact, or time frame of probably occurrence has become prominent enough to warrant additional mitigation activities (e.g. avoidance, transfer, immediate action, or deferred action).

A watch list thus serves as a constant reminder to reexamine risk factors that may become more serious as system implementation evolves. Project staffing, for example, might be sufficient for the next three months but a concern for future staffing needs might result in placing staffing issues on the watch list (i.e. a key person may be moving at the end of the children's' school year but it is not certain). If staffing issues have not been resolved as the time of need approaches an immediate action plan might be developed and implemented to replace the needed staff member.

Table 11.8 Example of an immediate action plan.

- *Action plan number and name*: AP#3, design tool training
- *Responsible party*: Sue Smith
- *Other responsible parties*: Joe Williams will set up the workstations and install the design tool on the server; an as-yet unidentified instructor will deliver the course.
- *Risk factor to be mitigated*: insufficient design tool skill
- *Actions to be completed*: training class and lab exercise for 10 mechanical engineers
- *Resources Needed*: instructor, classroom with workstations, release time for attendees
- *Duration of this action*: four weeks
- *Milestones for this action*:
  - *Week 1*: find instructor, reserve classroom, identify attendees
  - *Week 2*: load software on server, obtain/reproduce class materials
  - *Week 3*: conduct five-day class
  - *Week 4*: complete a design project (1/2 time, four days)
- *Success criteria for this action*: 9 of 10 attendees successfully complete the lab project

Source: Fairley 2009, table 9.9 modified (p. 385).

4. *Immediate action*

   Immediate actions are mitigation activities that are undertaken now to reduce the probability and/or loss if a potential problem (i.e. a risk factor) becomes a real problem in the near future. For example, an immediate action may be undertaken during the project initiation phase to train the mechanical engineers to effectively use a newly acquired design tool, thus reducing the probability and impact of design issues. An example of documenting an immediate action plan is illustrated in Table 11.8.

5. *Deferred action*

   Deferred action plans are prepared for potential problems that have significant probability and/or significant potential loss in the future but for which no immediate actions are currently warranted. If, for example, the I^3 iterative-increment development process is being used and lack of sufficient memory for an embedded microprocessor in a software-enabled physical system is of concern, a deferred action plan can be develop for activation later but, for now, there is sufficient memory. Deferred action risk factors are continually assessed and a deferred action plan is activated when (if) a problem trigger (an objective measure) crosses a predetermined threshold.

   Figure 11.7 provides an example of assessing memory allocated versus memory used in an embedded microprocessor. The risk factor is lack of sufficient memory to implement the required software. The software is being implemented in a series of weekly incremental builds. Memory is allocated to each successive build of the system and the cumulative amount of memory used for each demonstrated build is compared with the allocated amount. As indicated in Figure 11.3, 10% of memory is held in reserve. The software will fit in the available memory if the 10% margin is never exceeded.

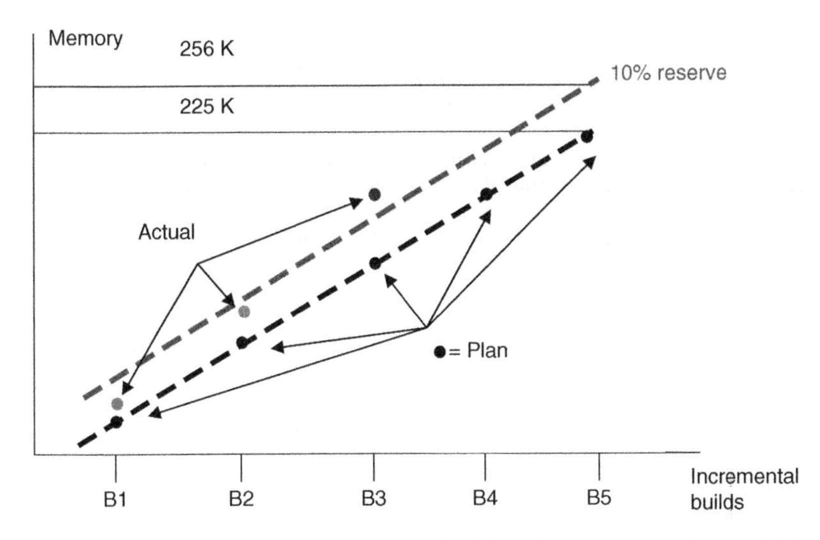

Figure 11.7 Illustrating a 10% memory threshold for risk management. Source: Fairley 2009, figure 9.3.

In Figure 11.7, actual memory used in incremental builds B1 and B2 exceed the allocated amount planned but not by more than 10%. Build B3, however, has caused actual memory used to exceed the 10% threshold. This is the trigger for invoking a deferred action plan.

One of the common failures of risk management is to "wait and see" whether the situation will improve without invoking the deferred action plan. Spontaneous improvement of a bad situation seldom happens; recall that the purpose of risk management is to identify potential problems with sufficient leadtime to avoid crisis situations. Although unused memory is still available after build B3, there is still a lot of remaining functionality to be implemented. The trend, based on builds B1, B2, and B3, indicates that successively more memory is being used than was planned on each build. Now is the time to invoke the deferred action plan to avoid the crisis of memory overrun.

Measuring attributes of a system to determine whether a system or some system element satisfies specified requirements is known as technical performance measurement (TPM). Measuring the amount of memory used versus the amount of memory allocated, as in Figure 11.7, is an example of TPM (Roedler and Jones 2005).

An example of a deferred action plan for the situation in Figure 11.7 is presented in Table 11.9. Exceeding the 10% margin provides the trigger for activating the plan. Table 11.9 indicates that the deferred action plan should be activated because incremental build B3 has exceeded the cumulative memory allocation for the build.

Table 11.9 Example of a deferred action plan.

- *Deferred action plan number and name*: DA #5 memory
- *Responsibly party*: John Smith
- *Risk factor to be mitigated*: lack of sufficient memory in the embedded microprocessor
- *Risk indicator to be measured*: planned versus actual memory used in successive incremental builds
- *Frequency of measurement*: weekly measurement of memory usage for successive incremental builds
- *Threshold DA value*: 10% over plan on any incremental build
- *Deferred action plan*:
    - *Actions to be completed*: reengineering of the software to fit within allocated memory
    - *Responsibilities*: Joe Williams and Sue Smith will attempt to rectify the memory overrun; they will be released from all other duties and receive overtime pay
    - *Resources needed*: an additional target machine will be flown in overnight from San Jose
    - *Milestones for this action*: no objective milestones; brief stand-up status meetings will be held at 11:00 a.m. and 6:00 p.m. each day
    - *Success criteria for this action*: memory usage reduced to not more than 5% over allocated amount
    - *Maximum duration of this action*: seven days

Source: Fairley 2009, table 9.11 (p. 387). Reproduced with permission of Wiley.

An unintended consequence of this deferred action plan may be that reengineering the software to compress it into the allocated memory will make it difficult to test or modify the software, or an already implemented capability may be degraded, or compressing the software may degrade quality attributes such a safety or security.

If the memory problem persists, an alternative action would be to eliminate some specified capabilities or to replace the microprocessor with one having more memory. But these approaches may not be feasible. Reducing the capabilities may not be possible because all capabilities allocated to the microprocessor are mission-critical. Replacing the microprocessor may not be possible because of the impact on cost, schedule, compatibility with other system elements, procurement delay, or sunk cost (i.e. the investment already made in the microprocessor and software development for it).

Development of a system or some part of a system enters crisis mode if the contingent actions do not achieve the success criteria specified in the plan within the specified duration, as in the example if the deferred actions do not resolve the problem within seven days.

6. *Crisis management*

A project enters crisis mode when an event or a situation stops progress. Crisis management is the process of clearing the roadblock so that progress can resume. Recall that the goal of risk management is to identify risk factors with sufficient leadtime to prevent crisis situations. Thus, a project would never enter crisis mode if risk management were 100% effective; but no process is 100% effective so the possibility of crisis situations must be acknowledged. There are several ways a project can get into crisis mode:

- Failing to systematically identify, prioritize, and mitigate potential problems;
- Identifying potential problems but taking no mitigating actions;
- Unforeseen and unforeseeable situations and events; and
- Failure of a risk mitigation strategy (i.e. a deferred action) to solve the problem in the established time frame.

It could be, for example, that no one foresaw that lack of sufficient memory in the microprocessor would be a problem, or perhaps the possibility of insufficient memory was discussed but no mitigating actions were taken (such as failing to include the risk factor on a watch list), or perhaps the contingent-action plan, as executed by Joe and Sue, failed to solve the problem within the seven days allocated to fixing the problem.

Guidelines for managing crisis situations include the following:

- Acknowledging the situation;
- Allocating sufficient resources;
- Seeking creative solutions;
- Reviewing status frequently; and
- Setting a "drop-dead" date.

If the crisis is overcome within the allocated time for crisis management, schedules, budgets, and work plans must be revised to account for the time and resources spent on crisis management.

The last guideline, setting a "drop-dead" date, is particularly important to force decisive action when a crisis cannot be overcome within a specified time frame; otherwise, a project may languish in crisis mode far beyond a reasonable decision point. The resulting decision at the drop-dead date will be to

1. Cancel the project; or
2. Significantly rescope it.

If the project is cancelled, a termination plan must be prepared and executed. The plan must include a redeployment plan for the project members and may require difficult negotiations with the customer, who may be internal or external to the organization. If the project is rescoped, the requirements and management plans must be developed for the (new) project.

In any case (successfully overcoming the crisis, canceling the project, or significantly rescoping it), the staff members who have made extraordinary contributions in attempting to overcome the crisis must be rewarded for their efforts.

The reward may take the form of a few days off to catch up on sleep and to see family members plus a letter or e-mail of appreciation (the recommended reward), or a dinner for the family, or conference travel, or some combination of these. Overtime pay should also be provided.

11.3 Assessing and Analyzing System Parameters

The first steps in assessing and analyzing system parameters for a software-enabled physical system is determining what parameters will be assessed and what measurement scales will be used to assess them. Technical parameters to be measured may include functional, behavioral, and quality attributes.

Functional parameters specify input/output relationships such as format, accuracy, and precision of input and output parameters; sequential or concurrent use of a function; precondition that must be satisfied before a function can be used correctly; and postconditions that will be satisfied by completion of a function, including the results of exception handling, if activated.

Behavioral parameters are concerned with system attributes that are manifest over time. Many of these attributes are indicators of system quality, such as reliability, availability, robustness, security, safety, and response time to various stimuli.

Measurement scales determine the form of assessment data to be collected and the operations that can be performed on the data. There are five commonly used measurement scales: nominal, ordinal, interval, ratio, and absolute. These scales provide a hierarchy of permissible operations on the measures. The hierarchy, based on the characteristics of each scale, is presented in Table 11.10.

A *nominal* scale assigns items to groups or categories. For example, a list of the number of engineers in each of several categories, analysis and design,

Table 11.10 A hierarchy of measurement scales.

Scale	Characteristics
Nominal	Frequency distributions among measurement categories
Ordinal	Ordering within categories; unspecified intervals among measures
Interval	Equal intervals among measures; arbitrarily determined zero element
Ratio	Equal intervals among measures; objectively determined zero element
Absolute	Similar to ratio but with uniqueness of measures

Source: Fairley 2009, table 7.1.

implementation, verification and validation, user training, and so forth, or a list of the number of installations of a system by region or country. The number of items in each category can be counted to provide frequency distributions among the categories but no ordering of the items within a category is implied. For example, a project might have 12 system developers and 5 testers; you can say there are 7 more developers than testers but nothing can be said about the rankings of skill levels of the implementers or the testers when using a nominal scale.

Measures based on symbols that form an ordered sequence, such as (low, medium, high), form *ordinal* scales. Skill levels or system complexity might be measured using the symbols (low, medium, high) with permitted transitive relational operations of less than, greater than, and equal to (<, >, =) defined on the sequence (low, medium, high). The intervals between adjacent symbols are not specified for ordinal measures and there is no objectively determined zero element; thus, we cannot say that a program of high complexity is 3 or 5 or 10 times more complex than a program of low complexity when using an ordinal scale. However, transitive relational operations can be applied if the symbols form an ordered sequence, so that a system of low complexity is less complex than medium, which is less complex than high, and two systems or subsystems of medium complexity are of comparable complexity. By the transitivity property, element A is less complex than element C if element A is less complex than element B and element B is less complex than element C or if module B is of equal complexity to module C (note the precedence among the "and" and "or" operators in this sentence). The value low, medium, or high might be determined for a system or a system element using subjective criteria.

When using an ordinal scale, elements within a category are ordered. Symbols higher in the ordering indicate larger values but the intervals between the symbols cannot be assumed to be equal. For example, a system developer rated high in skill is not necessarily three times more skilled than a system developer rated medium in skill. The zero point in an ordinal scale, if it exists, is chosen arbitrarily; for example, a scale of (1, 2, 3) equated to low, medium, high could equally be scaled as (0, 4, 6). Numeric values should not be used for ordinal measurements because numbers imply equal intervals between adjacent symbols.

Measures based on symbols that have equal intervals between any two adjacent symbols but having an arbitrarily determined zero element form *interval* scales. In an interval measurement scale, a unit of measure represents the same magnitude of a factor across the full range of the scale. For example, on the Celsius and Fahrenheit temperature scales, the difference between 30° and 40° is the same as the difference between 50° and 60° (10° in each case). However, the zero point on these scales does not denote the absence of temperature and ratios cannot be formed. Thus, a temperature of 60 °C or 60 °F is not twice as hot as a temperature of 30°.

Temperature

Daniel Fahrenheit invented the mercury thermometer in 1714 after discovering that mercury has a linear expansion/contraction factor over a wide range of temperatures, thus making it a suitable element for constructing glass thermometers with linear markings on the glass. Mr. Fahrenheit established three points on his measurement scale: 0 °F was determined as the temperature of a mixture of salt, ice, and water; 32 °F was determined as the freezing point of ice water without salt; and 96 °F was determined to be the bodily temperature of a healthy adult person. This point on the Fahrenheit scale later recalibrated the temperature of a healthy adult to be 98.6 °F. A fourth point on the scale, 212 °F, was later established as the boiling point of pure water at sea level. Mr. Fahrenheit could just as readily have assigned the number "10" to his zero element and added 20 to the other calibration points. Zero degrees Fahrenheit is thus an arbitrary value; i.e. 0 °F does not indicate the absence of temperature.

The numbers used as calibration points on the Celsius temperature scale are similarly arbitrary. Mr. Celsius initially established 0 °C as the boiling point of water and 100 °C as the freezing point. The measurement scale was later inverted to the now-familiar scale with 0 °C as the freezing point and 100 °C as the boiling point of pure water at sea level. The Fahrenheit and Celsius temperature scales are interval measurement scales because the intervals between measures are equidistant but the zero element does not denote the absence of the phenomenon being measured, namely, temperature.

In measurement theory, measures that have equal intervals between any two adjacent symbols and a zero element that denotes absence of the quantity being measured form a *ratio* scale. On measurement scales that use integer and real-number measures (for example) and that have objectively determined zero values, the relational operations and the arithmetic operations can be applied and ratios can be formed, as in measurement of the number of distinct software elements and physical elements in a system or subsystem. Because the measure is in integer units with equal intervals and with an objectively zero element (the absence of system elements), it can be said that a system having 100 elements is twice as large as a system that has 50 elements, the system having 50 elements has ½ the number of elements, and the combined systems have 150 elements.

Temperature measured in degrees Kelvin (°K) forms a ratio measurement scale because the zero element is an objectively determined value; temperature in degrees Kelvin is a measure of the kinetic energy associated with the motion of atoms and molecules. The 0 point of the Kelvin temperature scale is the temperature at which all movement at the atomic level ceases, i.e. the absence of temperature. Thus, 0 K is an objective measure and 200 K is twice as hot as 100 K (i.e. the kinetic energy of the atoms is twice as much).

An *absolute* measurement scale is one for which ratios are allowed (equal intervals plus an objectively determined zero element that denotes absence of the phenomenon) but the measures are unique. For example, measuring defect density in a computer program as an integer number of defects per line of code by measuring total number of defects and dividing by the number of lines of code forms a *ratio* measurement scale because defects per line of code can be converted to defects per thousand lines of code by the following transformation:

$$D/KLOC = (D/LOC) \times 1000.$$

If, for some reason, monotonic transformation of D/LOC were disallowed, the measurement scale would be an absolute one. Stated in another way, identity is the only transformation allowed for an absolute measurement scale, which isn't a transformation except perhaps for renaming.

Ordinal, ratio, and absolute scales are the most commonly used measurement scales in systems engineering and software engineering. System complexity, as measured by (low, medium, high) is an example of an ordinal measurement scale. Measuring software size as an integer number of lines of code is an example of a ratio scale. Forbidding transformations of a ratio measurement, such as disallowing defects per line of code to be transformed to defects per thousand lines of code, forms an absolute measurement scale.

Ordinal scales are used when there is no objective, agreed upon, method of determining the intervals between the measures being used. The evaluative measure of a low complexity subsystem combined with a high complexity subsystem, for example, will not result in a system of medium complexity (the average) because arithmetic operations are not allowed for ordinal data. The result for system complexity in this example is a high complexity rating, based on the transitive properties of the (low, medium, high) complexity measure. High complexity is more complex than medium or low complexity, but we cannot say by how much: the complexity rating is not 4 (i.e. 1 + 3).

11.3.1 Direct and Indirect Measures

Another way to categorize measures is direct versus indirect measurement.

A *direct measure* is obtained by directly applying measurements to the phenomenon of interest; for example, counting the lines of code in a computer program using well-defined counting rules. An *indirect measure* is obtained by combining direct measures using operations appropriate to that measurement scale. For example, the number of function points in a computer program is an indirect measure that is determined by applying the function point counting rules to determine the unique (integer) number of inputs, outputs, files, interfaces, and queries in the program; multiplying each by an integer complexity factor; and adding the results together. Tables 11.11 and 11.12 provide examples of direct and indirect measures used in systems engineering and software engineering.

Table 11.11 Some direct measures.

Measurements	Direct measures
System size	Number of use cases Number of system elements
Number of personnel by category	Number of developers Number of testers
Progress	Number of requirements traced to design elements Number of capabilities implemented Weeks taken to implement a system increment
Resource usage	CPU cycles used Memory bytes used Megawatts of power consumed
Time	System response time Weeks taken to implement a system increment
Quality	System response time A log of system failures

Source: Fairley 2009, table 7.3 (p. 275).

Table 11.12 Some indirect measures.

Measurement	Indirect measures
Size	Software function points
Productivity	System elements implemented per engineer-month
Production rate	System elements implemented per month
Testing rate	Tests conducted per staff-day
Defect density	Defects per thousand lines of code
Defect efficiency	Number of defects fixed per staff-day
Defect effectiveness	Number of defects detected/total defects
Requirements stability	Current number/initial number
Cost performance index	Actual cost/budgeted cost
Mean time between failures	Average of times noted between logged failures

Source: Fairley 2009, table 7.4.

Note, in Table 11.6, "weeks taken to implement a system increment" could be used to measure progress, time, or both and "system response time" could be categorized as a measure of time or quality. Categorizing a measure depends on the desired use of the measure and, as noted, a measure may fit into more than one category.

Indirect measures are based on direct measures using the operations permitted by the measurement scale.

11.3.2 Surrogate Measures

Surrogate measures comprise another category of measurement. Surrogates are used when direct or indirect measurements cannot be made. Surrogates for ease of system modification, for example, might include clearly defined and measured system structure, quality of interface documentation, a library of regression tests cases, automated testing tools, scripts for building system versions, a version control/configuration management tool, and documentation for previous system fixes and upgrades. These surrogates would indicate that a system can (or cannot) be sustained with efficient use of time and effort.

11.3.3 Technical Performance Measurement

Guidance for selecting technical parameters and developing a measurement program is provided in the report *Technical Measurement: A Collaborative Project of PSM, INCOSE, and Industry* (Roedler and Jones 2005). The Practical Software and Systems Measurement web site provides information about current and ongoing activities of practical measurement (PSM 2018).

As stated in the Roedler et al. report, technical measures include measures of effectiveness (MOEs), measures of performance (MOPs), TPMs, and key performance parameters (KPPs).

- MOEs are measures of success that are independent of the particular solution used to achieve the operational objectives. An objective measure used to determine the efficiency and effectiveness of a specified group of engineers when developing a specified kind of system, without regard to the methods and tools they are using, is an example of an MOE. TPMs are often derived from MOEs.
- MOPs provide insight into the performance of a specific system. Examples of MOPs are measurement of processing time and flow of material through a particular system.
- TPMs measure attributes of a system to determine how well a system or some elements of a system satisfy specified requirements. Measuring the amount

of memory used versus the amount of memory allocated, as in Figure 11.7, is an example of TPM.
- KPPs are a critical subset of the performance parameters representing the most critical capabilities or characteristics. MOPs that are related to essential capabilities and requirements are examples of KPPs.

11.4 Corrective Action

Controlling the technical work of a systems project is concerned with keeping the work activities within the bounds and constraints imposed by the systems engineering management plan (SEMP) while ensuring that the technical objectives of the project are being achieved. Assessment results are (or should be) periodically analyzed and corrective actions applied as necessary. The SEMP may be revised as a project evolves and quality assurance personnel may observe the work processes and provide recommendations, as described in Section 10.2.1.

11.4.1 Acceptable Corrective Action

Depending on the nature of the deviation from plan, acceptable corrective action may involve one or more of

- Extending the schedule;
- Adding more resources;
- Acquiring better resources;
- Improving one or more development processes;
- Improving the development environment, methods, or tools;
- Upgrading the technology base; and
- Descoping the system capabilities to be delivered.

Resources to be improved, added, or replaced might include people (e.g. adding an expert system designer for the domain of interest), software elements (e.g. reengineering an element to improve response time), hardware elements (e.g. replacing a black box element with a superior one), and development tools (e.g. acquiring and integrating a new design tool).

Descoping the system to be delivered can be gracefully accomplished (with customer agreement) if system capabilities are being realized in priority order. If the delivery date is constrained, it may be acceptable to deliver a subset of highest priority capabilities on schedule with delivery of a subsequent full-capability version at a later, negotiated date. As previously stated (Section 11.2.2), it is better to be 100% complete with 90% of the system capabilities than to be 90% complete with 100% of the capabilities.

11.4.2 Unacceptable Corrective Action

Unacceptable corrective actions include the following:

- Excessive overtime;
- Reducing or eliminating planned development activities;
- Failing to pay accumulated technical debt; and
- Falsely reporting progress.

Short bursts of self-imposed or management-imposed overtime is to be expected when developing a software-enabled physical system. Overtime becomes excessive when the short bursts require exceptional working hours (e.g. 12–15 hours per day for more than 7 days), or when short bursts become long bursts, or when short burst occur frequently (e.g. twice per month on a regular basis). Experience has shown that worker productivity and morale decrease with excessive overtime to an extent that shorter working hours would increase productivity and morale.

Eliminating or reducing planned development activities such as design reviews, maintaining traceability, or verification and validation will result in accumulation of excessive technical debt that must eventually be paid or the delivered system will not be acceptable.

Failing to acknowledge accumulated technical debt will provide a false sense of progress and failing to pursue needed corrective action is an unacceptable form of corrective action (i.e. the null case). Failure to pay technical debt in a timely manner may result in an unacceptable delivered system or project cancellation in an extreme case. Failure to pay technical debt can be thought of as mortgaging the future by deferring to a later date work that must be eventually accomplished at an increased cost in time and effort. Projects have been cancelled in the past because they were far behind schedule and no one knew how much technical debt had been accumulated and thus no one knew how much time, effort, and cost would be required to salvage the projects.

False reporting of progress is not, per se, an unacceptable corrective action (although it may be a legal offense in a contracted situation); rather, it is the result of failing to initiate corrective action, perhaps with the hope that an overly optimistic progress report will buy time to improve progress. However, this desperate approach is typically used after it is too late to make the necessary improvements. Another folk saying is

> It is too late to close the barn door after the horse has escaped.

11.4.3 Inhibitors of Corrective Action

Inhibitors of corrective action include the following:

- Contractual considerations;

- Unintended consequences;
- Failing to initiate corrective action in a timely manner;
- An ineffective assessment and analysis process; and
- Failure to analyze assessment results.

Contractual considerations that may inhibit needed corrective action include financial penalties for late delivery, failure to provide a mission-critical system when needed, stakeholder dissatisfaction with the delivered system, or loss of credibility or reputation with an important customer. This inhibitor is based on the belief, and perhaps the reality, that it is a lesser sin to deliver an inadequate system on time than to deliver it late.

Corrective actions may have unintended consequences that may or may not be recognized. Adding resources to improve the schedule may result in excessive cost. Adding development personnel or a new design tool may have an unacceptable impact on the schedule because of the additional time required to assimilate the change. Another folk saying is

> don't change horses in the middle of a river.

Perhaps the saying should be amended to add

> unless your horse is drowning.

Failing to initiate corrective action in a timely manner is based the following considerations:

- Assessment and analysis may indicate the need for corrective action but there may be a tendency to wait and see if the situation will improve during the current reporting period. But another well-known maxim is

  > Projects never improve without explicit corrective action.

- *Corrective action may not happen because of the organizational culture*: Bad news is not welcomed, we don't want the customer to know we are having issues, and no one wants to be the bearer of bad news in this environment.
- *Corrective action may not happen because the need for corrective action is not apparent*: Assessments based on wishful thinking, guesstimation, and excessive optimism rather than objectively obtained assessment data deny reality (see binary completion criteria and assessment of technical debt, as described in Sections 11.2.2 and 11.2.4).

Another inhibitor to corrective action is failure to analyze assessment data collected using objective measures. The need may be apparent but may be presented in a manner that does not reveal the issues or perhaps it is not carefully analyzed because of excessive work demands. Large software-enabled physical systems projects often have a staff member who collects and analyzes status data and reports results and recommended actions to the leaders of the technical work activities (i.e. a physical systems engineer and/or a software systems engineer).

Finally, it must be remembered that plans, policies, processes, and procedures are intended to provide a framework of guidance, whereby people can realize objectives and achieve desired goals. Perfect plans, assessments, analyses, and corrective actions will not achieve the intended results without enthusiasm and a sense of ownership by those who participate in developing a complex software-enabled system.

> Motivated and enthusiastic engineers may be able to overcome faulty processes but perfect processes cannot overcome low morale and lack of enthusiasm.

These issues are addressed in Chapter 12.

11.5 Key Points

- Clause 6.3 of ISO/IEC/IEEE Standards 15288 and 12207 provides guidance for managing technical work activities.
- The purposes of assessing and analyzing process parameters and system parameters are to
  - Provide frequent indications of progress;
  - Provide early warning of problems;
  - Permit analysis of trends;
  - Allow estimates of the final cost and completion date for the technical work; and
  - Build a data repository of histories for your organization.
- Process parameters to be assessed, analyzed, and controlled are
  - Effort;
  - Cost;
  - Schedule;
  - Progress;
  - Technical debt;
  - Earned value; and
  - Risk.
- Three categories of work are
  - Original work;
  - Evolutionary rework; and
  - Avoidable rework (retrospective and corrective).

- Binary assessment of progress provides accurate measurements and avoids the 95% complete syndrome.
- Technical parameters to be measured include functionality, behavior, quality attributes, and performance attributes.
- Different measurement scales permit different kinds of operations on the assessed attributes.
- Technical performance measures include Measures of Effectiveness (MOEs), Measures of Performance (MOPs), Technical Performance Measures (TPMs), and Key Performance Parameters (KPPs).
- Acceptable corrective actions include
  - Extending the schedule;
  - Adding more resources;
  - Acquiring better resources;
  - Improving one or more development process;
  - Improving the methods or tools; and
  - Descoping the system capabilities to be delivered.
- Unacceptable corrective actions include
  - Excessive overtime;
  - Reducing or eliminating planned development activities;
  - Failure to acknowledge or pay accumulated technical debt; and
  - Falsely reporting progress.
- Inhibitors to corrective action include
  - Contractual considerations;
  - Unintended consequences;
  - Failing to initiate corrective action in a timely manner;
  - An ineffective assessment and analysis process; and
  - Failure to analyze assessment results.

Perfect plans, assessments, analyses, and corrective actions will not achieve the intended results without enthusiasm and a sense of ownership by those who participate in developing complex software-enabled physical systems. These issues are addressed in Chapter 12.

Exercises

11.1. Implementation of smaller capabilities (i.e. smaller units of required effort) increases the accuracy of assessment but at the risks of micro-management. Briefly explain what is meant by "micro-management."

11.2. Briefly explain what is meant by binary assessment.

11.3. Briefly explain why capabilities should be sized for implementation during one to two weeks.

11.4. Suppose the technical work to implement a software-enabled physical system is estimated to require 425 engineer-months of effort. Suppose the work is to be completed in 18 months.
(a) What will be the average staffing level for the work to be done?
(b) How many teams of maximum size six members will be needed?

11.5. The calculations in Section 11.2.3 indicate that 44% of the work remains to be done and that the technical work could be completed in 11 months if four engineers were added. Identify and explain three factors that might prevent the completion of the technical work in 11 months.

11.6. Technical debt is incurred when some of the work needed to develop a system is deferred until later.
(a) Identify and explain three reasons some needed work might be deferred until later.
(b) Briefly explain why accumulating technical debt amounts to mortgaging the future.

11.7. Earned value (EV) is widely used to assess and analyze the progress of technical work for a systems project.
(a) Briefly explain what is meant by "earned value."
(b) Briefly explain how earned value is measured.
(c) Briefly explain how a cost performance index (CPI) is calculated, a schedule performance index (SPI) is calculated, and how the estimated actual cost (EAC) and the estimated completion date (ECD) are estimated. Do not merely write down the equations.
(d) Briefly explain three reasons earned value may not be accurately assessed and analyzed.

11.8. Briefly explain why individual performance data of engineers should never be revealed, except perhaps in private one-on-one meetings.

11.9. Choose two risk factors from Table 11.5. For each risk factor, describe how it could be mitigated using each of the five risk mitigation strategies in Table 11.7; that is, 10 answers are required.

11.10. Different measurement scales are useful for different purposes.
(a) Briefly describe a situation in your personal life where a nominal measurement scale was or would have been useful. Briefly explain how it was or could have been used.

(b) Briefly describe a different situation in your personal life where an ordinal measurement scale was or would have been useful. Briefly explain how it was or could have been used.
(c) Briefly describe a different situation in your personal life where an interval measurement scale was or would have been useful. Briefly explain how it was or could have been used.
(d) Briefly describe a different situation in your personal life where a ratio measurement scale was or would have been useful. Briefly explain how it was or could have been used.

References

DeMarco, T. (1986). *Controlling Software Projects.* Prentice Hall.

Fairley, R. (1985). *Software Engineering Concepts.* McGraw Hill.

Fairley, R. (2009). *Managing and Leading Software Projects.* Wiley.

Fairley, R. and Willshire, M. (2005). *Iterative Rework: The Good, The Bad, and The Ugly,* 54–61. IEEE Computer Society.

Fairley, R. and Willshire, M. (2017). *Better Now Than Later: Confronting Technical Debt in Systems Projects,* 81–87. IEEE Computer Society.

ISO (2015). ISO/IEC/IEEE 15288:2015, Systems and software engineering – System life cycle processes. ISO/IEEE, 2015.

ISO (2017). ISO/IEC/IEEE 12207:2017, Systems and software engineering – Software life cycle processes. ISO/IEEE, 2017.

PMBOK (2017). *A Guide to the Project Management Body of Knowledge,* 6e. Project Management Institute https://www.pmi.org/pmbok-guide-standards/foundational/pmbok (accessed 15 June 2018).

PSM (2018). Objective information for decision makers. Practical Software & Systems Measurement. http://www.psmsc.com/ (accessed 25 July 2018).

Roedler, G. and Jones, C. (2005). Technical Measurement: A Collaborative Report of PSM, INCOSE, and Industry. INCOSE-TP-2003-020-01, Version 1.1, 27 December 2005. http://www.psmsc.com/Downloads/TechnologyPapers/TechnicalMeasurementGuide_v1.0.pdf (accessed 30 January 2019).

12

Organizing, Leading, and Coordinating

12.1 Introduction

Three key assets for successful engineering projects are effective people, appropriate processes, and adequate technology. Successful projects have the correct number of engineers who have appropriate skills and are motivated to do their best work. Processes include procedures and techniques for planning, leading, and coordinating the work activities. Technology includes the infrastructure, hardware and software tools, and other equipment to be used by the people to accomplish the system development processes.

Effective people are the most important asset. People of outstanding skill, ability, and motivation can often overcome inadequate processes and technology but excellent processes and technology cannot compensate for lack of ability, motivation, and effective leadership.

Engineers who develop software-enabled systems are typically organized into multidisciplinary teams that are necessary to realize system capabilities being implemented in a concurrent manner. Teams are essential because the variety of skills possessed by different team members are needed and because it is impractical for 2 people to take 10 years to complete a 20 person-year project that could be completed by 10 people in 2 years.

And the synergy that results when the members of the team work together in a collaborative and cooperative manner often results in a system that is superior to the one that would have resulted from the efforts of engineers working in separate disciplinary groups. However, synergy seldom happens spontaneously so leadership is needed to plan and coordinate the work of multidisciplinary teams.

Systems engineers, both physical and software, have multiple responsibilities when developing complex software-enabled physical systems: accomplishing the technical work of analysis and design (both initially and ongoing); planning, assessing, and controlling the work processes and work products; and organizing, leading, and coordinating the work of the engineers and support

Systems Engineering of Software-Enabled Systems, First Edition. Richard E. Fairley.
© 2019 John Wiley & Sons, Inc. Published 2019 by John Wiley & Sons, Inc.

personnel who implement those systems. This chapter is concerned with the latter responsibilities.

Some of the material in this chapter is adapted from Chapters 10 and 11 of *Managing and Leading Software Projects*, with permission (Fairley 2009).

This chapter provides readers with the opportunity to learn how systems engineers can organize, lead, and coordinate the work of the engineers and support personnel who develop complex systems, including but not limited to software-enabled systems. The chapter concludes with 14 guidelines for organizing, leading, and coordinating the efforts of multidisciplinary engineering teams.

References and exercises provide additional opportunities to learn more about organizing, leading, and coordinating for development of complex software-enabled physical systems.

12.2 Managing Versus Leading

Some systems engineers are effective at and enjoy being both manager and leader of technical work but good managers are not necessarily good leaders and good leaders are not necessarily good managers. Managing is an analytical activity whereas leading involves human relations. Different skill sets and personality traits are required for managing and for leading.

Managing is concerned with making plans and estimates, assessing and analyzing process and product data, reporting progress, initiating and tracking corrective actions, controlling the development process and the work products, and identifying and managing technical risk. Leading is concerned with communicating with technical personnel; coordinating their work activities and reinforcing their morale; and, in addition, communicating and working with the project manager, other managers, and the project stakeholders. An effective leader is a good listener who listens for, hears, and responds to the subtext as well as the main text of a conversation; a facilitator who provides the catalyst for effective teamwork; a coach who provides guidance and encouragement; and an enthusiast who believes in the project team and the goals of the project.

Some systems engineers have aptitude for and enjoy both technical work and managing/leading. Other systems engineers specialize in the technical aspects of systems engineering and some specialize in managing and/or leading.

To be an effective technical manager and a technical leader, you must identify those activities of managing and leading for which you have an aptitude and that you enjoy doing. You must then find ways to compensate for the work activities for which you do not have the aptitude or that you do not enjoy doing. As my colleague Tom DeMarco said, in a private communication,

> find out what you are not good at and don't do it.

You may find, for example, that you are an excellent leader who easily establishes good working relationships with others and that others enjoy working with you, but that you do not enjoy analyzing assessment results and preparing earned value reports. In this case, you and your organization may find it cost-effective to delegate some of the analytical tasks to a designated person who will work with you on those tasks that, for you, are unappealing. That person might be a staff assistant working with you on a part-time or full-time basis, depending on the scope of your project, or that person might be someone you are mentoring and preparing to become a systems engineering technical manager and leader.

Or it may be, by dint of your personality and skill set, you are an excellent manager but less capable as a leader. You may find that the de facto technical leader is the person the development engineers look to for technical guidance, for example, the lead system architect. This can be a very effective arrangement, provided you and your technical leader have a close working relationship, a clear understanding of your realms of responsibility, and the ability to work out your differences in private, thus presenting a united front.

An alternative arrangement may be that you, the physical systems engineer, enjoy interacting with others and your counterpart software systems engineer enjoys the analytical tasks of managing, or vice versa.

You may find some managing and/or leading tasks are unpleasant because you do not have the training, tools, or organizational infrastructure to perform your tasks in an efficient and effective manner. One of the ironies of engineering organizations is that the best technical people are often selected to be technical managers without benefit of training, apprenticeship, or mentoring in management and leadership. If you are fortunate enough to receive proper support from your organization, you can become effective in areas you formerly regarded as your weaknesses.

Chapters 10 and 11 of this text provide information on managing (planning, estimating, assessing, analyzing, and controlling). The following sections of this chapter focus on organizing, leading, and coordinating work activities of the engineers and support personnel who develop complex, multidisciplinary systems.

12.3 The Influence of Corporate Culture

Corporate culture comprises the values, beliefs, and behavior patterns that exist within an organization. Corporate culture thus exerts a strong influence on individual performance. Cultural norms flow from the top down. Individuals look to their senior colleagues and managers for indicators of acceptable and unacceptable behaviors and to see "how should I try to fit in?" Managers, in turn, look to their managers for guidance up to the level of chief executive officer (CEO). From the individual's perspective, corporate culture provides the answer to the question "What does it feel like to work in this organization?" Some factors that determine cultural patterns within organizations are as follows:

- Dress code;
- Degree of formality;
- Fixed versus flexible working hours;
- Cooperation versus competition;
- Reward structure;
- Career progression;
- Conflict resolution;
- Disciplinary policies;
- Attitudes about quality;
- Customer relations;
- A vision statement;
- A mission statement; and
- Ethical behavior.

Different organizations have different dress codes, ranging from formal attire at all time (e.g. neckties, jackets, dresses) to formal when meeting with customers and informal otherwise to "business casual" to "dress-down" Fridays to jeans-and-tee-shirts at all times. In a similar manner, the degree of formality varies from first names throughout the organization to more respectful forms of address for managers and leaders.

Some organizations have flexible working hours and others have strictly enforced times to be at work. In the case of flexible hours, some organizations require everyone to be present between 10:00 a.m. and 3:00 p.m. so that meetings can be scheduled during times when everyone is available to attend. Strictly enforced working hours may be based on the desires of management for workers to be punctual or perhaps based on security considerations that dictate the facility must be locked down from 6:00 p.m. to 7:00 a.m. Monday through Friday and on weekends.

Some organizations encourage competition among individual contributors in the belief that the stress of competition improves productivity and quality of work. Other organizations encourage cooperation and teamwork in the belief that the synergy that results from cooperative teamwork improves productivity and quality of work. The reward structure in some organizations recognizes and rewards individual achievements while other organizations encourage and reward team efforts.

The approaches to conflict resolution and disciplinary action vary from laissez faire to intervention and amelioration. Disciplinary actions vary from hands-off to well-defined procedures that include documentation of unacceptable behaviors, counseling sessions, probationary periods, and dismissal policies.

Some organizations have well-defined career ladders, career development plans for each employee, qualifications and procedures for advancing up a career ladder, and a human relations department that works with employees to advance each employee's career. Career advancement in other organizations is on an ad hoc basis.

Customer relations, attitudes about quality, vision statements, mission statements, and ethical behavior are interrelated elements of organizational culture. In contrast to conventional wisdom, the customer is not always right and not all customers are the right customers for an organization. Some managers and leaders promulgate positive attitudes toward customers, quality, and ethics throughout an organization. Other organizations pay "lip service" to these issues, while others are silent. Regard for customers, attitudes toward quality, and mission and vision statements often instill or detract from ethical behavior.

Mission and vision statements serve distinct purposes and should be clearly differentiated.

A mission statement defines the purpose and goals of an organization. For example,

> We develop large scale electromechanical systems of highest quality for customers who value quality.

Organizational values and ethical behavior must be aligned with the mission statement. If providing large-scale electromechanical systems of highest quality is the mission, concerns for quality must be reinforced and supported throughout the organization and ethical considerations must prevent delivery of systems with known deficiencies. The phrase "for customers who value quality" indicates that the organization must be selective in the customers it deals with if the mission statement is to be fulfilled; i.e. customers who demand short schedules and low costs while sacrificing quality should not be pursued.

In contrast, a vision statement has specific objectives and a time frame for achieving them. An example of a vision statement is as follows:

> We will be one of the top three providers of large scale electromechanical systems throughout the world by 2024.

A mission statement and a vision statement, taken together, provide the basis for strategic planning and norms of organizational behavior.

Some organizations publicize their mission and vision statements to employees and the public and strive to live by them. Other organizations do not have mission and vision statements, or they have them but pay no attention to them.

Leaders, at all levels of an organization, should ensure that their members (including engineers) are aware of the mission and vision and the impact it has on the organizational culture.

12.4 Responsibility and Authority

Responsibility and authority are "two sides of the same coin." Each is considered in turn.

12.4.1 Responsibility

Responsibility is the obligation to perform the duties of your job position. When functioning as a manager and/or leader, the responsibilities of a systems engineer are to

- Prepare and update estimates and plans (see Chapter 10);
- Assess, analyze, and control the work processes and work products (see Chapter 11);
- Manage risk (see Section 11.2.6 and References); and
- Coordinate the work of and lead the system developers (this chapter).

When functioning as a system architect and system designer, your responsibilities are

- Clarifying and revising stakeholders' requirements, system capabilities, and system requirement, as needed;
- Developing alternative system architectures and system design options;
- Presenting the tradeoffs among them to decision makers;
- Refining and decomposing the chosen architecture and design;
- Organizing, leading, and coordinating the work activities of the implementation team or the leaders of the implementation teams;
- Communicating the system vision on a continuing basis; and
- Coordinating technical activities with other involved parties, as appropriate.

12.4.2 Authority

Authority is the power to make the decisions that must be made to fulfill one's responsibilities and the power to execute those decisions or to see that they are executed. A systems engineer, for example, must have the authority needed to make decisions that support the above listed responsibilities, in consultation with others as appropriate.

Authority can be delegated but responsibility cannot. You can, for example, delegate authority to your chief architect (if not you) to negotiate requirements with the customer. But you, as a system engineer, are responsible for delivering an acceptable system based on those requirements. If your architect fails to successfully negotiate the stakeholders' requirements and the resulting system is unsatisfactory, you will be responsible for the failure. And, of course, you deserve to share the credit for successful outcomes.

A common complaint is lack of authority to make the decisions needed to fulfill the responsibilities of your job. Sometimes this is the result of the ineptitude of a manager who delegates authority insufficiently, sometimes it is based on the desire of a manager to exert control over every aspect of the work for which he or she is responsible, and sometimes those who complain about lack of authority mistakenly think their responsibilities are larger than they are in fact.

Development of complex systems typically involves multidisciplinary engineering teams. Successful teams engage in teamwork.

12.5 Teams and Teamwork

A group of individuals engaged in a common endeavor are not necessarily a team. Teams are groups of individuals who work in a cooperative and coordinated manner to achieve shared goals and objectives. A group of people working together is not a team if they do not have a cooperative attitude and shared objectives and goals.

Coalescence of individuals into cohesive teams seldom happens spontaneously. Factors that contribute to team formation include the personalities of the individuals and the social and cultural norms in the organization. Some factors that contribute to efficient and effective engineering teams are the following:

- Appropriate number of engineers;
- Correct mixture of skills;
- Sufficient processes and tools;
- Adequate training;
- Respect for one another;
- Respect for managers and leaders;
- Willingness to be team members;
- Shared ownership of the work products;
- Good communication skills;
- Good communication channels;
- Good working environment; and
- Enjoyment of working together.

A primary responsibility of a systems engineering manager/leader is to facilitate the conditions listed above so that the disciplinary engineers will coalesce into a team, or teams. Project team(s) must have a sufficient number of engineers who have the necessary skills, tools, and training to achieve the project objectives within the constraints of schedule, budget, requirements, resources, and technology.

Gaining the respect and trust of engineers and support personnel is essential for effective leadership. Respect and trust, as is often said, must be earned. A systems engineer must work to earn the respect and the trust of team members by exhibiting

- Competence;
- Ability;
- Integrity;
- Follow-through; and
- Concern for their welfare.

Competence and ability include both technical and social attributes. Integrity involves staying true to a cause and those affected by it. Follow-through involves fulfilling agreements and commitments made to the team members and others. Concern for welfare is exhibited by expressing interest in the lives of team members outside of the work environment and the impact the work environment has on their lives. Comments such as "how was your weekend?" and "how is your sick daughter doing?" when expressed with sincerity rather than superficiality are expressions of concern.

A systems engineer may have positional authority over the engineers and others working on a systems project but collegial development of consensus is a better approach to fostering motivation and morale.

12.5.1 Lone Wolves

Some individuals may have no interest in being members of teams. These so-called lone wolves should, if possible, be removed from project teams. If this is not possible, tasks must be found that can be performed in relative isolation from the rest of the team. As the folk saying goes,

> one rotten apple can spoil the barrel.

One uncooperative individual can destroy a project team.

A key indicator of an effective team is shared ownership of the work products. While each individual is responsible for her or his work assignments and work products, shared ownership of work products is evident when team members willingly help one another. Conversely, dysfunctional groups are characterized by an unwillingness of individuals to help others, perhaps because of excessive schedule pressure, excessive overtime, and/or an organization that prizes competition over cooperation.

Good communication skills, good communication channels, and a pleasant physical environment are all conducive to team coalescence. Enjoyment of working together is the glue that can bond a team. Opportunities for casual interactions must be created. Simple expedients such as a communal break room, afternoon tea, and cookies on Friday afternoon (or soft drinks and beer) can be effective catalysts for developing team cohesion. Organizations have noted declines in team cohesion, productivity, and quality of work products when opportunities for casual interactions such as the communal coffee pot or the afternoon tea break are discontinued.

12.5.2 Teamicide

DeMarco and Lister (2013) in their book, *Peopleware,* introduced the concept of "teamicide," which is "sure fire ways to inhibit formation of teams and disrupt project sociology."

Teamicide techniques include the following:

- Defensive management;
- Mindless bureaucracy;
- Physical separation of team members;
- Fragmentation of time;
- Unrealistic schedules,
- Clique control;
- Lack of allotted time to produce quality work products; and
- Excessive overtime.

Avoiding teamicide is a necessary condition for building and maintaining cohesive teams. Table 12.1 lists some antidotes for teamicide.

As noted in Chapter 8, a "medical antidote" does not provide a cure but is used to counteract the adverse physiological effects of a poison. The antidotes listed in Table 12.1 will not cure the adverse psychological effect of teamicide but they can counteract the effects.

12.5.2.1 Defensive Management

Some managers/leaders practice defensive management, which is indicated by lack of trust in the engineers to fulfill their responsibilities; micro-managing and constant criticism are symptoms of defensive management. A defensive leader fixes blame for problems on others rather than confronting issues and resolving them in a timely manner. Unfortunately, personnel problems are sometimes allowed to persist to the detriment of teamwork when personnel problems are not confronted as they occur.

Other teamicide techniques listed in Table 12.1 include mindless bureaucracy and unrealistic deadlines.

Table 12.1 Some antidotes for teamicide.

Teamicide practice	Antidote
Defensive management	Trust team members until proven you can't
	Fix personnel problems as soon as they occur
	Don't micro-manage
Mindless bureaucracy	Use cost-effective meetings, procedures, and paperwork
	Demonstrate the benefit of the meetings, procedures, and paperwork to all involved parties
Unrealistic deadlines	Set progress milestones that have a reasonable probability of being met
Physical separation	Provide group workspaces and opportunities for casual interactions
Fragmentation of time	Assign people to one team at a time
	Avoid "firefighting" assignments
Clique control	Allow team members to work together for extended periods of time
Quality reduction	Don't compress schedules without de-scoping the requirements
	Don't add requirements without adjusting the schedule and resources
Excessive overtime	Avoid it!

12.5.2.2 Mindless Bureaucracy

Mindless bureaucracy is sometimes truly mindless but is often perceived as mindless based on lack of understanding because

> What is productive for a team or a project is not always viewed as productive by the individuals.

Recording time and effort spent on tasks, preparing reports, and attending meetings may detract from time available to fabricate or construct system elements or to develop and run test cases; however, this data is essential for effective assessment, analysis, and control of technical work (see Section 12.8.13).

It is important that each engineer see tangible results from the "overhead" that are of benefit to them, the project, the organization, and the customers. For example, analysis of reported defect data may result in better processes, methods, and/or tools that reduce rework and overtime. If the purpose, benefits, and value of time spent preparing requested data are not evident to them, the team members should not be asked to report it. Explaining to team members that accurate reporting of overtime work-hours can be of benefit them by providing the data needed to convince the project manager and the customer that, for example, the schedule needs to be extended or the requirements need to be descoped.

12.5.2.3 Unrealistic Deadlines

Some manager/leaders and customers set unrealistic deadlines in the belief that unrealistic deadlines will encourage engineers to achieve greater productivity in an effort to meet the deadlines. This is a wrong-headed, Machiavellian, Theory X approach to managing people (Machiavelli 1984; McGregor 1985). It is wrong-headed for several reasons:

- First, it engenders cynicism among team members;
- Second, it results in poor quality work products;
- Third, it results in job dissatisfaction because team members are not allowed to produce quality work products; and
- Fourth, it results in excessive overtime in attempting to meet unrealistic schedules.

Unrealistic deadlines set by managers and customers create pressure on engineers to engage in voluntary overtime. Unrealistic project milestones must be avoided.

12.5.2.4 Physical Separation

For many organizations, physical separation of teams and team members is a reality in the age of decentralization and globalization. In these cases, the work must be carefully partitioned so that geographically separated individuals and teams can pursue their work activities with relative autonomy and with ongoing communication with appropriate others.

As indicated in Table 12.1, physical separation is another aspect of teamicide to be avoided. Team members must be colocated whenever possible to facilitate teamwork. Engineers who are concurrently realizing the hardware and software elements of system capabilities must be in physical proximity to one another so that continuous and ongoing ad hoc communication is possible.

E-mail and video conference calls are useful adjuncts to communication but they are not substitutes for face-to-face discussions and meetings. Physical separation of team members is a major issue for geographically dispersed projects. These projects are most successful when work activities are partitioned in a manner that requires minimal communication among the partitioned teams. Learning about other cultures can be helpful, as can exchange of personnel among geographically disperse project teams. Investing in one or more face-to-face meetings, even when expensive travel is involved, is a worthwhile investment.

12.5.2.5 Fragmentation of Time

Fragmentation of time among multiple job assignments or projects is another teamicide technique to be avoided. Assignment to multiple projects prevents team members from working together over extended periods of time to learn one another's idiosyncrasies, regard the work products as shared artifacts, and build trust with one another. Matrixed organizations are beneficial, provided team members assigned to projects are allowed to stay together for the duration of a project, but a matrix may result in "firefighting" wherein engineers are shuttled from project to project to put out the latest fire emergency (Wirtz and Vantrappen 2016).

A related issue is disruption of an individual's concentrated flow of thought processes. *Flow* or "being in the zone" is a mental state in which a person is fully immersed in what he or she is doing; it is characterized by effortless enjoyment of an activity and a lost sense of time. The concept of flow was proposed by psychologist Mihály Csíkszentmihályi in 1975 and has been widely researched and referenced across a variety of fields (Csíkszentmihályi 2008).

According to Csíkszentmihályi, some of the factors that contribute to achieving a state of flow are focus of consciousness, self-assurance, and the appropriate environmental and social conditions. Flow occurs with a sense of purpose in a zone of interesting challenges and appropriate skills. The attributes of flow are loss of self-consciousness, the transformation of time, and a sense of enjoyment.

Many of the intellect-based work activities of developing software-intensive physical systems are best accomplished by concentrated mental effort that occurs when one is "in the flow" or "in the zone." In their text *Peopleware*, DeMarco and Lister's observations indicate that it takes 15–20 minutes for an engineer to enter the state of flow; if individuals are interrupted by phone calls and other distractions every 15 or 20 minutes, they will never enter the state of mental flow needed for full concentration on the task at hand.

DeMarco and Lister observe that small offices with doors that can be closed, perhaps shared by two or three people, provide better opportunities to perform concentrated work than do the ubiquitous carrels of work spaces. It is also productive to have small meeting rooms where people can meet to discuss mutual issues.

This author once worked in a large open room with desks and a phone on each desk. When a phone rang, everyone who was not at their desk would stop what they were doing to see if it was their phone and everyone looked to see who answered the phone. There was no private shared workspace for small meetings so people who needed to meet clustered around someone's desk to converse. The adjacent machine shop was noisy. Needless to say, flow never happened.

In another organization, I observed the effects of "quiet hours" from 1:30 p.m. to 3:30 p.m. each workday. During this time, phone calls were diverted and no meetings were scheduled. The resulting quiet atmosphere allowed individuals to enter the flow and perform concentrated mental work.

Other techniques include erecting a small flag or another signal that indicate "Don't bother me, I'm busy just now." The popularity of noise canceling headphones and working at home indicates the need for quietness when accomplishing intellect-intensive tasks.

12.5.2.6 Clique Control

Clique control is yet another teamicide technique sometimes used by insecure and distrustful managers who fear rebellions by united team members. These managers engage in practices that disrupt team cohesion, such as planting rumors and encouraging competition rather than cooperation. Detrimental clique control is sometimes the unintended consequence of rotating staff members among job assignments in the belief that constant job rotation broadens the skill sets of workers and injects new thinking into projects.

Human resource departments of enlightened organizations work with each engineer to develop and implement career plans that will provide the depth and breadth of skills that are beneficial to the organization and are in line with the individual's career goals. However, these plans should be implemented at intervals that do not disrupt continuity of a current job assignment or membership in a cohesive project team.

12.5.2.7 Quality Reduction

One of the major frustrations of engineers is not being allowed sufficient time to produce work products that have the features and quality attributes their personal creeds and senses of responsibility demand. Quality reduction occurs when a schedule is compressed without compensating actions such as descoping the requirements or increasing the resources. Quality reduction can also occur because of virtual schedule compression. This happens when new requirements are added and no compensating actions are taken; as a result, more work must be done in the allotted time.

Quality reduction must be avoided if team cohesiveness and individual morale are to be maintained. On the other hand, systems engineering leaders must guard against "gold plating," which occurs when engineers add features or performance attributes beyond what is needed to satisfy the requirements, users' needs, and customer's expectations.

Excessive overtime is the final item in Table 12.1.

Overtime is time worked beyond the obligated time commitment, which is typically eight work-hours per day, five days per week. Because system development and modification are intellect-intensive activities, it is not reasonable to expect that engineers can be intellectually productive on a continuing basis when they work more than eight hours per day, five days per week. Note that 10-hour workdays, on a routine basis, result in 25% overtime when 40-hour workweeks are the norm. The exception is four 10-hour workdays per week but this could be counterproductive in some instances when continuity of intellect-intensive work is disrupted for three days each week.

Excessive overtime results in mental fatigue and burnout, which demoralizes engineers and results in mistakes than show up as defects in the work products. Excessive overtime also leads to voluntary turnover of personnel, which is disruptive to progress; it is expensive and time consuming to replace qualified engineers.

Every systems engineer and disciplinary engineer knows there will be times when short bursts of overtime are required. However, more than 10% overtime per week (4 hours) or 10% per month (two 8-hour days) should be regarded as excessive. Short bursts of excessive overtime (i.e. one to two weeks of duration) must be compensated with time off to allow individuals to recharge physically, mentally, and emotionally.

To summarize the danger of teamicide, we quote DeMarco and Lister:

> Most organizations don't set out to consciously kill teams. They just act that way.

12.6 Maintaining Motivation and Morale

Another responsibility of a systems engineering leader is maintaining the morale of the engineering team. Morale is evident when a project team exhibits confidence, cheerfulness, discipline, and willingness to perform assigned tasks. Morale is the outward manifestation of motivation. Maintaining morale is thus largely a matter of providing the environment and conditions in which engineers are motivated to willingly perform their assigned tasks with confidence, cheerfulness, and discipline.

To motivate is to provide an incentive for performing an action. People can be motivated by fear and intimidation, which may take the form of fear of reprimand, fear of humiliation, or fear of losing one's job. This approach is not likely to produce the desired result of cheerfulness and willingness to confidently perform assigned tasks with discipline. A more positive approach is to create the conditions in which individuals can satisfy their psychological needs while pursuing their work activities. Individuals will thus derive satisfaction from their jobs and be motivated to do high-quality work in a timely manner.

Some of the ways people obtain psychological satisfaction at work are

- To believe their work is important;
- To have a continuing sense of achievement;
- To receive recognition for their contributions;
- To use a variety of skills;
- To perform well-defined tasks;

- To have profession growth opportunities;
- To have some autonomy; and
- To have pleasant social interactions.

It is probably true that most people, in order to derive psychological satisfaction from work, need to believe their work is important, have a continuing sense of achievement (i.e. "closure" in psychological terms), and receive recognition for their contributions. Other items listed vary in order of importance for different individuals. For example, the prevailing view is that those who work in marketing, sales, or human relations derive job satisfaction from social interactions, whereas engineers prefer autonomy above social interactions (i.e. to be allowed to do their jobs in their own ways). However, people are not so easily characterized. Some marketeers may prize autonomy to deal with customers in their own ways over pleasant social interactions with office mates, and some engineers may derive more satisfaction from performing well-defined tasks that contribute to their team's results than having a great deal of individual autonomy in performing those tasks.

In order to provide a positive work environment, a systems engineering leader must understand the motivational factors that are important to each engineer and attempt, to the extent possible, to create those conditions for each individual. Joe Tester may enjoy performing the well-defined tasks of testing more than interacting with users and Sue Analyst may enjoy interacting with users more than debugging software code. If circumstances allow, each team member should be assigned to the tasks that permit them to satisfy their psychological needs.

An anecdotal list of job satisfiers for engineers includes the following:

- A quiet place to work;
- Challenging technical problems;
- Autonomy to solve problems;
- Ability to control one's schedule;
- A chance to learn new things and try new ideas;
- Adequate infrastructure facilities and tools; and
- Competent technical leaders.

Engineers also prize interactions with peers via electronic mail, video conferencing, webinars, bulletin boards, news groups, blogs, and attendance at technical conferences.

It is not intended that the above lists be definitive but rather illustrative of factors that should be taken into consideration when creating a work environment in which workers can derive psychological satisfaction and thereby exhibit positive morale.

12.7 Can't Versus Won't

Andy Grove, a founder and former CEO of Intel Corporation, observed in his book *High Output Management*, that engineers who are not performing up to expectations either can't or won't (Grove 1995). If an engineer wants to do a good job but can't, it may be because she or he lacks training, skill, experience, tools, time, or basic ability to do the job. When a person has the necessary prerequisites to do a good job but won't, it is because he or she lacks motivation (or perhaps is perversely motivated to derail a project). Those who can't are unable and those who won't are unwilling. Table 12.2 lists the four combinations and the resulting situations.

As indicated, unable and unwilling is a realistic situation because the person who is not qualified to do a job is unwilling to do it and should not be assigned to that job. Unable and willing is a dangerous situation because the person will most likely make serious mistakes in attempting to do a job for which he or she is not qualified. Able and unwilling is the situation in which a person is qualified to perform a task but refuses to do it, or will do it grudgingly and with lack of enthusiasm. Able and willing is the most desirable situation; the person has the ability and is willing to perform assigned tasks.

Effective systems engineering leaders understand the personalities, skills, and motivations (or lack thereof) of individual engineers and respond as the situation requires. This approach is known as situational leadership (Hershey and Blanchard 1999). Each of the can't versus won't situations listed in Table 12.2 can be dealt with using the techniques of situational leadership indicated in Table 12.3.

Selling, as indicated in Table 12.3, is meant to be a positive method of motivating and not the negative methods of convincing people to buy something they don't want or need.

For those who are unable and unwilling, a teaching style is appropriate to enable them and a selling style is appropriate to motivate them. Teaching techniques include attending classes, reading papers, being mentored, and working with consultants. Selling techniques include anecdotes, testimonials of respected engineers, guest speakers, papers to read, and classes. Teaching

Table 12.2 Four combinations of can't and won't.

Combination	Manifestation	Result
Can't and won't	Unable and unwilling	A realistic situation
Can't but will	Unable but willing	A dangerous situation
Can but won't	Able but unwilling	Lack of motivation
Can and will	Able and willing	The best situation

Table 12.3 Four situations and leadership styles.

Can't versus won't	Leadership style
Unable and unwilling	Teaching plus selling
Unable but willing	Teaching plus reinforcing
Able but unwilling	Selling plus reinforcing
Able and willing	Reinforcing plus delegation

plus selling would be appropriate for individuals who do not have the training or experience to participate in design reviews, for example, and are skeptical of the value of reviews.

For those who are unable but willing, a teaching plus reinforcing style is appropriate; teaching to impart the skills and reinforcing to channel their efforts in the desired ways. Reinforcing techniques include a combination of teaching, selling, and other techniques, such as attending workshops, being coached, and apprenticeships to increase ability to do the job. Teaching plus reinforcing would be appropriate for those engineers who are willing to give design reviews a try but lack the necessary skill.

Conditions must be created for which engineers who are able but unwilling become willing to do the job with enthusiasm. Motivational techniques include selling and reinforcing, plus other techniques such as removing barriers that de-motivate the individual. Selling and reinforcing would be an appropriate approach for those individuals who have had bad experiences participating in poorly managed design reviews because the participants were not trained to do technical reviews. The demotivating barrier could be removed by providing training and coaching.

The best situation is when engineers are able and willing to do an assigned task. The appropriate leadership style is reinforcement and delegation. Reinforcement will provide confidence in their abilities and delegation will strengthen their motivation. Delegation techniques include working with individuals to set goals, giving them the authority and autonomy to do the job in the way they think best, and establishing procedures for reporting progress and problems. Individuals who are able and willing to perform design reviews, for example, are candidates to become design review moderators.

12.8 Fourteen Guidelines for Organizing and Leading Engineering Teams

This concluding section of the text summarizes many topics covered in the text and presents a few new ones.

As the size and complexity of software-enabled physical systems grows and as the demands for higher quality systems and shorter development cycles increase, the ability of individual engineers to work as members of teams and the ability of the systems engineers to effectively organize, coordinate, and lead the efforts of engineering teams becomes more critical to success. This section presents 14 guidelines for organizing, leading, and coordinating the efforts of engineering teams.

There are several reasons that teams are more effective than a collection of individuals working alone. Scheduling and skill sets are primary reasons. Customers won't wait five years for one person to develop or modify a software-intensive system requiring 60 staff-months of effort. At the other extreme, it is not realistic to assign 60 people for one month. We might scope a 60 staff-month project as a job for 5 people over 12 months or 6 people over 10 months, or by using the square root rule of 8 people for 8 months.

Teams are also needed to provide the variety of skills and aptitudes required to develop or modify a complex software-enabled physical system. In addition, the synergy that occurs when team members work together in a collaborative manner often results in a product superior to the one that would have resulted from the efforts of several individuals working in relative isolation.

Organizing and coordinating the activities of individuals engaged in intellect-intensive teamwork is a relatively new kind of human endeavor. Over time, humans have learned how to organize agricultural and manufacturing activities to utilize the skills of multiple individuals, but the corresponding organizational and leadership techniques for intellect-intensive work teams are not yet mature.

A simile for organizing the team effort required to develop a complex software-enabled system (or any other kind of complex system) is how this team effort is like writing a book as a team effort. "Book Writing" in Section 2.4.8 presents the simile.

Over time, this author has developed and observed a number of techniques that differentiate capable engineering teams from the less capable.

12.8.1 Use the Best People You Can Find

As previously stated, successful engineering projects depend on effective people, appropriate processes, and adequate technology. People are the most important asset. People of outstanding skill, ability, and motivation can often overcome inadequate processes and technology but excellent processes and technology cannot compensate for lack of skill, ability, or motivation.

For engineering teams, "best people" means engineers who have appropriate technical skills *and* sufficient interpersonal skills to effectively interact with other team members. Some engineers have outstanding technical skills but are neither interested in being nor psychologically suited to be members of cohesive teams. Too often, organizations are guilty of suboptimizing the productivity of a team by catering to the idiosyncrasies of technically skilled but socially inept individuals. In some cases, it may be necessary to remove an antisocial or asocial team member for the greater good of the team, the project, and the organization.

Hiring the best people means that more than the going rate of pay will have to paid for individuals who have outstanding skills within a given skill category. It has been repeatedly shown that software programmer productivity, for example, varies by factors of 10 : 1 or more among individual programmers who have similar backgrounds and experiences (DeMarco and Lister 2013). Simple economics would indicate that paying 10–20% more in return for a gain of 500% or 1000% is a bargain. Similarly, an expert system architect within the domain of interest is figuratively and sometime literally "worth his or her weight in gold."

12.8.1.1 Develop a Key Personnel Plan

Key personnel are those members of a system development group who have significant responsibilities and the skills, motivations, and authority to carry out their job assignments. For small projects (3–10 team members), the systems engineer manager/leader is project manager, systems engineer, requirements engineer, system architect, and leader of the system development team; every project member is a key person for the project. The systems engineer may assign some of the team members to assist him or her in playing some of the systems engineering roles. For medium-sized projects (10–50 team members), the systems engineer manager/leader will lead and coordinate the efforts of the team leaders. For large projects (more than 50 team members), the systems engineer manager/leader will lead and coordinate the efforts of the systems engineer managers/leaders of the subsystems.

Some key personnel roles for medium size systems projects are

- Systems engineer;
- Requirements engineer;
- System architect;
- Team leader;
- Interface coordinator;
- Verifier;
- Validator; and
- Configuration manager.

The team leaders are responsible for coordinating the work activities of their team members, who are in turn their key personnel.

The individual who plays the role of interface coordinator is a team member who works with other team members who are concurrently developing physical elements and software elements for a system capability; his or her role is to ensure that the interfaces among physical elements and software elements are compatible.

The verifier and validator roles are played by those who have responsibility for verifying and validating the new system capabilities developed during an incremental subphase of system implementation. The roles may be combined or distinct: the role players may be designated team members or the role players may rotate within a system implementation team.

The configuration manager should be a member of the development team, whether or not that person or persons report directly to the systems engineering manager/leader or to another manager; either arrangement is acceptable, provided the goals and procedures of configuration management are satisfied.

Note that process and product quality assurance and independent verification and validation (if involved) are not listed as key personnel roles. QA personnel are not project members, nor are the members of independent verification and validation groups (if any) because a systems engineer manager/leader does not have, and should not have, responsibility or authority to direct their work activities.

One of the first activities of a systems engineering manager/leader is to prepare a list of key personnel roles that includes the numbers and skills of personnel needed to play the roles, when they will be needed, and for how long. Some roles may require multiple personnel; for example, 10 software engineers proficient in using the Java programming language may be needed during the system implementation phase of system development. One person may play multiple roles, for example, the systems engineer for a small project who plays the roles of project manager, software architect, and team leader. One individual may play different roles at different times, for example, a hardware designer who later becomes a hardware integrator.

If, during initial planning, the names of those who will play some or all of the roles are known, their names can be entered in a personnel assignment matrix. If their names are not known, a personnel acquisition plan should be developed. The plan should state the roles to be played, the number of personnel needed to fill the roles, the job qualifications for the roles, and the dates by which the roles must be filled.

The project manager, the organization's human resources department, or some other organizational entity may be of assistance in filling the roles, or the systems engineering manager/leader may have full autonomy to recruit the technical staff members, depending on the infrastructure of the organization and possibly on the conditions stated in the contract between the customer's organization (i.e. the acquirer) and the development organization (i.e. the supplier). The personnel acquisition plan should be reviewed when developing the risk management plan to determine potential problems in acquiring the needed personnel.

12.8.2 Treat Engineers as Assets Rather than Costs

One of the first rules of business is to manage corporate assets to maximize return on investment in those assets; the second rule is to control costs. Unfortunately, many engineering organizations confuse the second rule with the first one and treat their engineers as costs rather than assets. Companies that regard their engineers as assets invest in the engineers: the engineers are adequately compensated, properly trained, and provided a work environment that is supportive of engineers and engineering work activities that enhances productivity and quality of work. The work environment for engineers includes the social, cultural, and intellectual work environment; the development infrastructure; and the physical workspace.

DeMarco and Lister (2013) have shown that the ability to divert phone calls and other interruptions, thereby preventing disruptions to thought patterns, is one of the most effective mechanisms for improving individual productivity and quality of work. Colocating engineering team members engaged in concurrent development of software and physical system elements are essential. Providing private breakout areas where two or three engineers can converse without disturbing others is another example of a workspace factor that can improve the efficiency and effectiveness of individuals and teams. A policy of quiet hours during part of each working day can improve the intellectual work environment of software engineering teams.

12.8.3 Provide a Balance Between Job Specialization and Job Variety

Many engineers are motivated by apparently conflicting needs: the need to be recognized for their expertise within a discipline, and the need to learn and apply new skills. In engineering, tasks that are initially challenging can quickly become repetitive and boring. On the other hand, organizations and projects often need highly skilled specialists in various arcane technologies. It is reasonable to assign tasks to those who are best qualified to perform those tasks; however, new and challenging job assignments must be provided so that individual contributors do not become technically stagnant. Short-term productivity may benefit from prolonged and concentrated specialization by individuals, but in the long term, the individual and the organization will both benefit from a judiciously chosen variety of job assignments for team members.

12.8.4 Keep Team Members Together

It takes time for team members to learn one another's work habits, skills, aptitudes, likes, and dislikes and for team members to become comfortable with their team environment. One of the potential problems of matrix organizations is lack of team cohesion; project members drawn from functional homes for short periods of time often have more allegiance to their functional managers and their functional colleagues than to their project manager and engineering colleagues. Keeping a team together over extended periods of time and using explicit team building techniques such as off-site planning and review meetings, team participation in training courses, and corporate-sponsored recreational activities are effective techniques for building a cohesive team. Weekly status meetings can be teambuilding experiences if properly conducted (see Guideline 10).

12.8.5 Limit the Size of Each Team

Small teams are best at accomplishing closely coordinated, intellect-intensive teamwork. For small projects, a systems engineer (either physical or software, as appropriate) will be the team leader but for larger projects, one or more systems engineers will coordinate the work of multiple team leaders. For very large projects, a team of systems engineers will coordinate the work of systems engineers for the subsystems, who will in turn coordinate the work of team leaders. This structure can be recursively decomposed to appropriate levels so that the lowest levels consist of collaborative small teams.

We have observed two types of cohesive engineering teams, the first being the more common. This team structure consists of three to six team members plus a team leader, which results in a maximum team size of seven. If a team grows to eight or more, it is split into two teams of three to seven members each plus two team leaders. When a team grows larger than six members plus a team leader, it is difficult for team members to coordinate their work activities with each of the other team members. It is also impossible for the team leader to provide the necessary level of planning, coordinating, and leading.

Teams larger than seven may be a symptom of inadequate partitioning of the requirements and insufficient decomposition of the system architecture. The system should be structured as a collection of loosely coupled, highly cohesive parts so that a small team can implement each part. The teams are then highly cohesive and loosely coupled.

The second team structure consists of 7–12 individuals plus a team leader. Within these larger teams, individual team members are more autonomous and tend to be more loosely coupled to other team members than three- to seven-person teams. To be effective, these larger teams must satisfy some special conditions; namely,

- Each team member must be a highly skilled and experienced professional;
- Each team member must have a well-defined functional role; and
- Everyone must have a clear understanding of his or her role, and the roles of the other team members.

In addition, each team member must have sufficient initiative and discipline to plan and organize his or her individual work activities and to communicate with the other team members and the team leader (see Guideline 8).

These larger cohesive teams have been observed in domains such as telecommunications, process control, and in teams of systems engineers for large complex systems. In these teams, team members are often highly skilled and experienced in their functional specialties, functional roles are clearly differentiated, and the role of each person is clearly understood by others. It must be emphasized that teams and projects are placed at risk when teams larger than seven (six members plus leader) are utilized without the prerequisites of individual skill, experience, job specialization, and initiative.

12.8.6 Differentiate the Role of Team Leader

In both types of teams described earlier, team leaders plays the pivotal roles of

- Planning, negotiating, and coordinating work activities of the team members;
- Setting performance goals for each team member;
- Tracking work progress of individuals and the team;
- Updating detailed plans;
- Validating the work products produced by team members; and
- Communicating with his or her leader: A physical systems engineer or a software systems engineer (as appropriate).

A team leader may assist a team member but should never take the initiative in generating work products – the job of a team leader is to plan and coordinate work activities, set performance goals, validate work products, monitor progress, advise and help team members, anticipate problems, and be the spokesperson for the team. Given these roles of planner, coordinator, progress monitor, communicator, and quality control agent (see Guideline 7), the leader of an engineer team is not "management overhead" but rather is the catalyst that causes a group of individuals to coalesce into a cohesive, productive team.

12.8.7 Each Team Leader Should Be the Team's Quality Control Agent

An important task for each team leader is specifying or tailoring the verification and validation criteria for work products and determining that the work products satisfy those criteria. In teams of three to six, the team leader is the moderator of technical peer reviews and determines that other quality control activities are conducted in an effective manner, for example, determining that test completion criteria are satisfied.

In larger teams (7–12 members), the team leader does not usually verify and validate all of the work products generated by all team members, but rather assigns validation tasks, such as moderator duties for technical peer reviews, to team members in a "round robin" manner so that everyone takes his or her turn.

This is possible because each member of the larger teams has sufficient skills and experience to lead peer reviews and apply objective acceptance criteria for the work products generated by other team members.

A team leader thus does "real work," stays in close contact with the efforts of each team member, and takes responsibility for the quality of work products generated by the team. The team leaders are thus the primary quality control agents for his or her team. The role of the quality assurance group is then to observe, advise the team leaders, analyze quality metrics data, recommend process improvements, and assure that the team leaders and their teams are fulfilling their responsibilities.

And for a small project (i.e. one team of seven or fewer team members), the team leader also plays the roles of systems engineer and system architect.

12.8.8 Safeguard Individual Productivity and Quality Data

A common area of concern among team members is how their productivity data and work-product quality data will be used. Data collected from individuals should always be aggregated for their small teams as needed for purposes of analysis and reporting. Team leaders and systems engineers must do everything possible to prevent disclosure of individual productivity and quality data. Nothing will kill trust and destroy team cohesion faster than disclosure or use of data that must be held in confidence.

When data indicates that an individual's performance is not up to expectations, the systems engineer leader should hold a private meeting with the individual. The goal and tone of the meeting should be to determine the conditions that are causing inadequate performance and to help the individual develop a plan of action that will remedy those conditions. Remedies may include one or more of

- Training;
- Mentorship;
- Better tools;
- Clarification of responsibilities;
- Reassignment of duties within the team; or
- Reassignment to another team, project, or department.

Poor performance may require that the systems engineer leader and the individual work with the human resources department to resolve the issue (perhaps you, the systems engineer, are unqualified or uncomfortable to deal with the issue, or perhaps, in the view of the individual, you are the issue to be resolved).

Systems engineers who are the leaders of small teams (four to six members) are responsible for monitoring the productivity and quality of output of their team members and for working with the members to apply corrective actions for deficiencies in performance. On larger projects, it may be necessary for the team leader and the supervising systems engineer to work privately with team members who are not performing or cannot perform up to expectations. Team members, team leaders, and others who cannot or are unwilling to improve must be removed from the engineering team.

12.8.9 Decompose Tasks into Manageable Units of Work

Lowest level tasks are sized to achieve a balance between micro-management of individual team members and macro-management of the entire technical effort. At the level of job assignments for individuals, we recommend the "one-to-two" rule: one to two persons, one to two weeks, but not to exceed 80 staff-hours per task. The one-to-two rule thus brackets work tasks in a range of 40 staff-hours to 80 staff-hours. A task assigned to one person for two weeks or two persons for one week would satisfy the one-to-two rule. Forty staff-hours of effort on the lower end avoids micro-management of individuals, who can plan and arrange their work activities week by week, perhaps on a flex-hour schedule. Eighty staff-hours on the upper end avoid macro-management by forcing attention to detailed planning and monitoring of progress by team leaders and systems engineers.

In teams of three to six, the one-to-two rule provides a manageable workload for the team leader. For a team of five engineers, each working on a 40 staff-hour task, the team leader will, on average, have one completed task to validate and one new task to initiate each day. A team of three, each working on 80 staff-hour tasks, represents one-and-a-half tasks to validate and initiate each day; however, the tasks are twice as large to initiate and the work products to be validated are larger and/or more complex than 40-staff hour tasks.

In engineering teams of 7–12 members, the team leader delegates some duties. Each team member is responsible for generating and documenting 40–80 hours of work plans, each of which generates a work product that is accepted by objective validation criteria, and for coordinating her or his plans with the team leader and the other team members. However, the team leader is still responsible for reviewing, approving, and coordinating plans; setting performance goals; monitoring progress; optimizing the allocation of team members and other resources; assuring that work products satisfy their validation criteria; and communicating with the coordinating systems engineer(s) while assuring that individual team members are meeting their performance goals (see Guideline 11).

12.8.10 Set Performance Goals for Each Team

In their book, *The Wisdom of Teams*, Katzenbach and Smith (1993) observe that the most effective way to build a cohesive team is to set challenging performance goals for the entire team *and* for each member of the team. The goals should be challenging but not impossible; DeMarco and Lister (2013) include impossible schedules (phony deadlines) as one of the "sure fire ways to inhibit formation of teams and disrupt project sociology." Other factors cited by Katzenbach and Smith that distinguish effective teams include having a meaningful purpose, a common approach, complementary skills, and mutual accountability.

It is important to set performance goals for each team member as well as for the entire team. This approach eliminates the possibility that collective team efforts might dilute individual accountability, initiative, and recognition. Elements of a performance goal include objective, measurable completion criteria and a time frame for achievement of the goal. For example, a team goal might be to reduce backlogged technical debt from 20% of the remaining work to be done to 10% during the next month (Fairley and Willshire 2017).

It is also important that team goals be discussed and negotiated in an open manner with the entire team and that individual goals are discussed, negotiated, and reviewed in private with each individual. Goals should be challenging but not impossible. Progress toward goals should be monitored and corrective action taken if extrapolation of the current trend indicates that a goal will probably not be met within the specified time frame. Failure to achieve an ambitious goal should be regarded as an opportunity to learn from the experience and to improve future individual and team performance.

Performance is reviewed periodically (e.g. monthly) with the team and privately with each team member. Achievement of ambitious goals is celebrated, problems are identified, and impediments to better performance are identified. Team goals, progress toward those goals, and team achievements are displayed in a public manner. Review of individual performance is a confidential matter between the individual and the team leader. Team members who consistently fail to meet agreed-to goals should be counseled to determine the reasons they are unable or unwilling to meet the goals and to develop courses of action that will enable the team member to improve his or her performance.

Setting of new goals is an ongoing process. Goal setting and performance measurement do not have to be, and should not be, Machiavellian (Machiavelli 1984). Attributes of cohesive teams include a collective sense of humor, healthy skepticism, and enjoyment of working together. These attributes facilitate collective setting of team goals.

12.8.11 Adopt a Contractual Model for Commitments

A contract is an agreement between two parties, for example, an agreement between a systems engineer and a team that will develop part or all of a system. Some contracts are formal and legally binding with consequences if the contract is broken. Formal contracts are typically used when subcontractors are employed.

Other contracts, as between a systems engineer and an in-house system development team, are informal agreements of commitment, similar to memos of understanding (see Section 6.2.2). A systems engineer, for example, might commit to providing needed resources within a specific time frame or to facilitate a meeting with a customer or a representative group of future system users. In return, the development team might commit to developing a prototype version of a system having specified features within a specific time frame. Team commitments should not be made without the consent and agreement of all team members.

12.8.12 Ensure Daily Interactions with Team Leaders and Team Members

It is important that a systems engineer manager/leader interact with the team leaders (if any), a team leader or systems engineer interact with each team member, and team members interact with one another on a daily basis. One mechanism for ensuring daily interaction at each level is a 15- to 20-minute "standup" meeting in which each team leader or team member briefly reports on work accomplished the previous day, work in progress for the current day, and any issues that should be brought to the attention of other team members. Issues to be resolved by two or three individuals should be noted but handled apart from the team meeting. Issues that require the attention of the entire team should be scheduled for later in the day or perhaps the following day, depending on the urgency of the situation.

A daily forum provides an opportunity for the systems engineer or team leader to communicate any new information, report on the status of work in progress, comment on the latest rumors, and provide advance notice of upcoming events. It also provides an opportunity for team leaders or team members to "pair-off" and discuss issues of mutual interest following the meeting.

Electronic mail, video conferencing, and groupware tools should never be used in place of daily "face-to-face" meetings unless they cannot be avoided. Electronic media can be effective communication mechanisms and should be fully utilized, but they should augment and not replace human contact among team members and between team members and the team leader. Team members who are on travel should speak, by phone or video, with their team leader on a daily basis; news of their activities should be communicated by the team leader to other team members during the daily standup meetings. An acting team leader should be designated when the official leader is absent and the acting and official team leaders should communicate, by phone or video, on a daily basis.

12.8.13 Conduct Weekly Status Review Meetings

Each team should meet each week (e.g. on Friday mornings) to review project status and prepare a status report. Each team member submits, by electronic mail, his or her individual weekly report to the team leader and includes a copy for the other team members on the afternoon before the weekly meeting (e.g. Thursday afternoon). Everyone comes to the weekly meeting on Friday afternoon having reviewed all individual reports. Issues that involve or impact the entire team are discussed and issues that involve one or two team members are resolved separately, perhaps in consultation with the team leader.

12.8.13.1 Maintain Weekly Top-*N* Lists at All Levels of a Systems Project

A weekly status report includes a "top-*N*" list of problems and risk factors, with $N \leq 10$ (Boehm 1981). The combined number of problems and risk factors should be 10 or fewer and the meeting should take one hour or less. A team and their work will be in serious trouble if there are more than 10 problems on the list and the meeting of the collective team takes more than one hour.

Each problem on the top-*N* list has an associated action item that specifies the nature of the problem, the actions to be pursued or being pursued, the responsible individual, and the scheduled closure date for the action item. Each risk factor is recorded with a description, probability and probable loss, and the time frame in which the potential problem (i.e. the risk factor) might become a real problem. Problems that cannot be resolved by a team are communicated to the systems engineer who is coordinating the work of the team for his or her weekly meeting with the team leaders (e.g. on Friday afternoon). As stated above, the top-*N* list should have 10 or fewer items and the meeting should take one hour or less; otherwise, the technical work is in serious trouble.

Each team leader sends his or her list of problems (and risk factors that may soon become problems) that cannot be resolved at the team level to the coordinating systems engineer. Each team leader reads the lists submitted by the other team leaders before the meeting. The purpose of the meeting is to develop the system engineer's top-*N* list that is developed by consensus of the team leaders. The systems engineer, in turn, prepares a weekly status report and initiates corrective actions for newly reported problems. He or she, in turn, sends a list of the problems that cannot be resolved at his or her level to the project manager. The project manager and the other systems engineers who are supervising one or more other teams (if any) attend a weekly meeting to help the project manager prepare a top-*N* list. The project manager, in turn, will resolve issues he or she can and meets with his or her manager to discuss issues that he or she cannot resolve.

There are thus top-*N* lists at all levels of a systems engineering project. In one organization familiar to the author, and perhaps in other organizations, the top-*N* lists are publicly displayed on bulletin boards for all to see (updated weekly). This creates an organizational culture where issues and resolutions of issues are openly discussed. This also helps to control rumors and overcome reluctance to candidly report problems and potential problems. Of course, there may be sensitive issues that cannot be publicly discussed but these should be limited to as few in number as possible.

Note that a top-N list is similar to a risk management watch list; however, top-N lists are used to proactively manage problems and mitigate risk factors, whereas a watch list is a means of periodically reviewing risk factors for possible mitigation.

A recommended agenda for weekly status meetings and follow-up activities is presented in Table 12.4. Some renegotiation of the contract with the team or teams may occur and involve trading off responsibilities among teams (with the team leaders' concurrence – see Section 12.8.11). In a structure of cohesive teams, it is quite common for those who are running ahead of schedule on their tasks to volunteer to help those who are running behind. Job sharing and willingness to help one another are key indicators of cohesive system engineering teams and cohesive members in each team.

It is important that team members agree on priorities among problems and risk factors and the associate action items, so that task priorities and allocations of resources are understood and supported by all.

Action items are entered into a tracking system and the status of each open action item is reviewed at each weekly status meeting. Action items are treated as discrepancy reports against the development process; as such they are tracked to closure in the same way problem reports are tracked against work products.

Table 12.4 Agenda and follow-up activities for weekly status meetings.

Meeting agenda (1 hr)
Review individual status reports
Review status of open action items
Generate a revised top-N problem list
Generate a revised and prioritized list of action items
Generate a revised and prioritized list of risk factors
Document planned corrective actions and mitigation strategies for newly identified problems and risk factors
Document action items that cannot, or should not, be handled at this level
Follow-up activities
Enter new action items into the tracking system
Reprioritize existing action items as necessary
Initiate new problem resolutions and risk mitigation strategies as necessary
Renegotiate contracts with teams as necessary
Revise schedule and resource allocations as necessary
Report upward action items that cannot be, or should not be, handled at this level

The top-*N* list should be included in a weekly status report. An effective technique for weekly status reporting is to adopt a standard format for the reports, for example,

Date:
Team leader:

- For each action item completed:
  Identifier:
  Responsible engineer:
  Problem description:
  Actions taken:
  Date initiated:
  Date closed:
- For ongoing action items:
  Identifier:
  Problem description:
  Responsible engineer:
  Current status:
  Current corrective actions:
  Alternative actions to be taken, if any:
  Estimated closure date:
- For newly identified problems:
  Identifier:
  Responsible engineer:
  Problem description:
  Date identified:
  Actions to be taken:
  Estimated closure date:
- For resurfaced problems:
  Old problem identifier:
    Previous actions taken:
    Date closed:
  New problem identifier:
    Responsible engineer:
    Actions to be taken:
    Responsible parties:
    Estimated closure date:
- For each risk factor:
  Description:
  Responsible engineer:
  Probability, potential impact, timing of likely occurrence:
  Recommended mitigation strategy, if needed:

Retaining the status reports provides a history of the evolving work processes and work products of a systems project. The status reports can provide information used to determine lessons learned and provide a basis for improvements in processes, methods, and tools.

12.8.13.2 Conduct Retrospective and Planning Meetings

Retrospective meeting should be held at the end of each incremental subphase of system development and planning meetings should be held at the beginning of the subsequent subphase (see Chapter 5).

The purpose of a retrospective meeting is to review the newly completed incremental subphase of system development, identify what went well and what didn't go well, and determine corrective actions needed in the work processes and/or work products for the next subphase.

A planning meeting occurs at the start of each subphase of the I^3 implementation phase (Phase 5), as depicted in Figure 5.2. The purpose of a planning meeting is to identify, analyze, and revise, as necessary, the system capabilities and system requirements to be realized during the subphase; initiate any corrective actions from the preceding retrospective meeting; reach agreement concerning tasks assignments among the team members; and negotiate a contract with the supervising systems engineer for the incremental subphase.

In some cases, the retrospective and planning meetings may be combined into one meeting and that meeting may be scheduled as one of the weekly status meetings. In other cases, the retrospective and planning meetings may be separately conducted, for example, when there are significant issues to be addressed in one or both of the meetings or when there is a break in continuity between the end of one incremental subphase and the start of the subsequent one.

For small projects, each meeting should require not more than two hours. For large complex projects, each meeting may require one or two days. Excessive meeting times may be indicative of deeper issues that the project team cannot solve alone.

12.8.14 Structure Large Projects as Collections of Highly Cohesive, Loosely Coupled Small Projects

Projects requiring more than six team members plus a team leader (or 7–12 members in certain circumstances; see Guideline 5) are best accomplished using multiple teams. Figure 12.1 illustrates a project structure that can be tailored to fit the needs of projects ranging from three or four team members to projects having 100 or more system developers.

In the case of a small system development team, as illustrated in the figure, the systems engineer may play the roles of customer contact, project manager, system architect and system designer, and leader/coordinator of the development teams. The structure in Figure 12.1 can be recursively decomposed

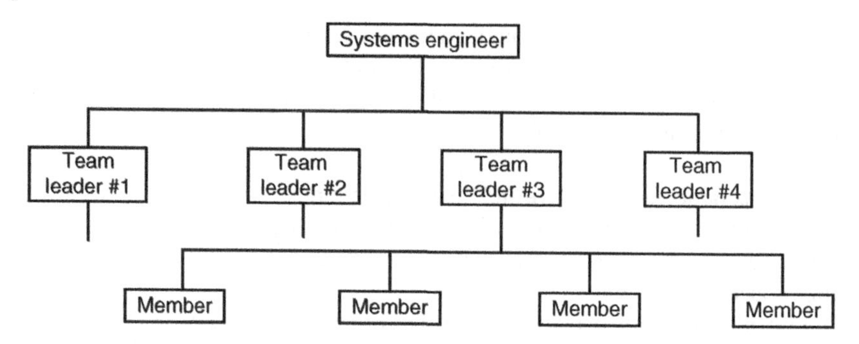

Figure 12.1 Structure of a small engineering team.

as needed – the team leaders could become or be replaced by subsystem manager/leaders with subordinate team leaders, who could, in turn, become sub-subsystem manager/leaders.

On projects of medium size (say, 10–50 members), multiple team leaders (six or fewer members per team) report to a physical systems engineer and a software systems engineer; in addition, the systems engineers are leaders or members of the software architect design team. On large projects (more than 50 members), the team leaders depicted in Figure 12.1 are systems engineering managers/leaders of subsystems' development; each subsystem will have multiple teams and team leaders. Depending on the size and complexity of a subsystem, the systems engineering manager/leader for the subsystem may also be the software architect for that subsystem or may be assisted by a subsystem architect. On a large project, the systems engineering manager/leader may be assisted by one or more staff members and the chief architect may head a design team consisting of the subsystem architects.

Different subsystem groups may use different methods and tools, as appropriate within the overall framework of the project and the system domain. Some subsystem developers might use an incremental development strategy with weekly milestones and internal demos; some teams might use a state-based approach to design and implementation, while others might be using functional decomposition development techniques.

12.8.14.1 A Rule of Thumb

The rule of thumb for structuring engineered teams, as with systems, is to design a collection of highly cohesive, loosely coupled elements. In a 1968 article, Mel Conway (1968) made an observation that has become known as "Conway's law":

> organizations which design systems … are constrained to produce designs which are copies of the communication structures of these organizations.

Like many profound statements, this one is obvious once pointed out: the protocols and conventions worked out among team members and between teams become the interfaces and protocols between the system elements and among the subsystems developed by the team members and teams.

Conway's law implies "Fairley's corollary":

> To maximize efficiency and effectiveness, task assignments for system development teams and for individual team members must be structured to mirror the decomposed system architecture.

One technique for assuring adherence to Fairley's corollary is to embed the architectural structure of a system or subsystem in a work breakdown structure (WBS) by partitioning the development tasks in the WBS in one-to-one correspondence with the system or subsystem architecture.

Figure 10.9 illustrates a partial WBS for developing the hardware and software elements of an automated teller machine (ATM) in which the product structure is denoted by the work products that will result from completion of the corresponding elements of the WBS.

The fourth-level elements (i.e. 3.x.y.z) in the WBS are tasks to be performed when building the validator, processor, recorder, and terminator. Note that each element of the WBS is a verb phrase denoting an activity or task to be accomplished. Needed hardware and software can be developed concurrently. Using the I^3 system development model, separate hardware engineer/software engineer teams could be assigned to realize the capabilities for each of the four major system elements. For larger projects, each of the four major elements in Figure 10.9 would be a subsystem for which a separate WBS would be developed. Depending on size and scope of the system, one or more systems engineers and development teams might be assigned to each subsystem.

By Conway's law and Fairley's corollary, the WBS/embedded architecture approach will result in a system with loosely coupled interfaces between the subsystems, loose coupling among the elements of each subsystem, and high cohesion within the elements generated by each small team. Equally important, it is possible for each individual contributor and each team leader to know the names and faces of all other individuals working on the subsystem, which provides individual identification with the larger effort, thus avoiding the feeling of anonymity in a large bureaucracy.

Perhaps most important, this approach allows the team leaders and their team members to function as a small team that reports to the systems engineer who is coordinating development of a system or subsystem. The subsystems' systems engineers can coordinate and lead the team leaders using the same techniques the team leaders use to manage their teams, including individual and collective performance goals, WBSs, schedule networks, contractual models, binary acceptance criteria for work products, weekly status meetings, top-*N* problem lists, action item tracking, risk management, and earned value reporting.

In turn, the various subsystems' systems engineers are a team that reports to their leader systems engineers and the subsystem architects are members of the system architect's design team. Contractual commitments and performance goals flow downward through the hierarchy; negotiated commitments, work products, and performance metrics flow upward. In this way, the team techniques described herein can be generalized from individuals to projects of arbitrary size.

The text *Work Breakdown Structures* and the Project Management Institute (PMI) practice standard for WBSs provide guidance for developing and using WBS (Norman et al. 2008; Perrin 2015).

Section 11.6 of *Managing and Leading Software Projects* provides 15 guidelines for designing WBSs (Fairley 2009).

12.9 Summary of the Guidelines

As stated in the introduction to this chapter: three key assets for successful engineering projects are effective people, appropriate processes, and adequate technology. In systems engineering and software engineering, people are the most important element. Competent engineers and cohesive teams can overcome weak processes and poor technology but excellent processes and outstanding technology can never compensate for inadequate skills or dysfunctional teams.

Fred Brooks' famous observation is by now a cliché, albeit an important one:

> There are no silver bullets for software engineering.
>
> (Brooks 1987)

His statement applies equally to engineering projects of all sizes and shapes; there are no silver bullets that will slay the complexity demon. However, groups of engineers, working in small, well-organized, and well-led teams are the "silver-plated" bullets of systems engineering. The leadership skills of systems engineers and team leaders, coupled with the technical skills of individual engineers and the ability of the engineers to participate, as members of cohesive teams engaged in intellect-intensive teamwork, are the keys to success.

The 14 guidelines for organizing and leading systems engineering teams are itemized as follows:

1. Use the best people you can find.
2. Treat engineers as assets rather than costs.

3. Provide a balance between job specialization and job variety.
4. Keep team members together.
5. Limit the size of each team.
6. Differentiate the role of team leader.
7. Make each team leader the team's quality control agent.
8. Safeguard individual productivity and quality data.
9. Decompose tasks in manageable units of work.
10. Set performance goals for each team.
11. Adopt a negotiated contractual model for team commitments.
12. Ensure daily interactions with team leaders and team members.
13. Conduct weekly status review meetings.
14. Structure large projects as collections of highly cohesive, loosely coupled small projects.

These guidelines can be, and should be, tailored to fit each particular situation. However, most of the guidelines are applicable to most kinds and sizes of systems projects. But do not be misled by these "Fourteen easy steps to organizing, coordinating, and leading" engineering teams. The guidelines presented here are by no means complete or comprehensive, nor are they foolproof.

There are no physical laws or mathematical theories for building and maintaining cohesive engineering teams. Remember that systems engineers, team leaders, and team members performing in a cooperative and cohesive manner are the most important asset for project success. Interpersonal skills and goodwill are key ingredients of successful teams. Good intentions, alone, are not sufficient. The techniques presented in this chapter, when applied with common sense and within a supportive organization, can produce gratifying results.

12.10 Key Points

- Corporate culture comprises the beliefs, values, and behavior patterns that exist within an organization.
- A mission statement defines the purpose and goals of an organization.
- A vision statement has specific objectives and a time frame for achieving them.
- The primary assets of an engineering organization are the skills and abilities of the project managers, the systems engineers, the team leaders, the disciplinary engineers, and support personnel.
- The first rule of business is to manage corporate assets to maximize return on investment in those assets; the second rule is to control costs.
- Unfortunately, many software organizations confuse the second rule with the first one and treat their engineers as costs rather than assets.

- Key personnel are those project members who are assigned responsibilities and are given the authority to carry out those assignments by the project manager and the systems engineers who are managers/leaders of the technical work.
- Responsibilities are obligations to perform assigned tasks; they are (or should be) documented in job descriptions.
- Authority is the power to make the decisions that must be made in fulfilling one's responsibilities and the power to implement those decisions.
- Authority can be delegated; responsibility cannot.
- The 14 guidelines for organizing and leading engineering teams are by no means complete or comprehensive, nor are they foolproof.
- There are no physical laws or mathematical theories for building and maintaining cohesive engineering teams.
- However, the 14 guidelines, when applied with common sense and within a supportive organization, can produce gratifying results.

Exercises

12.1. Choose three of the factors of corporate culture listed in Section 12.3.
 (a) Briefly explain how those factors apply to your work environment or your school environment.
 (b) Briefly explain the positive or negative influences of each of your three factors on your attitude about your work environment or school environment.

12.2. Briefly explain the difference between a mission statement and a vision statement.

12.3. Briefly explain why authority can be delegated but responsibility cannot be delegated.

12.4. Teams are necessary for developing large systems.
 (a) Briefly explain why multiple small teams are preferable to one large team.
 (b) Briefly explain some negative aspects of multiple small teams.

12.5. Briefly explain why leaders of small teams are not excessive "management overhead."

12.6. Select three teamicide practices in Table 12.1 and briefly explain why they are detrimental to effective teamwork.

12.7. Briefly describe a situation where you have experienced being "in the flow."

12.8. Section 12.6 lists several ways people obtain psychological satisfaction at work. Select three of the factors that are most important to you and briefly explain why they are important to you.

12.9. Section 12.6 also lists several job satisfiers for engineers. Select three of the job satisfiers that are most important to you and briefly explain why they are important to you.

12.10. Section 12.8 includes 14 guidelines for organizing and leading systems engineering teams.
(a) Choose three guidelines you think would be easiest to implement in systems engineering organizations. Briefly explain why you think they would be easiest to implement.
(b) Choose three guidelines you think would be hardest to implement in systems engineering organizations. Briefly explain why you think they would be hardest to implement.

References

Boehm, B. (1981). *Software Engineering Economics*. Prentice Hall.

Brooks, F. (1987). No silver bullets: essence and accidents of software engineering. *IEEE Computer* 20 (4): 10–19.

Conway, M. (1968). How do committees invent? *Datamation* 14 (5): 28–31.

Csíkszentmihályi, M. (2008). *Flow: The Psychology of Optimal Experience*. Harper Perennial Modern Classics.

DeMarco, T. and Lister, T. (2013). *Peopleware: Productive Projects and Teams*, 3e. Addison-Wesley.

Fairley, R. (2009). *Managing and Leading Software Projects*. Wiley.

Fairley, R. and Willshire, M. (2017). Better now than later: managing technical debt in systems development. *IEEE Computer* 50: 80–87.

Grove, A. (1995). *High Output Management*, 2e. Vintage.

Hershey, P. and Blanchard, K. (1999). *Leadership and the One Minute Manager*. William Morrow.

Katzenbach, R. and Smith, D. (1993). *The Wisdom of Teams: Creating the High-Performance Organization*. McKinsey and Company, Inc.

Machiavelli, N. (1984). *The Prince*. Bantam Classics.

McGregor, D. (1985). *The Human Side of Enterprise*. McGraw-Hill.

Norman, E., Brotherton, S., and Fried, R. (2008). *Work Breakdown Structures: The Foundation for Project Management Excellence*. Wiley.
Perrin, R. (2015). *Practice Standard for Work Breakdown Structures*. Project Management Institute.
Wirtz, F. and Vantrappen, H. (2016). Making matrix organizations actually work. *Harvard Business Review*, H02OXC-PDF-ENG (1 March).

Appendix A

The Northwest Hydroelectric System

A.1 Background

This case study presents the Northwest Hydroelectric System (NwHS) as an example of a complex software-enabled system that includes natural elements and engineered elements. NwHS is situated in the northwestern United States (the Northwest). NwHS encompasses the Columbia River and its tributaries. The headwaters of the Columbia are in the Rocky Mountains of British Columbia, Canada. The river flows south into the United States and then east to west; it forms the north–south border of the states of Washington and Oregon and empties into the Pacific Ocean near Astoria, Oregon. The Columbia River drainage basin (Columbia and tributaries) is roughly the size of France and extends across seven U.S. states and British Columbia, as illustrated in Figure A.1 (CRITFC 2019).

NwHS includes a large number of loosely coupled autonomous elements (hydroelectric dams). It is instructive to note that the NwHS is characterized as a project that is concerned with planning, developing, and maintaining a hydroelectric system that has evolved, and continues to evolve, over time. Each of the hydroelectric dams within the NwHS is also referred to as a project, which indicates that the individual elements of the NwHS are also evolving over time (FWEE 2018).

As shown in this case study, NwHS is an adaptive and reconfigurable system that exists within a complex environment of policies, rules, regulations, and agreements. The NwHS is presented using each of the four provisioning paradigms in Part IV of the *Guide to the Systems Engineering Body of Knowledge* (SEBoK): product system engineering, service system engineering, enterprise system engineering, and system of systems engineering (sebokwiki.org).

The Columbia River has 14 hydroelectric dams (hydro dams) on its main stem. In total, there are more than 250 hydro dams in the Columbia River Basin, each of which has a generating capacity of five or more megawatts of electricity. The NwHS produces approximately one-third of all hydroelectric power generated in the United States – more than any other North American

Systems Engineering of Software-Enabled Systems, First Edition. Richard E. Fairley.
© 2019 John Wiley & Sons, Inc. Published 2019 by John Wiley & Sons, Inc.

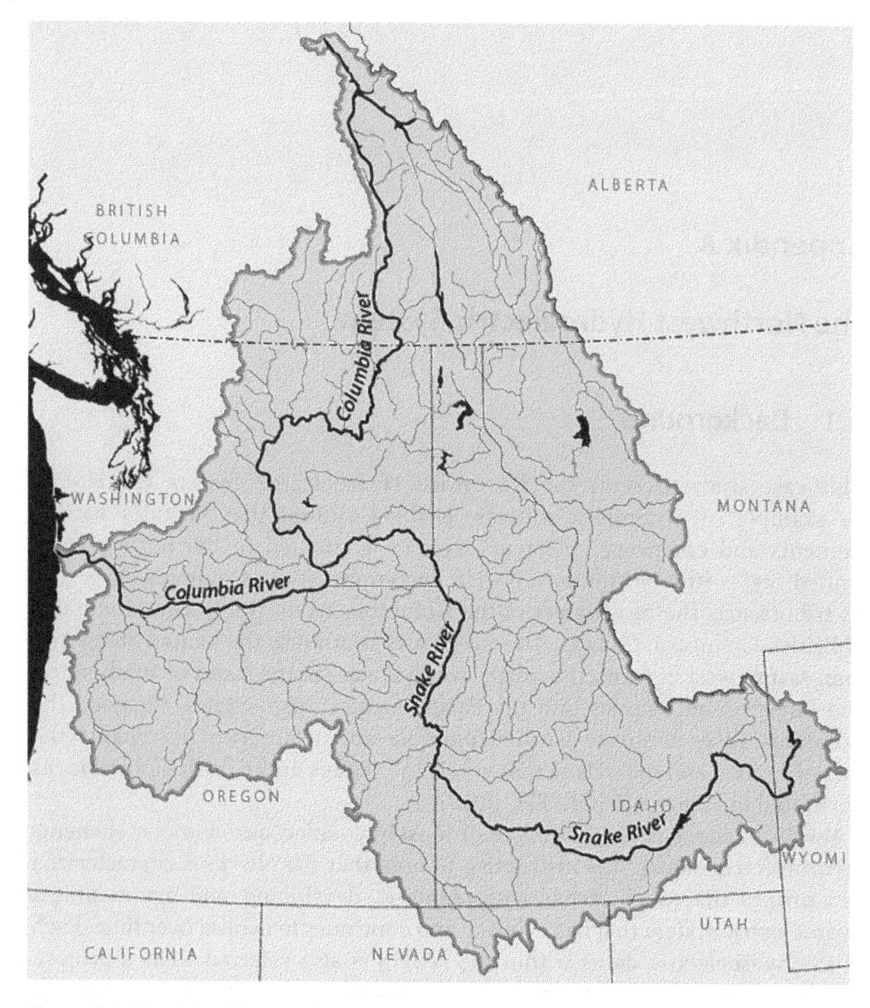

Figure A.1 The Columbia river drainage basin. Source: CRITFC 2019.

hydroelectric system. The amount of electrical power generated by the NwHS fluctuates between 50% and 75% of all electricity used in the Northwest; other sources include coal, natural gas, nuclear, wind, and solar. Excess electrical energy generated by NwHS is sold to other electrical grids.

There are more small hydro dams than large dams. Small hydro dams have a generating capacity of 100 kW to 30 MW; large hydro dams have a capacity of more than 30 MW. In addition, there are numerous micro dams that each can generate fewer than 100 kW. Most micro dams provide power to isolated homes or small communities but some are elements of the NwHS and sell power to utilities (DOE 2018). Utility companies own some of the dams, some are locally and privately owned, and some are owned and operated by federal agencies.

The Bonneville Power Administration (BPA) provides about one-third of the electrical power generated by NwHS and used in the Northwest; BPA is a federal nonprofit agency – it is part of the U.S. Department of Energy. It is self-funded and covers its costs by selling electrical power generated by 31 federal hydro dams in the Columbia River Basin. The U.S. Army Corps of Engineers and the Bureau of Reclamation operate the BPA dams.

The Bureau of Reclamation also operates a network of software-enabled hydrologic and meteorologic monitoring stations located throughout the Northwest. This network and its associated communications and computer systems are collectively called Hydromet. Remote data collection platforms transmit water data and environmental data via radio and satellite to provide near-real-time water management capabilities. Other information, as available, is integrated with Hydromet data to provide timely water supply status for river and reservoir operations (USBR 2016).

Distinguishing characteristics of individual hydro dams are their capacity to generate electrical energy and their physical structure. Several factors contribute to the differences in generating capacity of hydro dams: the number of turbines/generators, the flow rate of a river or tributary, the amount of elevation that water falls in order to spin the turbines, and environmental factors such as providing for fish passage and regulating water flow to provide downstream irrigation water and maintain downstream ecosystems.

Dam structures include impoundment, diversion, and pumped storage dams (DOE 2018). An impoundment dam retains water in a reservoir for release through the dam. A diversion dam (also known as a run-of-river dam) converts the potential energy in a river into kinetic energy as water flows through the dam but the dam does not impede the water flow. Pumped storage dams use electricity to pump water back up to a storage reservoir during times of low demand for use during times of high demand. Pumped storage dams are not common in the NwHS because they may not be able to meet their energy commitments when rivers are lower than anticipated and flowing slowly.

An impoundment dam and a diversion dam are illustrated in Figures A.2 and A.3.

A software-enabled control room for a hydroelectric plant is illustrated in Figure A.4.

A.2 Purpose

NwHS has three primary goals (NwHS 2018):

(1) To provide most of the Northwest's "firm-energy" needs;

Figure A.2 The Grand Coulee impoundment dam. Image obtained from: https://www.usbr.gov/pn/grandcoulee/about/index.html.

Figure A.3 The Safe Harbor diversion dam. Image obtained from: https://www.pgc.pa.gov/Wildlife/WildlifeSpecies/BaldEagles/BaldEagleWatching/SoutheasternPA/Pages/LowerSusquehannaRiver.aspx.

Figure A.4 Control room for the Itaipu hydroelectric dam. Image obtained from: https://www.powerengineeringint.com/articles/print/volume-24/issue-7/features/maintaining-engineering-expertise-in-hydro-plants.html.

(2) To maximize "non-firm" energy production; and
(3) To maintain the ecological environment.

Firm energy is the amount of electricity the Northwest will need each year. Planners rely on NwHS and other sources to produce enough firm energy to ensure that sufficient electricity will be generated to meet estimated needs (energy sources include hydroelectric, nuclear, coal, natural gas, solar, and wind). NwHS "firm energy" is the amount of electricity that can be generated by NwHS when the amount of available water is at a historical low, thus guaranteeing the amount of energy the NwHS can provide.

Non-firm energy is the electricity generated when the annual hydrologic cycle makes more water available for power generation than in a historically low-water year. Non-firm electricity generated by hydro dams is generally sold at a lower price than the alternatives of electricity generated by nuclear, coal, or natural gas, thus making it attractive to customers. Excess non-firm electricity is also sold to interconnected software-controlled regional grids when the demand on those grids exceeds supply.

Other goals for NwHS include flood control, navigation, irrigation, and maintaining the water levels of all reservoirs.

A.3 Challenges

The NwHS is a large, complex software-enabled system that has many challenges to be met.

NwHS hydro dams have varying design details (e.g. the types of turbines, generators, control systems, and fish passage facilities used). This makes routine maintenance, retrofitting, and other sustainment issues unique for each dam.

Safety and security (both physical and cyber) are common challenges for all dams; cyber security is a growing concern. Smaller dams, having fewer resources, may be more susceptible to cyberattacks than larger ones. It has been reported that on 12 occasions in the past decade hackers gained top-level access to key power networks (Cyber 2015).

The ways in which electricity generated by a hydro dam is transmitted and sold to utilities and large industrial customers varies widely. For instance, the BPA transmits and sells power generated by federal dams. Nonfederal dam operators must manage transmission and sale of power produced by their dams. Depending on factors such as size, structure, location, and ownership of each dam, a large number of policies, regulations, and agreements have dramatically different effects on how dams are operated.

Some of the most contentious environmental issues are associated with maintaining the ecology of the rivers, preserving salmon and other fish, and providing sufficient irrigation water while preserving sufficient reservoir water to meet firm-energy demands (Impact1 2018).

Preserving the salmon population endangered by dams is a continuing challenge. Salmon populations have been depleted because dams impede the return of salmon to upstream spawning beds. Native Americans advocate for their traditional fishing rights, which conflict with governmental policies intended to maintain healthy salmon populations in the Columbia River Basin (Impact2 2018). The National Marine Fisheries Service has recently declared that salmon recovery is a higher priority than all other purposes except flood control at 14 federal dams.

A.4 Systems Engineering Practices

The Northwest Hydro System is a large, complex software-enabled system composed of loosely coupled autonomous elements; each dam operates semi-independently within a large network of similar entities and contextual constraints. The NwHS evolves over time: some dams have been retrofitted to increase power generation capability or to reconfigure connections to electrical transmission lines, new dams have been constructed, and some existing dams have been decommissioned and removed. Systems engineering practices are used to facilitate and coordinate the many diverse and complex endeavors of the NwHS.

The needs of and processes used by the human elements of the NwHS are also considered for operators, maintainers, inspectors, and regulators. Others stakeholders are suppliers to NwHS (vendors and contractors), some are users of the electricity generated by NwHS (businesses and home owners), and some are stakeholders who depend on and are impacted by the NwHS (utilities, large industries, farmers, ranchers, ecology advocates, Native Americans, and towns).

The systems engineering context of NwHS includes natural elements (rivers, terrain, weather systems, fish); elements purposefully built by humans (transmission lines, electrical grids); cyber connections (both wired and Internet); and rules, regulations, and agreements at the federal, regional, state, and local levels.

Given the complexity of the NwHS and its context, it is instructive to analyze the NwHS by applying each of the four application paradigms of systems engineering presented in Part 4 of SEBoK: the product, service, enterprise, and system-of-systems application paradigms of systems engineering (sebokwiki.org).

A.4.1 Product Provisioning

Product provisioning applies systems engineering processes, methods, tools, and techniques to conceive, develop, and sustain the purposefully developed elements of a system (e.g. a hydro dam or a hydro system). In addition, some of the naturally occurring physical elements of a system may be shaped and configured (e.g. a river channel).

Major elements of a hydro dam include the physical structure of the dam (including the spillway used to divert excess reservoir water), the penstock (used to direct water into the turbines), and the generating plant (i.e. the turbines used to turn generator rotors, generator stators and rotors that generate the electricity, step-up transformers used to increase the voltage level of electricity produced by the generators, and connections to transmission lines).

Software elements sense, measure, regulate, and control water flow, power generation, safety, and security and measure the structural integrity of a dam. Some turbines, for example, have adjustable vanes that are software controlled to harvest maximum energy from the water, depending on the flow rate, power demand, and other factors. Other software elements include supporting software for computing devices (operating systems, databases, spreadsheets), data management software (collection, analysis, reporting), application software (displays of monitored status and interfaces for controlling operation of a dam), and communication interfaces to hard wired links and Internet-enabled links. In addition, software support is provided for the analog and digital devices needed to sense, measure, regulate, and control the purposefully built and naturally occurring elements of a dam and its environment.

Product provisioning is also concerned with other issues that apply to individual dams, elements of dams, and the overall NwHS. They include issues such as manufacturability/producibility; logistics and distribution; product quality; product disposal; conformance to policies, laws, regulations, agreements, and standards; value added for stakeholders; and meeting customer's expectations.

Many different technologies and engineering disciplines are needed to develop and sustain a hydro dam and the overall Northwest Hydro System. Software-enabled product provisioning can also provide the coordination and control of systems engineering processes, methods, and techniques needed to develop, reconfigure, adapt, analyze, and sustain the hydro dams and the overall NwHS.

A.4.2 Service Provisioning

A service is an activity performed by an entity to help or assist one or more other entities. Service provisioning can be applied within the various contexts of services provided by the NwHS to meet stakeholders' requirements, users' needs, and system interactions with operators, users, and maintainers, plus the interactions with the contextual elements that determined services provided by the NwHS in the social, business, regulatory, and physical environments.

The NwHS provides electricity to a grid that serves commercial, industrial, governmental, and domestic customers. Stakeholders in addition to customers served include those who affect or will be affected by development, operation, and sustainment of a dam. Downstream stakeholders served include Native Americans, farmers and ranchers, and communities that receive the service of water released by the dam.

Additional service attributes include the services provided to enable operators and maintainers to efficiently and effectively operate and maintain the physical and software elements of a dam; release water from the dam in a manner that services the upstream and downstream ecosystems; manage sharing of electrical power with other regional grids; provide emergency responses to power demands that result from electrical brownouts, blackouts, and overloads; and handle system failures that might be caused by earthquakes, terrorist attacks, and other catastrophic events.

A.4.3 Enterprise Provisioning

An enterprise, such as the NwHS, consists of one or more organizations with shared mission, goals, and objectives to provide a product or service. The mission, goals, and objectives of the NwHS are to provide most of the Northwest's firm energy needs and to maximize non-firm energy production while serving stakeholders and preserving affected environmental ecosystems. To meet those goals, the Northwest Hydroelectric Association (NWHA) coordinates the planning, design, improvement, and operation of the hydro dams that constitute the NwHS enterprise.

NWHA members represent all segments of the hydropower industry – independent developers and energy producers; public and private utilities; manufacturers and distributors; and local, state, and regional governments including water and irrigation districts. Other NWHA members include contractors, Native American tribes, and consultants: engineers, financiers, environmental scientists, attorneys, and others (NWHA 2018).

An enterprise typically consists of multiple semiautonomous organizations engaged in a common endeavor. The NwHS is a large complex enterprise that has many constituent organizations, namely, the organizations that own and operate the hydro dams and the other stakeholder members of the NWHA. Differences in ownership, structure, location, and size of hydro dams, the special interests of various NWHA members, and a complicated regulatory process are some of the distinguishing characteristics of the NwHS that can be analyzed by enterprise provisioning.

A.4.4 System-of-Systems Provisioning

Many large composite systems are composed of autonomous systems that are combined to provide increased capabilities that cannot be provided by the individual systems operating in isolation. The Northwest Hydro System is a system of systems composed of autonomous hydro dams that have different owners, different operators, different stakeholders, and different regulators. The autonomous hydro dams could not provide the NwHS capabilities without the overall coordination and control that can be managed by applying system-of-systems provisioning.

A.5 Lessons Learned

NwHS is a collection of many interrelated ongoing projects that have shared common goals and shared constraints. The unique characteristics of NwHS make it a useful case study to illustrate how the four provisioning paradigms in SEBoK provide essential viewpoints for analyzing large complex software-enabled systems composed of loosely coupled, autonomous elements.

Product engineering allows the collection of natural and purposefully built NwHS elements and their interconnections to be analyzed by applying product systems engineering processes and methods.

Systems engineering of services supports analysis of the NwHS services provided to customers, users, farmers, ranchers, Native Americans, and other stakeholders who rely on NwHS for those services.

Enterprise systems engineering considers the broad scope and impact of the NwHS enterprise, both positive and negative, on the northwestern United States within the context of economic, social, physical, and regulatory environments.

System engineering of system-of-systems applies the principles of planning, coordination, and operation to a collection of semiautonomous hydro dams that form the Northwest Hydro System. The complexity of adding new dams as well as modifying and decommissioning existing dams in a seamless manner can best be understood by applying system-of-systems engineering processes and methods.

Taken together, the four provisioning paradigms in Part 4 of SEBoK present a comprehensive view of a very large complex software-enabled system whose many dimensions would be otherwise difficult, if not impossible, to comprehend when the NwHS is examined using only one of the paradigms.

References

CRITFC (2019). Columbia river drainage basin. http://plan.critfc.org/vol1/tribal-restoration-plan/biological-perspective/habitat-of-anadromous-fish/ (accessed 30 January 2019).

Cyber (2015). US in fear of new cyber attack. http://www.independent.co.uk/news/world/americas/bowman-avenue-dam-us-in-fear-of-new-cyber-attack-

as-dam-breach-by-iranian-hackers-is-revealed-a6782081.html (accessed 5 September 2018).
DOE (2018). Types of hydropower plants. https://www.energy.gov/eere/water/types-hydropower-plants (accessed 5 September 2018).
FWEE (2018). Overview of hydropower in the Northwest. http://fwee.org/education/the-nature-of-water-power/overview-of-hydropower-in-the-northwest/ (accessed 5 September 2018).
Impact1 (2018). How a hydroelectric project can affect a river. http://fwee.org/environment/how-a-hydroelectric-project-can-affect-a-river/ (accessed 5 September 2018).
Impact2 (2018). The impact of the Bonneville Dam on Native American Culture. https://www2.kenyon.edu/projects/Dams/bsc02yogg.html (accessed 5 September 2018).
NWHA (2018). About NWHA. Northwest Hydroelectric Association. http://www.nwhydro.org/nwha/about/ (accessed 5 September 2018).
NwHS (2018). How the Northwest Hydro System Works. http://fwee.org/nw-hydro-tours/how-the-northwest-hydro-system-works/ (accessed 5 September 2018).
USBR (2016). Hydromet. https://www.usbr.gov/pn/hydromet/select.html/ (accessed 5 September 2018).

Appendix B

Automobile Embedded Real-Time Systems

B.1 Introduction

This appendix presents a case study of the software-enabled real-time systems embedded in modern automobiles. The case study is small enough to be presented in detail and is representative of the methods and techniques used to develop complex embedded real-time systems using systems engineering processes, methods, and techniques (both physical and software).

Modern automobiles are software-enabled systems, as illustrated in Figure B.1.

Modern automobiles have been termed mobile software platforms. They may include 100 or more electronic control units (ECUs) connected by electrical cables that form embedded real-time networks. The first automotive ECUs were developed to control emissions generated by gasoline engines and the ECUs were termed engine control units. Over time, ECU has come to mean electronic control unit; automotive electronics have evolved until almost every element of a modern automobile is sensed and controlled by an ECU. The growth of ECUs in luxury automobiles is illustrated in Figure B.2 (Crolla et al. 2015).

As many as 2500–3000 different digital signals and messages are generated and shared to control the responses of the ECUs in an ECU network, including simple information such as the speed of the vehicle and complex information for engine and transmission performance parameters (Navet and Simonot-Lion 2013).

Vehicles that have electronic systems similar to those described in this case study include electric and hybrid automobiles, trucks, buses, trains, boats, and airplanes. This case study is restricted to ECUs and networks for modern gasoline-powered automobiles. The characteristics of ECUs, ECU domains, and ECU real-time communication networks are presented here.

© 2019 John Wiley & Sons, Inc. Published 2019 by John Wiley & Sons, Inc.

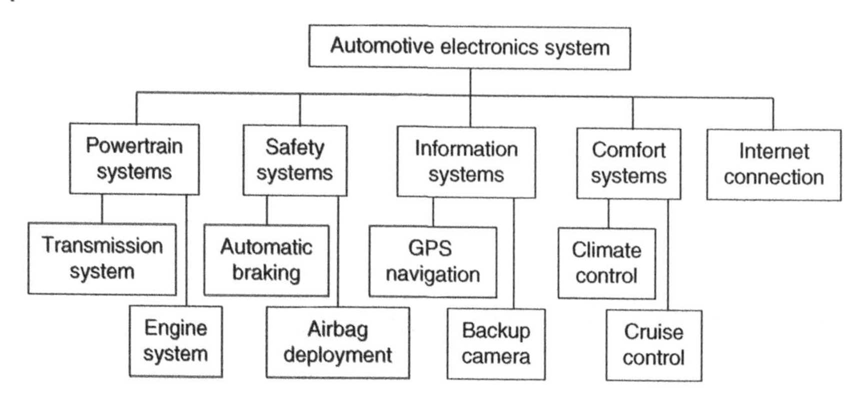

Figure B.1 Partial structure of a complex software-enabled engineered system.

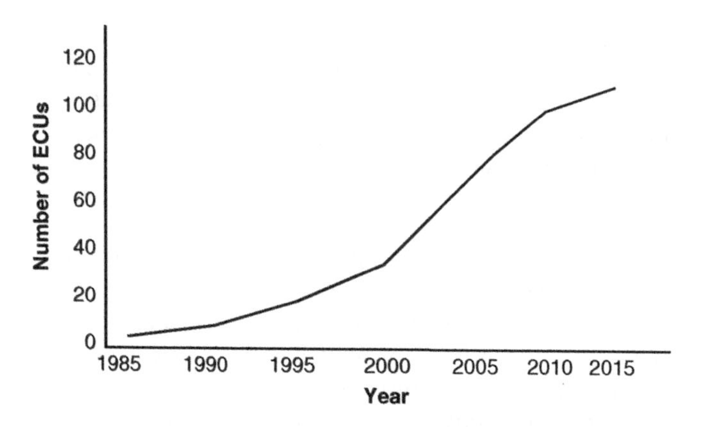

Figure B.2 Growth of ECUs in luxury vehicles. Source: Crolla et al. (2015).

B.2 Electronic Control Units

Automotive ECUs are embedded microprocessors that have software to receive input signals from sensors and apply control algorithms to generate output actuator signals. ECUs operate in sensor-control-actuator loops, as illustrated in Figure B.3.

In addition, ECUs receive messages from other ECUs and send messages generated by the controller to other ECUs across one or more ECU networks.

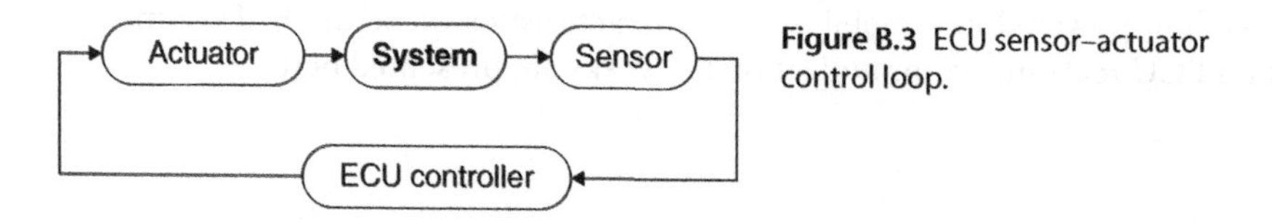

Figure B.3 ECU sensor–actuator control loop.

The ECU hardware is a microcontroller, which is a microprocessor designed for embedded applications, in contrast to microprocessors used in personal computers, cell phones, and other consumer products. Hardware elements of an ECU microcontroller include a central processing unit (i.e. a CPU that includes an instruction processor and memory), an input connector for power delivery, and input connectors for receiving messages and sensor signals. Programmable output connectors are provided to connect ECUs to various kinds of actuators such as relays for controlling door locks and controllers for engine injector fuel/air ratios. Sensors and actuators are connected to microcontrollers by electrical wires. Connectors are also provided to connect ECUs to ECU network cables. ECU hardware converts analog input signals to digital form for processing, as needed.

ECU microcontrollers are usually packaged on a single electronic board. The physical structure of an ECU is a small "black box" that is built to withstand extremes of temperature, humidity, shock, vibration, and electromagnetic interference. ECUs are typically $6\times 4\times 2$ in and weigh ~1 pound (i.e. $147\times 105\times 40\,\text{mm}^3$; weight 500 g).

An ECU is illustrated in Figure B.4.

Some microcontrollers are inexpensive microcomputers that use four-bit words and operate at comparatively slow clock speeds with low power consumption. The software architecture in these microcontrollers consists of two

Figure B.4 A Ficosa telematic control unit. Source: https://www.ficosa.com/multimedia/images/products/telematic-control-unit.

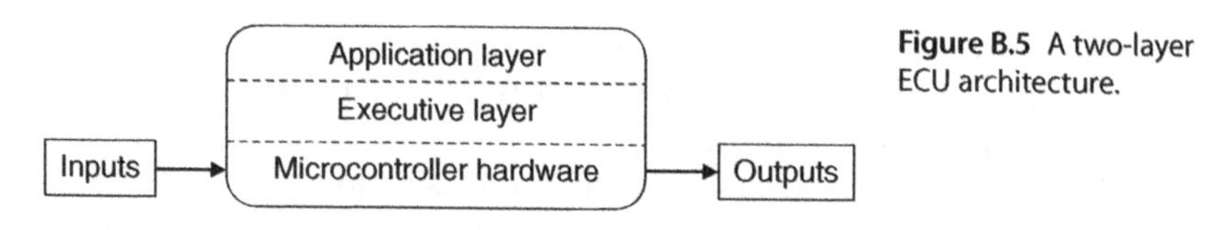

Figure B.5 A two-layer ECU architecture.

layers: an executive layer and the application control algorithms, as illustrated in Figure B.5.

The executive layer coordinates arrival of signals and messages, stores them in hardware memory, and notifies the application layer. The application layer processes the received signals and messages, and then stores processed data in the hardware memory and signals the executive layer, which coordinates transmission to actuators and an ECU network. The software control algorithms in these ECUs may be relatively simple-state machines for operation of devices such as raising and lowering windows with pressure sensing feedback to stop raising a window when it is fully raised or lowered, or an obstacle such as a child's hand is encountered. Other ECUs control digital radios that have AM, FM, and Internet connections; station seeking; settable tuning buttons; and steering wheel control buttons. Simple control algorithms may be implemented in microcontroller software or in firmware.

Microcontrollers for performance-critical applications such as engine and transmission control require more powerful CPUs, more and different memory, faster clock speeds, and correspondingly higher power consumption. Software architectures for more powerful ECU microcontrollers typically have three layers, as illustrated in Figure B.6.

The middleware layer provides a bridge between the executive layer and the application layer. It provides application program interfaces (APIs), support for cellphones and streaming video, and interfaces to the operating system to support multiple application programs that are running concurrently. The executive layer is a "real-time operating system" (RTOS). The RTOS provides the software interfaces to the ECU hardware's input and output ports and interrupt subsystem and provides interfaces used by the application layer. The RTOS and the middleware layer also provide operations such as receiving signals and messages, unpacking them and relaying them to the application layer, formatting and packaging outgoing signals and messages, sending the outgoing signals to actuators controlled by the ECU, and placing outgoing messages on an ECU network.

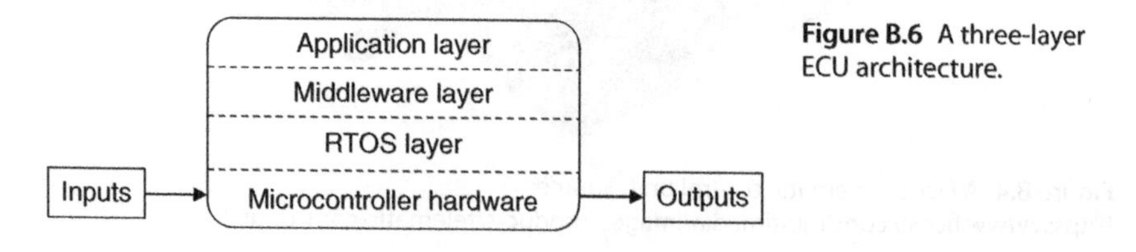

Figure B.6 A three-layer ECU architecture.

Additional capabilities of a microcontroller and a three-layer software architecture include advanced driver assistance systems (ADASs), often camera and radar based, for braking and parking assistance, lane departure detection, night vision assistance for pedestrian recognition, and adjustment of the automobile suspension system by scanning the road surface. ADAS features are increasingly found in midrange and luxury automobiles.

An Example of an ECU and the Related System Elements

A US patent provides an example of an ECU and the related system elements for an electronic acceleration system that includes a method of processing malfunctions of the acceleration sensor (Hattori et al. 1989). This activity occurs in a closed control loop, as illustrated in Figure B.7.

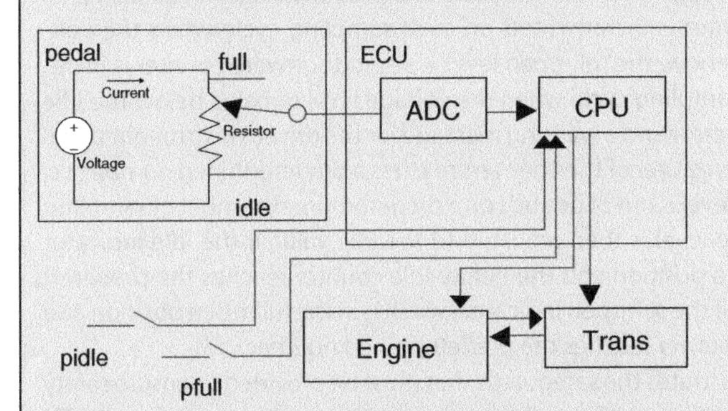

Figure B.7 Schematic for an electronic acceleration sensor. Source: Figure 2 of U.S. Patent 4843555. https://patents.google.com/patent/US4843555.

The ECU acceleration sensor receives an electrical signal generated by a potentiometer connected to the gas pedal. The analog voltage level varies with the amount of pedal depression; the voltage is converted into a digital signal by an analog–digital converter (ADC) in the ECU. The voltage provided by the potentiometer varies in a nonlinear manner when the accelerator pedal is depressed in a linear manner so a polynomial algorithm is applied to convert the nonlinear voltage level to a corresponding linear value, which is delivered to the acceleration actuators of the engine and transmission.

The ECU software also stores the acceleration voltage level at the idling position of the accelerator pedal (idle) to provide a calibration value that is used as a ratio variable for generating actual acceleration signals obtained from the processed accelerator pedal input signal. The voltage level for the fully depressed

(Continued)

(Continued)

accelerator pedal (full) is stored for managing malfunctions of the acceleration system.

The method of processing a malfunction of the acceleration sensor involves storing the idle value of the pedal voltage when the accelerator pedal is not depressed (i.e. the "idle" value in Figure B.7) and the value of the pedal voltage when the accelerator pedal is fully depressed (i.e. the "full" value in Figure B.7). The voltage level is sampled in continuous cycles to generate engine acceleration signals. A possible malfunction is detected when the voltage level falls outside the acceptable range by being higher than the full value or lower than the idle value. Signals from the physical indicator switches (pidle and pfull) are examined to determine whether the accelerator pedal is currently in the idle position or the full-open position (the pidle and pfull switches in Figure B.7).

A software counter is incremented on each sampling cycle when the voltage level remains above the full-open level; a second software counter is incremented on each sampling cycle when the voltage level remains below the idle level. A diagnostic error and a warning message for the driver's instrument panel are generated when either of the counters reaches a predetermined number, i.e. when the voltage level is out of bounds on a predetermined number of sampling cycles. The voltage level is then adjusted to the idle value if the idle indicator switch is in the idle position and the below-idle counter reaches the predetermined number, or if the full open indicator switch is in the full-open position and the above-open counter reaches the predetermined number.

This example illustrates the safeguards that must be provided to ensure safety of the passengers, the vehicle, and the surrounding environment for all ECU states, even for this simple ECU.

B.3 ECU Domains

The ECUs in modern automobiles are typically categorized by domain; different domains have different requirements for network communication bandwidth and performance, as well as different requirements for reliability, safety, and security of the ECUs and the network or networks. One commonly used categorization of automotive domains includes three domains that are concerned with real-time control and safety of a vehicle's behavior: the engine, transmission, and chassis domains. The engine and transmission domains are sometimes combined in a powertrain domain.

Other domains include the automobile body domain, which is concerned with passenger comfort and convenience, and the infotainment domain, which integrates entertainment and wireless communication devices. An emerging domain is "X-by-wire," where X is a generic term for replacement of mechanical parts with mechatronics devices and associated ECUs, sensors, and actuators, such as steering-by-wire.

An on-board diagnostics (OBD) computer provides self-diagnostics and reports vehicle status information for an automobile's various subsystems. OBD information is used to trigger warning lights and alarms for the driver and is used by service technicians to diagnose vehicle problems. Most countries require that all automobiles sold in that country must include OBD capabilities.

ECU microcontrollers for the engine, transmission, and chassis domains are designed to operate in never-ending cycles, periodically sampling sensor signals and messages, applying (sometimes complex) algorithms to generate control signals for actuators, and to send messages to other ECUs as necessary on each cycle. Provisioning of these domains is of the highest priority and requires the most powerful microcontrollers and complex software in an automobile; the engine ECU (hardware and software) is typically the largest and most complex.

B.4 The Powertrain Domain (Engine and Transmission)

Currently, the powertrain ECUs for gasoline-engine automobiles monitor 20 or more sensors and have 20 or more actuators to control the ignition system, cooling system, oil pressure, air/fuel ratio, fuel injection sequencing and rate, and other parameters.

Engine control for a gasoline engine is realized by computations, based on several complex control equations that perform computations on cycles having sampling periods of a few milliseconds, which is sufficient for the speed at which a gasoline engine rotates. The complex computations may require software multitasking with stringent time constraints on scheduling of the computational tasks. Frequent data exchanges with the transmission ECU and ECUs in other domains, such as the antilock braking system (ABS) in the chassis domain and the cruise control and other driver's indicators and controls in the body domain, are also required. This requires software that includes robust middleware and operating system layers.

A transmission ECU controls selection and shifting of gears. The transmission ECU must sense incoming signals and messages from the engine, the braking system, the cruise control system, driver controls, and other messages and signals in the chassis and body domains; perform the necessary calculations; and generate transmission control signals. Some transmissions have manual clutches, some have both fully automatic and semiautomatic clutches, and some have fully automatic clutches. Manual transmissions have simple ECUs whose primary functions are to notify the engine ECU and the cruise control ECU when the clutch pedal or brake pedal is depressed. A semiautomatic clutch allows manual up-shifting and down-shifting that is restricted to revolutions-per-minute (RPM) ranges monitored by an ECU; downshifting also occurs by ECU control at specified RPM levels. Some automobiles have continuously variable transmissions (CVTs); the clutch ECU signals the engine ECU to adjust RPMs while smoothly varying the ratio of RPMs to drivetrain rotations, thus maintaining constant speed when the cruise control system is engaged, or to control engine RPMs for acceleration or deceleration when the accelerator pedal is pressed or released.

B.5 The Chassis Domain

The ECUs for the chassis domain are similar to those for the powertrain but they have a stronger impact on the vehicle's stability and dynamics and are thus critical for safety of the passengers, the vehicle, and the surrounding environment. Braking is an important subdomain of the chassis domain. The braking ECUs provide control of the automatic braking system (ABS), the traction control system (TCS), electronic brake distribution (EBD), and an electronic stability program (ESP). The braking ECUs control each of these systems individually and provide coordinated operation among them plus communication to and from the engine, transmission, and individual wheel ECUs.

Other chassis elements include ECUs for sensors that sense acceleration/deceleration of collisions and rollovers and signal ECUs that trigger air bag deployment and safety belt pretensioners and inflators. Other ECUs may provide for electronic power steering, control of a rear camera, forward and reverse warning sensors, and a blind spot detector; lane drift warning and lane change assistance; and parking assistance. Advanced chassis ECUs may also include a predictive emergency braking system, which senses the degree of pressure on the brake pedal and under strong pressure reduces the throttle setting to idle and applies maximum antilock braking until the brake pressure is lessened. Another feature is autonomous cruise control, which automatically adjusts cruise control speed to maintain a safe distance behind vehicles using transmitter signals for radar or laser sensors. Selected cruising speed is resumed when it is safe.

B.6 The Body Domain

The automobile body domain includes ECUs and communication links that provide passenger comfort and convenience. The body domain typically includes ECUs for controlling elements such as windows and doors; mirror adjustment; seat adjustment, remote entry, automatic sensing and controlling of headlights on/off; headlight and rear-view mirror dimming and brightening; ventilation, heating, and air conditioning; cruise control; and warning indicators for oil pressure, tire pressure, battery, windshield washer level, and others.

Some automobiles include ECUs for additional conveniences such as memory for steering wheel, seat heating, and seat adjustments. Multiple functions may be provided by a single ECU, for example, a comfort ECU that integrates ventilation, heating, and air conditioning and that may include temperature sensors and controllers for separate zones. Other functions may have dedicated ECUs, for example, an ECU for manual door and window locking and a separate ECU for automatic door locking controlled by sensing the speed sensor in a chassis ECU.

B.7 The Infotainment Domain

Automotive information and entertainment functions include radio, CD and DVD, hands-free phone, GPS navigation systems, rear seat entertainment, remote vehicle diagnostics, and others. Infotainment functions require wide communication bandwidth within the vehicle and for communication with the external world using wireless technology. Emphasis in the infotainment domain is on processing multimedia data streams, bandwidth sharing, and multimedia quality of service. An external telecommunication link provides an entry point that must be protected to guard against hackers who can potentially modify ECU software including door lock and ignition key codes or engine and braking ECUs.

B.8 An Emerging Domain

X-by-wire is an emerging automotive domain (at the time of writing), where X is a term that refers to replacement of mechanical and hydraulic systems by mechatronic ones, similar to the "fly-by-wire" systems in modern aircraft (Balajee et al. 2015). Mechatronic elements may replace mechanical elements such as the steering column and steering connections, pumps, vacuum-powered actuators, and the hydraulic brake master cylinder. Human–machine interfaces provide simulated feel of elements such as steering and braking. Other X-by-wire elements include complex elements such as control of suspension and simple ones such as control of the windshield wiping speed based on sensing of vehicle speed and the intensity of rainfall or snowfall. The sidebar example in this case study is a "throttle-by-wire" system that eliminates mechanical linkages. Safety and acceptable response times (for safe operation and human perception) are major concerns for X-by-wire systems.

B.9 The ECU Network

An ECU network enables exchange of messages among ECUs and ECU domains, which is necessary because some processing functions involve multiple ECUs and the status of some ECUs must be repetitively communicated to other ECUs. For example, vehicle speed, as determined by the engine-transmission ECUs or a wheel rotation sensor, is used to display the speed, adjust electronic power steering, control suspension and traction, and perhaps adjust the windshield wiping speed. The primary goal for the network is predictable performance. Safety is paramount; failure to deliver a message in a predictable, timely manner can place the vehicle, the passengers, and the physical environment in an unsafe situation.

The first ECU networks were point-to-point connections among ECUs but this soon became unfeasible because the number of wired connections grew as the square of the number of ECUs if each node is connected to all other nodes. The number of necessary wires and connectors negatively impacted weight, cost, complexity, and reliability. The result is network domains that use multiple subnetworks within domains and multiplexed communication across domains.

B.10 Automotive Network Domains

An ECU network is a complex distributed system that treats each ECU as a network node. Network quality of service issues include bandwidth, response time, timing jitter of signals, redundant channels for tolerating transmission errors, efficiency of the error detection mechanisms, retransmit mechanisms, and other similar issues.

Different ECU domains have different bandwidth, response time, and predictability requirements. The drivetrain and chassis domains have the most stringent requirements. The body and infotainment domains do not need the cost and complexity of the networks needed by the drivetrain and chassis domains because vehicle occupants trigger the functions in those domains. The body domain has the least stringent network requirements and the infotainment domain has the greatest bandwidth requirements.

In 1994, the Society for Automotive Engineers[1] (SAE) defined a classification for automotive communication networks, still used, that is based on data transmission speed and network performance (SAE j3131).

Class A networks have a data rate lower than 10 kbit/s. They are used to transmit simple control data with low-cost technology and are mainly used in the

1 SAE International was initially established as the Society of Automotive Engineers; it is a U.S.-based, global professional association and standards organization for engineering professionals.

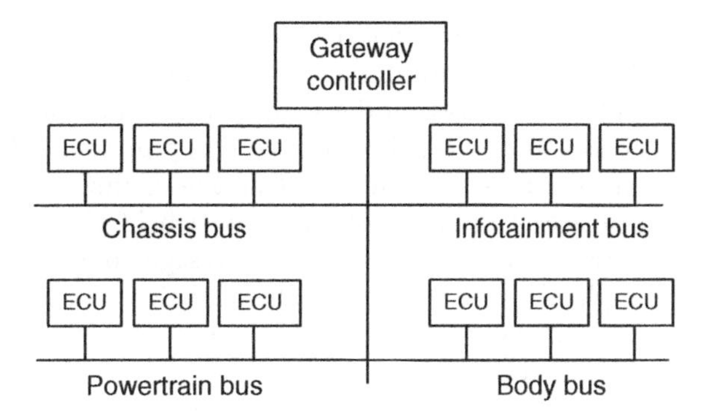

Figure B.8 ECU networks with a gateway controller.

body domain (seat control, door lock, lighting, trunk release, rain sensor, and so forth).

Class B networks are dedicated to sharing information among ECUs in order to reduce the number of sensors needed. They operate from 10 to 125 kbit/s.

Class C networks support applications that require real-time communication at speeds of 125 kbit/s to 1 Mbit/s. Class C networks are used for the powertrain and chassis domains.

Class D networks (speeds over 1 Mbit/s) are used for the infotainment domain.

The increasing diversity of needed bandwidth, performance, and dependability requirements plus cost considerations has resulted in multiple interconnected networks in modern automobiles. Current network architectures typically include different types of networks interconnected by a gateway, as illustrated in Figure B.8.

B.11 Network Protocols

Network protocols provide the rules and conventions used to manage network traffic, including scheduling of message transmission. The following discussion of network protocols presents the main concepts while avoiding the complexities of the many variations and enhancements.

Network messages are transmitted in fixed format "frames" that contain the sending ECU's address, error detecting and correcting bits, the message, and other information. The ECU software unpacks and processes incoming frames and constructs outgoing frames. The ECU operating system receives and sends the messages. Scheduling of messages is usually accomplished using a time-triggered protocol; the alternative is an event-triggered protocol.

In time-triggered scheduling, each ECU message frame receives a dedicated time slot in each network communication cycle; a master clock synchronizes scheduling of the time slots. Time-triggered scheduling allows communication among ECUs without central coordination of a host computer.

Some critical functions may receive more than one time slot per processing cycle while less critical ones receive only one time slot. Each message frame includes the address of the sending ECU. All other ECUs check the sender's address and process the message if appropriate; otherwise, the other ECUs ignore the message. One disadvantage of time-triggered scheduling is inefficient use of the network because each ECU transmits a message on each processing cycle even when the message contains no new information; for example, the ABS is still idle. The inefficiency of time-triggered networks can be somewhat compensated by using multiple networks connected by gateways, as described above.

The primary advantage of time-triggered scheduling is the ability to verify that messages will be transmitted and received at specified times when designing the time slots for the ECU messages. Guaranteed timing is critical because the calculations used to update control signals may become unstable if the incoming data samples are not received in the required time frames. Also, it is easy to determine that an ECU is offline because the message frame for that ECU will be blank. For these reasons, time-triggered scheduling is the most widely used protocol in automotive vehicle networks. However, scheduling for as many of 2500 different kinds of messages for 70 or 80 ECUs with guaranteed message delivery times is a complex problem that is solved using algorithms and heuristics in an iterative trial-and-error simulation process.

Time-triggered scheduling is a synchronous process; the processing cycles continue indefinitely with fixed time slots. In contrast, event-triggered scheduling is asynchronous; messages are transmitted only when an event occurs that requires notification of another ECU; for example, the brake pedal has been depressed, which results in a notification message that will be processed by the engine, cruise control, and other relevant ECUs. However, verifying guaranteed timing of messages is difficult, or impossible, when using an event-triggered protocol; this discourages use of event triggering.

Some automotive networks use hybrid protocols that combine time-triggered and event-triggered protocols by leaving blank time slots in the time-triggered schedule for use by event-triggered alarms such as deployment of an air bag.

B.12 Summary

Modern automobiles are software-enabled physical systems that can be characterized as mobile electronic networks having interconnected subnetworks of ECUs used to sense and control all aspects of the vehicles, as illustrated in Figure B.1. The basic units are ECUs that interact in complex ways using various communication protocols.

Complex interactions among the ECUs and connections to the Internet create the risk of both internal and external unsafe and insecure events.

Self-driving automobiles and other land-based modes of transportation will soon be consumer items. ECUs, interconnected subsystems, and the interactions among them will continue to be core elements of these advanced systems.

References

Balajee, S., Balajai, P., Shrejas, J. et al. (2015). An overview of X-by-wire systems. *First National Conference on Trends in Automotive Technology, International Journal of Innovative Research in Science, Engineering, and Technology*, Velammal Engineering College, Chennai, India (20 March 2015) (Vol. 4, No. 3, March 2015). http://ijirset.com/upload/2015/etat/21_ET140-2.pdf (accessed on 5 September 2018).

Crolla, D. et al. (2015). *Encyclopedia of Automotive Engineering*, Chapter 14, Figure 2, 2015. Wiley.

Hattori, T. et al. (1989). Signal processing system for vehicular acceleration sensor. U.S. Patent 4,843,555. https://www.google.com/patents/US4843555 (accessed 6 September 2018).

Navet, N. and Simonot-Lion, F. (2013). In-vehicle communication networks – a historical perspective and review. In: *Industrial Communication Technology Handbook*, 2e. CRC Press, Taylor & Francis.

Glossary of Terms

Application domain A systems engineering field of application such as aerospace, defense, finance, or health care.

Acquirer An organization or individual that obtains a product, service provisions, or enterprise enablers. An acquirer may be a distinct entity or may, in addition, be a customer, a system user, or any other system stakeholder. See also *Stakeholder* and *Supplier*.

Aggregate A designated collection of system elements. See also *Containment, Composition, Strong aggregation,* and *Weak aggregation*.

Black box An approach to specifying and activating a system element based on observable outputs that result from stated inputs. No implementation details are considered. See also *White box*.

Capability The ability of a system, subsystem, or system element to display a behavior or assume a state of being.

Composition A relationship in which system elements in a designated collection do not exist independent of the composer. See also *Containment* and *Strong aggregation*.

Concurrent development Simultaneous development of hardware elements, software elements, and the interfaces among them. See also I^3.

Constraint A limitation placed on a process, method, technique, schedule, resources, budget, requirement, system, system element, or any other aspect of a system project that will affect system development, deployment, or operation. See also *Stakeholder requirement, System requirement,* and *Design constraint*.

Construction The process of realizing (i.e. making real) the software elements of a system. See also *Fabrication*.

Containment A relationship in which system elements in a designated collection exist independent of the composer. See also *Composition* and *Weak aggregation*.

Customer An organization or an individual that provides the requirements; supplies agreed upon resources (including financial resources); and accepts the resulting product, service enabler, or enterprise provisioning. A

Systems Engineering of Software-Enabled Systems, First Edition. Richard E. Fairley.
© 2019 John Wiley & Sons, Inc. Published 2019 by John Wiley & Sons, Inc.

customer may also be a user, an acquirer, or both. See also *Acquirer* and *System user.*

Derived requirement A system requirement added to clarify or augment a primary system requirement. See also *System requirement* and *Primary system requirement.*

Design constraint A limitation placed on system requirements, architecture definition, or design definition of a system. See also *System requirement.*

Design goal The period after *System requirement* has an underbar.

Deterministic behavior System behavior that is completely determined by the current state of the system and the current stimulus. See also *Nondeterministic behavior.*

Emergent behavior Unexpected behavior that results from adding or modifying one or more system elements during system development or system operation. An emergent behavior may increase or decrease the value of a system.

Engineered system A system for which humans have purposefully developed the system elements, the relationships among them, and the connections to the system environment. See also *Natural system.*

Estimation A prediction of future events based on historical and current data and adjustments made to account for differences between the historical plus current data and what is known or assumed about the future.

Fabrication The process of realizing (i.e. making real) the physical elements of a system. See also *Construction.*

Flow A mental state in which a person is fully immersed in what he or she is doing, characterized by effortless enjoyment of an activity and a lost sense of time (i.e. being in the zone).

Hardware Physical elements of a system that are purposefully engineered and realized, in contrast to naturally occurring physical system elements and software elements; may include analog and digital elements. See also *Physical system element* and *Software-enabled physical system.*

Holism An approach to understanding or analyzing a system, or a proposed system, by understanding the purpose of the system, relationships among the system elements, interactions with the system environment, and impacts of the system on the operational context. See also *Reductionism.*

I^3 The integrated-iterative-incremental system development model.

Implementation The process of realizing a system increment or a system by fabricating, constructing, and procuring system elements, as in "to implement." See also *Fabrication, Construction,* and *Procurement.*

Incremental development A segmented development process wherein system implementation is accomplished in a sequence of processes that each consists of realization, verification, and validation processes. See also *Implementation, Realization, Verification,* and *Validation.*

Iterative development Repeated application of a system or software development process wherein each application of the process produces additional tangible results. Iterative development processes include evolutionary, agile, and spiral processes. See also *Incremental development.*

Model-based systems engineering (MBSE) An approach to system development that uses visual models to supplement and replace the documents used to specify system requirements and system design; used to facilitate analysis, verification, and validation. See also *System modeling language (SysML).*

Modeling and simulation-based systems engineering (M&SBSE) An approach to system development that uses executable MBSE models to simulate system architecture and system elements. See also *Model-based systems engineering (MBSE).*

Natural system A system that has formed as a result of natural forces and processes. See also *Engineered system.*

Nondeterministic behavior System behaviors that occur in different unpredictable ways in response to identical system states and stimuli.

Operational feature An observable system function, system behavior, or quality attribute exhibited during operation of a system. Also known as a user feature.

Physical system element A system element that may be naturally occurring, or is purposefully engineered and composed of physical materials. See also *System element* and *Software-enabled physical system.*

Physical systems engineer (PhSE) A systems engineer whose background and expertise is in engineering of physical elements and physical systems. See also *Software systems engineer (SwSE).*

Primary system requirement A system requirement based directly on a stakeholder requirement. See also *System requirement* and *Stakeholder requirement.*

Procurement The process of obtaining one or more system elements by contracting for, purchasing, leasing, or otherwise obtaining the elements from an external source. See also *Reuse.*

Quality assurance An engineering management process used to determine the degree to which regulations, policies, procedures, guidelines, and plans that impact the quality of work products are being followed by each project and to recommend improvements to decision makers. See also *Quality management* and *Quality control.*

Quality control Application of processes, methods, tools, and techniques to assess, analyze, and control the quality of work products being generated by a systems project. See also *Quality management, Quality assurance, Verification,* and *Validation.*

Quality management The process of defining and implementing, at the organizational level, policies and procedures for quality assurance and quality control processes. See also *Quality assurance* and *Quality control.*

Realization The process of fabricating physical elements, constructing software elements, and procuring and reusing system elements (i.e. to "make real"). Also used to denote the process of making a real system, as in "system realization."

Reductionism An approach to understanding or analyzing a system, or a proposed system, by decomposing the system into system elements and analyzing the purpose and functionality of each element. See also *Holism.*

Reuse The process of using a system element or any other work product originally developed for another use. See also *Procurement.*

Rework The process of modifying a work product to add value (evolutionary rework), add necessary attributes that should have been included earlier (retrospective rework), or fix defects in a current or previous version of a system element (corrective rework). See also *Technical debt.*

Software-enabled physical system A system composed of physical elements and software elements. Software elements enable the physical elements' functionality, behavior, and quality attributes; coordinate interactions among the physical elements; and provide connections to the system environment. Software-enabled systems include systems termed as "software-intensive," "cyber-physical," "embedded," and "Internet of things." See also *Physical system element.*

Software performance measurement Used to measure the degree to which a software element or a software subsystem achieves performance parameters such as rate of throughput, response time, or memory usage. See also *System performance measurement.*

Software systems engineer (SwSE) A systems engineer whose background and expertise is in engineering of software elements and software systems. See also *Physical systems engineer (PhSE).*

Specialty engineering Application of engineering skills needed to evaluate and improve critical performance parameters; typically applied to quality requirements such as safety, security, and reliability.

Stakeholder Stakeholders include system users, operators, maintainers, and others who interact with (or will interact with) a system in a hands-on manner; other stakeholders include those who will not directly interact with a system but will affect or be affected by development, and/or deployment, and/or operation of a system. See also *Stakeholder requirement.*

Stakeholder requirement A statement of a system feature, quality attribute, or design constraint provided by a system stakeholder. See also *Stakeholder* and *Design constraint.*

Strong aggregation A relationship in which a designated systems element does not exist independent of the aggregator. See also *Composition* and *Weak aggregation*.

Supplier An organization or an individual that enters into an agreement with an acquirer to provide a product, enable a service, or provision an organization. The agreement may be formal (e.g. an SOW) or informal (e.g. an MOU). See also *Acquirer*.

Synergy The positive effect on productivity and quality of work that results when members of interdisciplinary engineering teams work together in a collaborative and cooperative manner, as compared with working in separate disciplinary groups.

System element A discrete part of a system (physical or software) that satisfies the requirements allocated to it.

System modeling language (SysML) A set of graphical notations used to model system requirements, architecture, system design, and the relationships among them.

System-of-systems (SOS) A system composed of a collection of independently developed systems that provides capabilities the independently developed systems cannot provide when operating independently.

System performance measurement Used to measure the degree to which a system satisfies various attributes of the entire performance envelope, such as adaptability, quality, quantity, readiness, resilience, and robustness. See also *Software performance measurement*.

System requirement Definition of a primary requirement, derived requirement, design constraint, or design goal derived from a stakeholder requirement. See also *Stakeholder requirement*.

System user An individual who interacts with a system in a hands-on manner. Users include system operators, maintainers, and those who use (or will use) a system to enable or facilitate their job functions or leisurely activities.

Technical debt A measure of work that should have been done previously and must be done now. See also *Rework*.

Technical management Management of the technical work activities needed to develop a complex software-enabled physical system.

User feature See *Operational feature*.

Verification The process of determining the degree to which a work product satisfies the requirements, conditions, and constraints placed on it by other work products, organizational policies and procedures, and regulatory and legal considerations.

Validation The process of determining the degree to which a work product is suitable for use in the intended ways by the intended users in the intended context and environment.

Weak aggregation A relationship in which a designated system element exists independent of the aggregator. See also *Strong aggregation* and *Containment*.

White box An approach to specifying and testing a system element by considering the internal structure of the element (also known as glass box). See also *Black box*.

Work product Any artifact generated during development of a system, including requirements, design documents, system elements, traceability matrices, work plans, test plans and test results, risk management and corrective action plans and strategies, assessment results, and so forth.

Index

Systems Engineering of Software-Enabled Systems, First Edition. Richard E. Fairley.
© 2019 John Wiley & Sons, Inc. Published 2019 by John Wiley & Sons, Inc.

Printed and bound by CPI Group (UK) Ltd, Croydon, CR0 4YY
16/04/2025
14658581-0004